The Birth of NASA

The Work of the Space Task Group, America's First True Space Pioneers

Other Springer-Praxis books by Manfred "Dutch" von Ehrenfried

Stratonauts: Pioneers Venturing into the Stratosphere
2014
ISBN 978-3-319-02900-9

Manfred "Dutch" von Ehrenfried

The Birth of NASA

The Work of the Space Task Group, America's First True Space Pioneers

 Springer

Published in association with
Praxis Publishing
Chichester, UK

Manfred "Dutch" von Ehrenfried
Lago Vista, TX, USA

SPRINGER-PRAXIS BOOKS IN SPACE EXPLORATION

ISBN 978-3-319-28426-2   ISBN 978-3-319-28428-6   (eBook)
DOI 10.1007/978-3-319-28428-6

Library of Congress Control Number: 2015960756

Springer Cham Heidelberg New York Dordrecht London

Cover design: Jim Wilkie
Project Editor: David M. Harland

Printed on acid-free paper

Praxis is a brand of Springer
Springer International Publishing AG Switzerland is part of Springer Science+Business Media (www.springer.com)

Contents

Frontispiece.. viii

Dedication ix

Acknowledgments xi

Preface.. xiii

Part I   Setting the Stage

1   **Introduction**.. 2

2   **The Sputnik Reaction** 5

3   **The President**... 7
    3.1   A Secret Conference with the President.................................... 7
    3.2   The Press Conference Opening Remarks.................................. 10
    3.3   The President's Remarks 11
    3.4   Dr. Killian's Secret Memorandum.. 16

4   **The National Security Council**.. 21

5   **The Congress** 24

6   **The Department of Defense and Other Agencies** 26
    6.1   Advanced Research Projects Agency 27
    6.2   Army Ballistic Missile Agency 28
    6.3   U.S. Air Force 29
    6.4   U.S. Navy 30
    6.5   Department of State.. 31

Part II   Creating the Space Team

7   **Creation of the Space Task Group**.. 35
    7.1   NACA – From Aeronautics to Astronautics............................. 35
    7.2   The Core Team 36

7.3	The Lewis Contribution	40
7.4	The Goddard Contribution	43
7.5	The Ames Contribution	46
7.6	Wallops Island Contribution	47
7.7	The High Speed Flight Station's Contribution	50
7.8	White Sands Missile Range Contribution	52
7.9	The Arnold Engineering Contribution	53
7.10	Marshall Space Flight Center	53
8	**The AVRO Canadians**	**54**
8.1	History	54
8.2	The STG Captures the Talent	55
8.3	The AVRO Contribution to the STG	57
9	**The STG Organization**	**59**
9.1	The Directive	59
9.2	Staff Offices	59
9.3	Flight Systems Division	65
9.4	Operations Division	69
9.5	Engineering and Contracts Division	76
10	**Representatives and Contractors**	**79**
10.1	Military	79
10.2	STG Contractors	80
10.3	Mercury Control Center Contractors	80
11	**The Need for More People**	**83**
11.1	Langley Support to Project Mercury	83
11.2	STG Hiring	85
12	**The End of the Space Task Group**	**89**
12.1	The Decision	89
12.2	The Move	92
13	**Some Key Project Mercury Decisions and Lessons Learned**	**95**
13.1	Management	95
13.2	Engineering	101
13.3	Operations	120
13.4	Scientific	133
13.5	Medical	137
Part III	**Achievements**	
14	**Facilities Created for Project Mercury**	**143**
14.1	Mercury Control Center	143
14.2	Bermuda Control Center	151
14.3	Mercury/Manned Space Flight Network	153
14.4	Mercury Procedures Trainers	155

15 Mission Designs and Concepts.. 158
  15.1 Mission Rules... 158
  15.2 Operational Procedures 160
  15.3 Simulations... 161
  15.4 Spacecraft Designs 164
  15.5 Launch Vehicle Designs 165
  15.6 Mercury Full Pressure Suit.. 165
  15.7 Mission Analysis and Trajectory Planning........................... 166

16 The Impact of NASA and the STG on History................................... 169
  16.1 Organizational Excellence.. 169
  16.2 Mercury Mission Accomplished 171
  16.3 Future Programs 171
  16.4 Technology Transfer.. 175
  16.5 National Pride... 176
  16.6 Generational Impact 177

Appendix 1: STG Organization Lists, Charts and Manning............................... 179

Appendix 2: Biographies 197

Appendix 3: STG Technology 295

Appendix 4: Some Photos.. 301

Appendix 5: Quotes.. 309

Appendix 6: Stories and Trivia 319

Appendix 7: Author's STG Experience 329

References 335

Credits 337

Glossary 340

About the Author 344

The Front Cover.. 347

Index... 348

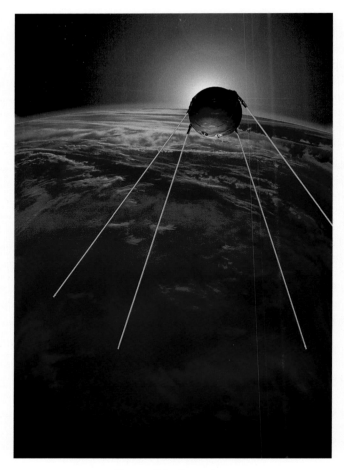

Dawn of the Space Age

Created by Gregory R. Todd to mark the 50th anniversary of the launch of Sputnik.
This is the little ball that started it all!
(Photo courtesy of Wikimedia Commons)

*When I first started this book, I wanted to dedicate
it to the approximately 750 people who were
at the Space Task Group (STG) at the Langley
Research Center between 1958 and 1961,
many of whom then stayed into 1962. I also
wanted to include the small group at the NACA/
NASA Headquarters who met at the Dolley
Madison House to kindle the sparks of a new
spaceflight organization and program.
These were the people that I thought
of as America's first true space pioneers.
But then I realized that hundreds of Langley
Research Center scientists, engineers, technicians,
tradesmen, secretaries, clerks, and others
had also been working hard to support
the STG without formally being part
of the STG organization. I also realized
that many of the men and women's families
were also heavily involved. Everyone supporting
Project Mercury worked very long hours
and took the work home to the dinner table
and often burned the "midnight oil."*

*Then I also realized that the entire Hampton,
Virginia community and surrounding towns
and villages were involved as well. They provided
the food, the cars, the gasoline, the schools,
the shops, and the entire infrastructure to support
those working at Langley Field to establish
a new space program. Then there were hundreds
of contractors across the Nation supporting
the project and, later, thousands of civilian
and military people providing launch*

*and recovery support. Also, there were people
all around the world at tracking stations
and in other support roles. How can I dedicate
the book to only 750 people?*

*The initial effort took its toll on men, women,
and children. Years later, the Project Mercury
Director Robert Gilruth pined about the good old
days at Langley, saying that he couldn't do it
again; it was a young man's job. And many
of us were young and "wet behind the ears."
Many of us were just or recently out of college.
In our "20-something" eyes, our managers
were what we thought of as "older" men;
why, we thought, they must be in their late thirties
or forties! At the time, I couldn't imagine
how smart, indeed brilliant if not geniuses,
these men and women were. It's only now
that I have the experience of old age
that I realize what a unique gathering
of eagles came to alight in a nest called
Hampton, Virginia.*

*If you were part of Project Mercury
in any capacity, in any location, doing
any support work, then this book is dedicated
to you. You are a space pioneer because you were
there at the very beginning! You made it happen!
That was over half a century ago. Many, probably
most, are now gone. Only we "20-somethings"
and a handful of the "older men" are left.
So this book is also dedicated to our prodigy
and the next several generations of space
enthusiasts and workers. You could be the ones
to be "planetary pioneers." But we "Mercurians"
were the original space pioneers!
Forge ahead; it's your turn!*

Acknowledgments

One cannot write a history of such a unique organization as the NASA Space Task Group (STG) without a lot of help. Fortunately, some of the members of this group participated in the Oral History program at the Johnson Space Center (JSC), formerly the Manned Spacecraft Center (MSC) where most of them moved to after the STG was merged into the MSC. Unfortunately, many of the members didn't participate in this effort and their histories are lost except for the memories of others and perhaps archive files at universities or NASA History and Personnel offices.

My thanks to Rebecca Wright of the JSC History Office for her work on the history project and help in answering questions about former STG people. Thanks also to Jennifer Ross-Nazzal for her research assistance.

Similarly, Anne K. Mills of the Glenn Research Center (formerly the Lewis Research Center) provided input about their role in Project Mercury and sent me some reference books and a video disk. April D. Gage of the Ames Research Center also provided support, as did Christian Glezer and Curtis Peebles from the Armstrong Flight Research Center (formerly the High Speed Flight Station and the Dryden Flight Research Center). Thanks also to Travis Kinchen for his support.

Mary E. Gainer from the Langley Research Center was most helpful and even started another page on their website to include the STG. It now includes cross links to the JSC histories as well as photographs of some original members. She also helped me with building locations, telephone books, and the confirmation of members. The Langley Alumni Association is aware of this effort to recognize those at Langley who contributed to the success of Project Mercury.

Thanks to Bill Barry and Connie Moore of the NASA Headquarters History Office for their contribution of an early phone book that helped me to identify who was at the Dolley Madison House and at nearby offices.

Many thanks to some of the original members of the STG who provided input and comments on my drafts. In this role I must particularly thank Arnold "Arnie" Aldrich, who was one of the first persons I met at Langley. He reviewed nearly all the chapters in the book. Arnie could not have had a more distinguished career in spaceflight. He rose to the highest levels in NASA and later in the aerospace industry. I'm grateful for his knowledge and insight.

Other STG and former NASA people who provided input are (in alphabetical order): Peter Armitage, Harold Beck, Norm Chaffee of the JSC Alumni League, Henry "Pete" Clements, Jack Cohen, Bill Davidson, Bryan Erb, Dennis Fielder, Tom Gallagher, George Harris, Clay Hicks, John Hodge, Gene Kranz, Charles Lewis, Glynn Lunney, C. Frederick Matthews, Duncan McIver of the Langley Alumni League, Harold Miller, William Pratt, John C. Stonesifer, Kenneth Suit, Robert Thompson, and Dr. Robert Voas. They supplied input into those sections about their work and/or details of their biographies. My thanks also to William H. Taylor of DB Consulting at JSC for some photos and identification of people.

While I obtained a lot of information from various NASA websites, they say very little about the STG people from nearly 60 years ago except for the key managers and the astronauts. I made good use of the internet search engines like Wikipedia, Bing and Google. The many books that I used for reference are included at the back of this volume.

My thanks to Maury Solomon and Nora Rawn of Springer in New York, Clive Horwood of Praxis Publishing in the United Kingdom and, of course, my editor extraordinaire, David M. Harland in Glasgow, Scotland. In the process of getting approval for the book, my concept and outline were reviewed by referees who provided various comments. I nominated Gene Kranz, Arnold Aldrich, Clay Hicks and John Hodge. There were others nominated by Springer-Praxis for a total of seven referees. I wasn't privy to who actually responded but I would like to thank them and acknowledge their inputs, which I definitely considered in developing the book.

Preface

This book should have been written about half a century earlier! For such a great period in space history, more can be said about the personal contributions and stories of the early space pioneers who scrambled after the surprise of Sputnik to start the American space program. While I knew many of the people in the Space Task Group (STG) at the Langley Research Center in Hampton, Virginia, there are hundreds I didn't know. Even then, as a young man, I had little knowledge of their backgrounds and experiences. I was what they now call a "newbie." In those more formal days of the 1950s and early 1960s, we might be addressed as "young man." There were many of us in our twenties. Our managers were, for the most part, in their late twenties to middle thirties.

I've learned more about the STG people in writing this book than I ever knew at the time. It is difficult even now to find some of their names, let alone their contributions. The NASA History Offices at the Johnson Space Center (then the Manned Spacecraft Center) and the Glenn Space Center (then the Lewis Research Center) have, over the years, obtained oral histories from many of the Project Mercury people. The Langley History Office recently added a Space Task Group webpage with links to the Johnson oral histories. I have read most of them. Unfortunately many people didn't participate in the Oral History Project, with the result that their contributions are essentially lost. Some of the histories aren't available online but are VHS tapes held in storage somewhere. It is sad that the contributions of some very key people are not recorded anywhere that I could find.

In many cases, when I read the oral histories the individual says very little about their early STG career, focusing more on their later contributions to major programs like Apollo and the Space Shuttle. While I find these oral histories very interesting, the average reader today might view them as rather rambling and sometimes incoherent memories. To get an overall sense of what was going on, you would have to read a lot of them. I wanted to capture what these early Mercury space pioneers accomplished.

During 2015, in researching this book, I talked with many STG people who are now in their twilight years – as indeed am I. It seems easy for them to recall special events such as the spaceflights, but not the day-to-day particulars of their work over half a century ago.

They remember only some of their co-workers. Some of them have kept in touch, but most drifted apart over the years. To my great delight, I heard from one man who is now 93 years old and is able to recall events in great detail.

The STG only existed for three years. Almost immediately after NASA was itself formed on October 1, 1958 the STG was formally organized on November 3, 1958. Only three years later on November 1, 1961, the STG staff was formally declared part of the new Manned Spacecraft Center which didn't even physically exist. Everyone's badges changed, but it had little effect on those preparing for John Glenn's flight. Over the next eight months, people relocated to the as-yet-unbuilt Manned Spacecraft Center in Houston, Texas. They were temporarily housed in a variety of rented office buildings in Houston. We all wondered why we were leaving beautiful Virginia for what we considered the "Wild West." After John Glenn's flight, I took a trip to the proposed site and found cows in a big pasture. A now-famous photo of those cows is included later just to show you how things were in those days. It was hard to believe that out of 20 cities evaluated to host the Manned Spacecraft Center, Houston was chosen, especially considering its distance from the launch site and control center at Cape Canaveral in Florida. I have included a discussion of that decision.

When NASA was first established there was great organizational upheaval, with some people transferring to NASA Headquarters, some from one Center or Laboratory to another, and some to various aerospace contractors. A new agency was being pieced together to lead the Nation's new civilian space program. This involved bringing together people from many locations and organizations to tackle an unprecedented technical challenge. To express it in the context of the title of this book, it was a rather sudden and difficult birth!

I have made an attempt to write the story about the birth of NASA and the STG in three parts. The first part, "Setting the Stage," discusses the beginning of America's space program ranging from Sputnik to the creation of NASA out of many existing organizations. Then "Creating the Space Team" begins with the creation of the STG organization, explaining where people came from and where they ended up in the organization. This part ends with the decision to disband the STG and establish the Manned Spacecraft Center, but it lists some of the key decisions and lessons learned in management, engineering, operations, science, and spaceflight medicine. The third part, "Achievements," lists the major accomplishments of the STG and the Project Mercury team. This includes the facilities that were specifically created as well as the unique and creative mission designs, operational concepts, and methodologies. The story is wrapped up with some philosophical thoughts on the impact of this experience on future spaceflights, management of complex systems, political will, and national pride. I also predict the date of the first landing of humans on Mars.

These three parts are supplemented with many appendices that give more detail, including a significant number of biographical profiles that describe where these space pioneers came from and the work that they did, both in the STG and subsequently.

I describe the Mercury missions from operational, science, and medical perspectives. The astronauts were part of the STG and many of us worked with them as part of their daily work routines. Most of their time was spent on training and a variety of engineering and operational assignments. Only two astronauts flew during the three years of the

STG. In fact, more animals than astronauts flew during this period. The lives and contributions of the astronauts of Project Mercury have been well covered by many historians. Excellent books are referenced at the end of this volume.

In summary, the intent for this book is to capture as much as possible, the roles of America's first true space pioneers. Most are now in their twilight years. Many of those that feature in the history books are long gone, having taken the ultimate spaceflight. So the intent of this book is to chronicle as much as possible the Space Task Group's contributions to history; if not for the participants themselves then for their children and grandchildren.

Lago Vista, TX, USA   Manfred "Dutch" von Ehrenfried
Winter of 2015

Part I
Setting the Stage

1

Introduction

This is the story of the men and women who were America's first true space pioneers. History books have often focused on the National Aeronautics and Space Administration (NASA) from the point of view of the early Russian space achievements and the need to organize a U.S. space capability. The Nation's response to Sputnik in 1957 and its military implications was to "Wake Up and Catch Up."

Even before Sputnik, the National Advisory Committee on Aeronautics (NACA) was already considering moving into astronautics, and it had studies of capsule design and re-entry heating underway. The Air Force was working on the Atlas intercontinental ballistic missile (ICBM) and had ideas on manned spaceflight of their own. The Army Ballistic Missile Agency had its Redstone and Jupiter ballistic missiles. And the Navy had its Naval Research Laboratory and Project Vanguard, which was a rocket designed for civilian scientific use.

A new national manned spaceflight effort would require Presidential and Administration policies and directives as well as a new Congressional Law. These efforts led to Congressional hearings and special committees to discuss a future space program, most notably the President's Scientific Advisory Committee and the Joint NASA-ARPA Panel. By July 29, 1958, President Dwight Eisenhower signed the National Aeronautics and Space Act. During that same summer, even before NASA was created, a select group of people from NACA, ARPA, and many from various laboratories, met in the Dolley Madison House near the White House to discuss how to proceed and organize a space program. These approximately two dozen people could arguable be considered the "Founding Fathers" of the space program. Some would go on to be the leading administrators, managers, and engineers of Project Mercury and even follow-on programs like Gemini and Apollo.

NASA was established on October 1, 1958, just one year after Sputnik, from three NACA research laboratories, namely the Langley Memorial Aeronautical Laboratory at Langley Field, Virginia, the Lewis Flight Propulsion Laboratory at Cleveland, Ohio, and the Ames Aeronautical Laboratory at Moffett Field, California. The Muroc Flight Test Unit at Edwards Air Force Base, the NACA Wallops Island Station in Virginia, and selected elements from the Army and Navy flight test programs were also included.

© Springer International Publishing Switzerland 2016
M. von Ehrenfried, *The Birth of NASA*, Springer Praxis Books,
DOI 10.1007/978-3-319-28428-6_1

But suddenly, and by decree, on October 1 all NACA employees across the country became NASA employees and the challenge of forming a manned spaceflight organization fell to the Space Task Group (STG) at the newly named Langley Research Center.

The history books have well documented these events and dates, and I have certainly made good use of some of them (as listed at the end of this volume). In history books, emphasis is devoted to why this group was created, and the roles of the first astronauts and key managers; less is said about the scientists, engineers, mathematicians, technicians, and administrative people who were also part of this first great and historical team. History chronicles what was accomplished by this unique group by describing the first manned space missions during this period; mostly from the perspectives of the astronauts and the Nation's role in space. In those days the national press focused its attention on the astronauts and the launches, and apart from some of the top managers, less on the many other people involved.

The STG was a relatively small group of people. It only formally existed as an organization for three years from November 3, 1958 to November 1, 1961, at which time it was folded into the Manned Spacecraft Center (MSC), even though that did not yet physically exist. The STG employees suddenly got MSC badges. The new MSC was now being built to house not only the STG, but many more organizations. The completion of the relocation of the STG from Langley Field, Virginia to MSC in Houston, Texas was completed by July 1, 1962. Many of the STG employees associated with flight operations relocated to Houston in the wake of John Glenn's historic flight in February 1962, after being temporarily based at Cape Canaveral, Florida and deployed around the world to man the remote tracking stations.

This first group, led by Robert R. Gilruth, began with 36 people (counting himself) from the now named Langley Research Center and 10 from the now named Lewis Research Center in Cleveland, Ohio. This total of 46 included 37 engineers; 27 from Langley and 10 from Lewis. It also included 8 women, some of them secretaries and others operating mechanical calculators (in those days referred to as "computers"), plus one male file clerk.

Shortly thereafter, NASA offered jobs to 32 engineers from Canada who were victims of the cancellation of the AVRO (A. V. Roe Company) CF-105 program on February 20, 1959. Seven declined but the remaining 25 joined NASA. The CF-105 was to have been Canada's first and most advanced supersonic interceptor, and the company employed Canada's best and brightest engineers.

By the end of 1959 the STG staff had grown to approximately 287 in all capacities, ranging from astronauts, doctors, and life support engineers, training people, flight systems engineers, operations people, mission planning and analysis people, mathematicians, engineers, contracts people, to mission recovery people. Now secretaries, accounting, travel people, and security personnel were also needed. The STG swelled with the need to staff up for spaceflight. As a result, by the end of 1961 the total was approaching 750; not all of whom moved to Houston. It also included military personnel and contractors assigned to the STG.

During this short three year period, the STG was also planning a world-wide tracking system capability, designing and constructing the Mercury Control Center in Florida and the Bermuda Control Center, integrating the space capsules to military missiles, setting up to use the military missile ranges, and planning for follow-on programs including Gemini and Apollo. In just three years, NACA/NASA had gone from focusing only on aeronautics to embracing aeronautics and astronautics.

This book looks more closely at the people of the Space Task Group, and is a tribute to their Herculean efforts. They were at the right place at the right time, and became the original NASA space pioneers. Many became rather famous; some are now legends in the annals of spaceflight. But most have remained in the shadows, until now. This book will list their names and provide summary details on as many as it was possible to find after more than nearly six decades. Many are long gone and most are in their seventies, eighties, and even their nineties. It is "altogether fitting and proper" that we should attempt to document their efforts before they are *all* gone, in order that the history of the STG will be more complete and later generations will know of their personal achievements and their impact on the development of spaceflight.

2

The Sputnik Reaction

There is no question that the American reaction to the launch and orbit of Sputnik on October 4, 1957 was more than unnerving; it even caused fear and foreboding in some. While the satellite was just a 23-inch diameter sphere weighing 184 lbs. that simply went "Beep Beep," it was the first satellite in orbit. More disturbing was the fact that the upper stage of the R-7 booster rocket weighing over 7 tons was also in orbit. If the Soviets could put that much mass into orbit they could clearly launch a nuclear weapon. If you didn't believe the news, you could go outside and see it (the upper stage, not the satellite) moving across the night sky; a phenomenon that no one had ever witnessed before.

While the United States was trying to digest what had just happened, the Soviets launched a second Sputnik on November 3rd in what was effectively "a slap in the face" or at least a "take that." This spacecraft weighed 1,120 lbs. and carried a dog named "Laika." That indicated the vehicle must possess a life support system; albeit just for a dog. President Eisenhower tried to "spin" the event by saying that our satellite program was not being conducted as a race against other nations. The Soviets, however, had considered it a race for at least two years. Terms like "missile gap," "arms race," and "space race" were now everywhere in the media.

To add to the Nation's embarrassment, the first attempt to launch the Vanguard rocket and a "grapefruit" size payload from Cape Canaveral on December 6 in front of a world press and on TV ended in an ignominious explosion. These three launches took on a new meaning within the Washington bureaucracy, within the Department of Defense, and within the missile contractor industry. There are books written about this period of space history. Here is what subsequently happened within NACA/NASA and the Space Task Group.

The Air Force made overtures to NACA Director Dr. Hugh L. Dryden to collaborate on their Dyna-Soar program. This seemed only natural to the Air Force, as they had worked with NACA for 40 years on aeronautical issues. But Dr. Dryden knew that NACA Langley wanted to work on a manned "capsule" of their own. He also knew that only the Air Force and the Army could provide the requisite launch vehicles. NACA wanted to add astronautics to their traditional role of aeronautics. The last "A" in NACA is "Aeronautics," but

© Springer International Publishing Switzerland 2016
M. von Ehrenfried, *The Birth of NASA*, Springer Praxis Books,
DOI 10.1007/978-3-319-28428-6_2

NACA wanted a leadership role in the new field of manned spaceflight. NACA engineers weren't waiting for approval, they had been working on aspects of aerodynamic flight that were also applicable to spaceflight.

During the last three months of 1957, there were scores of committees from all the federal agencies concerned, discussing what should be done and who should undertake it. There were meetings in the Pentagon, in Congress, at NACA Headquarters and its field laboratories, in the National Academy of Sciences, in the National Science Foundation, in universities, and within industrial corporations. Even the American Rocket Society was ready to offer input.

But, as concerns the STG, it was the NACA Lewis Flight Propulsion Laboratory, headed by Associate Director Abe Silverstein, that produced a bold plan called "A Program for Expansion of NACA Research in Space Flight Technology." The impact of this report will feature in later chapters.

3

The President

President Dwight Eisenhower was dealing with a lot of major issues even prior to the events of October 4, 1957. This was the period of the Hungarian Uprising, the Suez Crisis, the McCarthy "witch hunts," schoolchildren practicing "Duck & Cover" air raid drills, and the riots in Little Rock Central High School.

For us, nowadays, to judge Eisenhower's response to Sputnik, I think it is important that the reader have a good understanding of his position at the time, and in particular what his advisors were telling him. This will facilitate an appreciation of why first NASA and then the STG were created.

3.1 A SECRET CONFERENCE WITH THE PRESIDENT

Ever since the launch of Sputnik, the President had meetings almost every day with his advisors, sometimes many meetings. As you might expect, he selected very senior people for his staff. One was the highly decorated WW-II combat veteran Brig. Gen. Andrew Jackson Goodpaster, who also had an M.S. in engineering and a Ph.D. in international affairs from Princeton. Eisenhower had appointed him to be his Staff Secretary and Defense Liaison Officer.

Also present at a meeting on October 8, 1957 was the Assistant Secretary of Defense, Donald Aubrey Quarles. He had served as Secretary of the Air Force, President of Sandia Laboratories, and Vice President of both Western Electric and Bell Labs, as well as being assigned to NACA. In addition he had an honorary doctorate in engineering.

The meeting included many other distinguished and knowledgeable advisors to the President, all of whom were concerned about the Sputnik event.

The following SECRET memorandum (declassified 11/17/1971) provided the President with information that helped him to prepare for the press conference on October 9. This would be his opportunity to tell the concerned public what his position was concerning Sputnik and his views concerning the Nation's response.

© Springer International Publishing Switzerland 2016
M. von Ehrenfried, *The Birth of NASA*, Springer Praxis Books,
DOI 10.1007/978-3-319-28428-6_3

Although we don't know the details of the undocumented discussions, it is evident from this memorandum that the Army had ambitions in space and there was a reconnaissance program in the works that was clearly military, not part of a civilian space program. It is also clear that we weren't about to share technology with the Soviets.

SECRET,

October 9, 1957

MEMORANDUM OF CONFERENCE WITH THE PRESIDENT
October 8, 1957, 8:30 AM

Others present: Secretary Quarles
                      Dr. Waterman
                      Mr. Hagen
                      Mr. Holaday
                      Governor Adams
                      General Persons
                      Mr. Hagerty
                      Governor Pyle
                      Mr. Harlow
                      General Cutler
                      General Goodpaster

Secretary Quarles began by reviewing a memorandum prepared in Defense for the President on the subject of the earth satellite (dated October 7, 1957).   He left a copy with the President. He reported that the Soviet launching on October 4th had apparently been highly successful.

The President asked Secretary Quarles about the report that had come to his attention to the effect that Redstone could have been used and could have placed a satellite in orbit many months ago. Secretary Quarles said there was no doubt that the Redstone, had it been used, could have orbited a satellite a year or more ago. The Science Advisory Committee had felt, however, that it was better to have the earth satellite proceed separately from military development.   One reason was to stress the peaceful character of the effort, and a second was to avoid the inclusion of materiel, to which foreign scientists might be given access, which is used in our own military rockets.   He said that the Army feels it could erect a satellite four months from now if given the order -- this would still be one month prior to the estimated date for the Vanguard.   The President said that when this information reaches the Congress, they are bound to ask why this action was not taken.   He recalled,

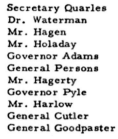

DECLASSIFIED
E.O. 11652, Sec. 11
MR 76-49   DOCUMENT #65
By ____ . Date 11-17-71

SECRET

SECRET

- 2 -

however, that timing was never given too much importance in our own program, which was tied to the IGY and confirmed that, in order for all scientists to be able to look at the instrument, it had to be kept away from military secrets. Secretary Quarles pointed out that the Army plan would require some modification of the instrumentation in the missile.

He went on to add that the Russians have in fact done us a good turn, unintentionally, in establishing the concept of freedom of international space -- this seems to be generally accepted as orbital space, in which the missile is making an inoffensive passage.

The President asked what kind of information could be conveyed by the signals reaching us from the Russian satellite. Secretary Quarles said the Soviets say that it is simply a pulse to permit location of the missile through radar direction finders. Following the meeting, Dr. Waterman indicated that there is some kind of modulation on the signals, which may mean that some coding is being done, although it might conceivably be accidental.

The President asked the group to look ahead five years, and asked about a reconnaissance vehicle. Secretary Quarles said the Air Force has a research program in this area and gave a general description of the project.

Governor Adams recalled that Dr. Pusey had said that we had never thought of this as a crash program, as the Russians apparently did. We were working simply to develop and transmit scientific knowledge. The President thought that to make a sudden shift in our approach now would be to belie the attitude we have had all along. Secretary Quarles said that such a shift would create service tensions in the Pentagon. Mr. Holaday said he planned to study with the Army the back up of the Navy program with the Redstone, adapting it to the instrumentation.

There was some discussion concerning the Soviet request as to whether we would like to put instruments of ours aboard one of their satellites. He said our instruments would be ready for this. Several present pointed out that our instruments contain parts which, if made available to the Russians, would give them substantial technological information.

A. J. Goodpaster
Brigadier General, USA

SECRET

That same day, Eisenhower also met with the President of the National Academy Sciences, Dr. Detlev Wulf Bronk, and several others to review Eisenhower's proposed statement about Sputnik at the White House press conference to be held the next day. Only a few changes were made to the speech, with some emphasis added.

On October 9, White House Press Secretary James Hagerty held a press conference called a "Summary of Important Facts in the Development by the United States of an Earth Satellite" to provide background information for the Q & A that would immediately follow the President's statement.

However, what the following transcript (reproduced verbatim) suggests, in retrospect, is that the science and defense people were following the programmatic direction they were previously given by the President and the National Security Council. In particular, they weren't integrating the satellite program with the missile program. After reading the President's remarks, you will understand the reason why the Soviets beat us into orbit.

3.2   THE PRESS CONFERENCE OPENING REMARKS

It is important to quote Mr. Hagerty's introduction in order to set the stage for the President's appearance.

The first serious discussion of an Earth satellite as a scientific experiment to be incorporated in the program for the International Geophysical Year took place at a meeting of the International Council of Scientific Unions in Rome in October 1954. At this meeting, at which Soviet scientists were present, a resolution was adopted by the scientists of the world recommending that in view of the advanced state of present rocket techniques, thought be given to the launching of small satellite vehicles.

Following this International Council meeting, the United States National Committee for International Geophysical Year, working under the sponsorship of the National Academy of Sciences, recommended that the United States institute a scientific satellite program. It was determined by the Administration that this program would be carried out as part of the United States' contribution to the International Geophysical Year.

Responsibility within the Government for scientific aspects of the program was assigned to the National Science Foundation, working in close cooperation with the United States National Committee for the International Geophysical Year. The Department of Defense was made responsible for supplying the rocketry needed to place a satellite in orbit without interfering with the top priority ballistic missile program. In line with the recommendations of a group of United States scientists advising the Department of Defense, the satellite project was assigned to the Naval Research Laboratory as Project Vanguard.

On July 29, 1955, at a White House press conference, participated in by representatives of the National Science Foundation and the National Academy of Sciences, it was announced that plans are going forward for the launching of small, unmanned Earth circling satellites as part of the United States participation in the International Geophysical Year, which takes place between July 1957 and December 1958.

At this press conference it was specifically stated that the data which will be collected from this program will be made available to all scientists throughout the world. The National Science Foundation, it was also announced, would work with the United States National Committee for the International Geophysical Year to formulate plans for the satellite and its instrumentation as well as plans for the preparation and deployment of the ground observer equipment required for the program.

In May of 1957, those charged with the United States satellite program determined that small satellite spheres would be launched as test vehicles during 1957 to check the rocketry, instrumentation, and ground stations and that the first fully-instrumented satellite vehicle would be launched in March of 1958. The first of these test vehicles is planned to be launched in December of this year.

As to the Soviet satellite, we congratulate Soviet scientists upon putting a satellite into orbit.

The United States satellite program has been designed from its inception for maximum results in scientific research. The scheduling of this program has been described to, and closely coordinated with, the International Geophysical Year scientists of all countries. As a result of passing full information on our project to the scientists of the world, immediate tracking of the United States satellite will be possible, and the world's scientists will know at once its orbit and the appropriate time for observation.

The rocketry employed by our Naval Research Laboratory for launching our Vanguard has been deliberately separated from our ballistic missile efforts in order, first, to accent the scientific purposes of the satellite and, second, to avoid interference with top priority missile programs. Merging of this scientific effort with military programs could have produced an orbiting United States satellite before now, but to the detriment of scientific goals and military progress.

Vanguard, for the reasons indicated, has not had equal priority with that accorded our ballistic missile work. Speed of progress in the satellite project cannot be taken as an index of our progress in ballistic missile work.

Our satellite program has never been conducted as a race with other nations. Rather, it has been carefully scheduled as part of the scientific work of the International Geophysical Year.

I consider our country's satellite program well designed and properly scheduled to achieve the scientific purposes for which it was initiated. We are, therefore, carrying the program forward in keeping with our arrangements with the international scientific community.

In retrospect this statement indicates what the U.S. was focusing upon, as compared to the Soviets.

3.3   THE PRESIDENT'S REMARKS

Following Mr. Hagerty's introduction, President Eisenhower restated his position until the press (whose names have been omitted) pushed for answers.

Good morning ladies and gentlemen. Do you have any questions you would like to ask me?

(Question) Mr. President, Russia has launched an Earth satellite. They also claim to have had a successful firing of an intercontinental ballistic missile, none of which this country has done. I ask you sir, what are we going to do about it?

The President: Well, let's take, first, the Earth satellite as opposed to the missile, because they are related only indirectly in the physical sense, and in our case not at all.

The first mention that was made of an intercontinental – of an Earth satellite that I know of, was about the spring of 1955 – I mean the first mention to me – following upon a conference in Rome where plans were being laid for the working out of the things to be done in the International Geophysical Year. Our people came back and with studying a recommendation of that conference that we now undertake, the world undertake, the launching of a small Earth satellite, and somewhere in, I think May or June of 1955, it was recommended to me, through the Committee of or by the Committee for the International Geophysical Year, and through the National Science Foundation, that we undertake this project with a satellite to be launched somewhere during the Geophysical Year, which was from June 1957 until December 1958.

The sum asked for to launch a missile was $22 million and it was approved.

For the government, the National Science Foundation was made the monitor of the work, for the simple reason that from the beginning the whole American purpose and design in this effort has been to produce the maximum in scientific information. The project was sold to me on this basis.

My question was: What does mankind hope to learn? And the answer of the scientists was we don't exactly know, and that is the reason we want to do it, but we hope to learn lots of things about outer space that will be valuable to the scientific world.

They did mention such things as temperatures, radiation, ionization, pressures, I believe residual pressures, from such air as would be at the altitude where successful orbiting was possible. That is the kind of information the scientists were looking for, and which they hoped to obtain from this project.

Now, in the first instance, they thought they would merely put up a satellite, and very quickly they found they thought they could put up a satellite with a considerable instrumentation to get, even during the Geophysical Year, the kind of information to which I have just referred.

So they came back, said they needed some more money. This time they went up to $66 million and we said all right, in view of the fact we are conducting this basic research this seems logical. So we did that.

Then they came back, and I forget which one of the steps it came along, and they realized when you put this machine in the air, you had to have some very specially equipped observation stations, so the money, the sum of money, again went up to provide for these observation stations; and so the final sum approved, I think about a year ago, something of that kind, was $110 million, with notice that they might have to go up even still more.

There never has been one nickel asked for accelerating the program. Never has it been considered as a race; merely an engagement on our part to put up a vehicle of this kind during the period that I have already mentioned.

Again emphasizing the non-military character of the effort, we have kept the Geophysical Year Committees of other nations fully informed all the time – as, for example, the frequencies we would use when we put this in the air so that everybody, all nations, could from the beginning track it exactly – know exactly where it was. And I believe it was 108 megacycles we were to use, and that was agreed throughout the world.

We are still going ahead on this program to make certain that before the end of the calendar year 1958, we have put a vehicle in the air with the maximum ability that we can devise for obtaining the kind of scientific information that I have stated.

Now, every scientist that I have talked to since this occurred – I recalled some of them and asked them – every one of them has spoken in most congratulatory terms about the capabilities of the Russian scientists in putting this in the air. They expressed themselves as pleased rather than chagrined, because at least the Soviets have proved the first part of it, that this thing will successfully orbit. But there are a lot of other things in the scientific inquiry that are not yet answered, and which we are pushing ahead to answer. Now that is the story on the satellite. It is supplemented by a statement that we prepared this morning that has some of the basic facts to include the sequence of events.

As to their firing of an intercontinental missile, we have not been told anything about the details of that firing.

They have proved again and, indeed, this launching of the satellite proves, that they can hurl an object a considerable distance.

They also said, as I recall that announcement, that it landed in the target area, which could be anywhere, because you can make a target area the size you please, and they also said it was a successful re-entry into the – to the atmosphere, and landing at or near the target.

Now that is a great accomplishment, if done. I have talked to you in the past about our own development in this regard as far as security considerations permit, and I can say this: It – the ICBM, the IRBM – we call them, we are still going ahead on those projects on top priority within the government, but incidentally a priority which was never accorded to the satellite program. The satellite program, having an entirely different purpose, even the scientists did not even think of it as a defense – or security instrument, and the only way that the Defense Department is in it at all is because one of them, the Navy, was called upon as the agency to have the sites and the mechanisms for putting it in the air.

(Question): Mr. President, Khrushchev claims we are now entering a period when conventional planes, bombers, and fighters, will be confined to museums because they are outmoded by the missiles which Russia claims she has now perfected; and Khrushchev's remarks would seem to indicate he wants us to believe that our Strategic Air Command is now outmoded. Do you think that SAC is outmoded?

The President: No, I believe it would be dangerous to predict what science is going to do in the next twenty years, but it is going to be a very considerable time in this realm, just as in any other, before the old is completely replaced by the new, and even then it will be a question of comparative costs and accuracy of methods of delivery.

It is going to be a long-term. It is not revolutionary, a revolutionary process that will take place in the re-equipping of defense forces, it will be an evolutionary.

(Question): Mr. President, do you think our scientists made a mistake in not recognizing that we were, in effect, with Russia – in a race with Russia in launching this satellite, and not asking you for top priority and more money to speed up the program?

The President: Well, no I don't, because as – even yet, let's remember this: The value of that satellite around the Earth, going around the Earth, is still problematical, and you must remember the evolution that our people went through and the evolution that the others went through.

From 1945, when the Russians captured all of the German scientists in Peenemünde, which was their great laboratory and experimental grounds for the production of the ballistic missiles they used in WW-II, they have centered their attention on the ballistic missile.

Originally, our people seemed to be more interested in the aerodynamic missile, and we have a history of – going back for quite a ways – in modest research in the intercontinental ballistic missile, but until there were very great developments in the atomic bomb, it did not look profitable and economical to pursue that course very much, and our people did not go into it very earnestly until somewhere along about 1953, I think.

Now, so far as this satellite itself is concerned, if we were doing it for science and not for security, which we were doing, I don't know of any reasons why the scientists should have come in and urged that we do this before anybody else could.

Now, quite naturally, you will say, "Well, the Soviets gained a great psychological advantage throughout the world," and I think in the political sense that is possibly true. But in the scientific sense it is not true, except for the proof of the one thing, that they have got the propellants and the projectors that will put these things in the air.

(Question): Mr. President, could you give the public any assurance that our own satellite program will be brought up to par with Russia, or possibly improve on it?

The President: Well now, let's get this straight: I am not a scientist. I go to such men as Dr. Waterman, Dr. Bronk, Dr. Lawrence, all of the great scientists of this country, and they assured me back in the spring, I think it was, of 1955, this could be done, and they asked for a very modest sum of money compared to the sums we were spending on other research. So, in view of the fact that, as I said before, this was basic research, I approved it.

Now, the satellite that we are planning to put in the air will certainly provide much more information, if it operates successfully throughout, according to plan, it will provide much more information than this one can.

(Question): Mr. President, you have spoken of the scientific aspects of the satellite. Do you think that it has immense significance, the satellite, immense significance in surveillance of other countries, and leading to space platforms which could be used for rockets?

The President: Not at this time, No, there is no – there is – suddenly all America seems to become scientists, and I am hearing many, many ideas. (Laughter) And I think that within time, given time, satellites will be able to transmit to the Earth some kind of information with respect to what they see on the Earth or what they find on the Earth. But I think that that period is a long ways off when you stop to consider that even now, and apparently they have, the Russians, under a dictatorial society, where they had some of the finest scientists in the world, who have for many years been working on it, apparently from what they say, they have put one small ball in the air.

I don't – I wouldn't believe that at this moment you have to fear the intelligence aspects of this.

(Question): Mr. President, considering what we know of Russia's progress in the missile field —

The President: Yes?

(Question): — are you satisfied with our own progress in that field, or do you feel there have been unnecessary delays in our development of missiles?

The President: I can't say there has been unnecessary delay. I know that from the time that I came here and got into the thing earnestly, we have done everything I can think of and know… I will say this: Generally speaking, they have – more than one scientist – has told me we were actually spending some money where it was doing no good.

Now the great reason for spending more money is because of the number of strings you put on you bow. In almost every field we have had several types and kinds working ahead to find which would be the more successful, so I can't say that I am dissatisfied.

I can say this: I wish we were further ahead and knew more as to accuracy and to the erosion and to the heat-resistant qualities of metals and all the other things we have to know about. I wish we knew more about it at this moment.

(Question): Is there some way that could have been done, something that could have been done that wasn't done?

The President: Well, I'll tell you, shortly after I came here, I immediately assembled a group of scientists, through the Defense Secretary, to study the whole thing, and to give us something on which we could proceed with confidence, or at least pursuing the greatest possibilities according to scientific conclusions.

That we have done, and I think we have done it very earnestly, with a great deal of expense, a great deal of time and effort, and I don't know what we could have done more.

(Question): Mr. President, could you give us, sir, the American story, that is, this government's version of the incident that Mr. Khrushchev described to Mr. Reston in his interview when the Soviet government put forth a feeler as to whether or not Marshal Zhukov would be welcomed in this country, and according to Mr. Khrushchev were rebuffed?

The President: Well, I will say this about the rebuff, I know nothing. If there was any committed, I am sure it was unintentional.

Now, what happened: You will recall somebody in one of these meetings asked me whether I thought that a meeting between Mr. Wilson and Marshal Zhukov might produce anything useful, and I said it might, and that later – I was talking to the Secretary about it, and he said it was a hypothetical question and got a hypothetical answer.

I don't know whether it would do any good or not; and he said, "Well, there's this one thing about it, we have got to beware" – and of course this we all know – "of bilateral talks when you have allies and comrades in very great ventures like we have in NATO, and so on." And at that moment talks were going on in Britain on the disarmament business on multilateral basis, and it would have probably had a very bad interpretation in the world if such things at that time had taken place.

The only follow up that I know of, was somebody asked the State Department – it may have been the ambassador, I don't know, somebody asked the State Department – well, was this a serious thing? Was this an invitation? And he said exactly what I just told you, it was merely a hypothetical answer to a hypothetical question.

So far as I know, there has never been any additional activity in connection with it.

At this point the reporters turned to other issues and shortly thereafter the conference was concluded.

3.4 DR. KILLIAN'S SECRET MEMORANDUM

The following SECRET (declassified 2/13/1974) memorandum dated December 28, 1957 to the President by Dr. James Rhyne Killian illustrates how Eisenhower utilized his scientific advisors following the Sputnik incident.

In 1956 Dr. Killian, who had been the President of MIT since 1948, was appointed by Eisenhower to chair his Foreign Intelligence Advisory Board. The following year he became Eisenhower's Special Assistant for Science and Technology, in effect making him the first true Presidential Science Advisor. It is also significant to note that the memorandum mentioned five extraordinary men who assisted Dr. Killian with his analysis and conclusions, namely:

- Dr. James Brown Fisk – an atomic physicist from Bell Labs and developer of high frequency radar during the war.
- Dr. Herbert Frank York – a Manhattan Project physicist and Director of the Lawrence Livermore Laboratory (at age 28). He was the first Chief of ARPA.
- Dr. George B. Kistiakowsky – a Manhattan Project physical chemist and developer of the explosive lens for the atomic bomb. He later took over from Dr. Killian as the Chairman of the President's Scientific Advisory Committee (PSAC).
- Dr. James McRae – Army Scientific Advisory Panel.
- Dr. Emanuel Piore – physicist and former head of Naval Research. He was the first director of the IBM Research Center.

There were about two dozen distinguished people on the full Killian committee. Others not mentioned in the following memorandum include: Dr. Edward Purcell, a Harvard physicist and Nobel laureate, General James H. Doolittle, who was then chairman of NACA, and Dr. Edwin Land of the Polaroid Corporation.

Although the memorandum seems to focus on the military, the implications for the satellite programs were obvious in that those programs relied on military missiles to obtain orbit. It was clear that the panel favored the Jupiter C, which was a variant of the Army's Redstone, rather than the Vanguard missile.

It is important to bear in mind that this briefing occurred two months after Sputnik and nine months prior to the creation of NASA. The President believed that the civilian space program should be separate from the military program, even though the civilian program was reliant on military missiles.

Dr. Killian was a science administrator, not a scientist, although he had been awarded many doctorates. He laid the groundwork for the creation of what he defined as "a civilian-directed and civilian-oriented space science and exploration program providing research support for military aeronautics and space programs." It was largely through the Killian committee that the outlines of a civilian space program were defined. After meeting with the President to explore ideas, he formerly made a recommendation that the civilian agency be built around the National Advisory Committee for Aeronautics (NACA). The President approved the recommendation, and the due legislation was drafted. Hence it can justifiably be said that Dr. Killian was an early architect of the civilian space agency now known as NASA.

SECRET

THE WHITE HOUSE
WASHINGTON

December 28, 1957

MEMORANDUM FOR THE PRESIDENT

I have the honor to present herewith a brief progress report on
the U. S. missile and satellite programs. This report reflects the
judgment and views of panels of scientists and engineers which have
reviewed the programs at my request. The conclusions which they
have reached are concurred in by Dr. J. B. Fisk, who is a consultant
to my office, and myself.

The Missile Programs

It is our judgment that <u>technically</u> our missile development is
proceeding in a satisfactory manner. Although it is probably true
that we are at present behind the Soviets, we are in this position
largely because we started much later and not because of inferior
technology. Our technological progress in the missile field, in fact,
has been impressive.

The so-called failures of flight test vehicles, to which much publicity has been given, are normal and unavoidable occurrences in the
development of complex mechanisms, many functions of which can
be tested only in flight. A flight test, which to a casual observer
appears to have been a failure, provides a great deal of necessary
information to the test crew. We shall continue to have such occa-
sional "failures" as long as we pursue a vigorous search for more
advanced missiles.

At present, the development programs of the IRBM are moving
ahead very rapidly. There have been flights of both the JUPITER and
the THOR which were complete technical successes. The regular
production of IRBM's is soon to begin. In the immediate future a
report will be prepared on POLARIS, the ship-based IRBM. In the
development of ICBM's, the progress is also good and the recent
successful flight test of the ATLAS gives confidence in the future of
this missile. Another more advanced ICBM, the TITAN, is reaching
the stage where initial flight testing will begin in 1958. We are confident that the U. S. has ample technical competence in our ballistic
missile technical groups to achieve satisfactory operational missile
systems at an early date.

Along with the testing and production of the missiles currently
programmed, we must devote major efforts to achieving improved
missiles for the future. During the next few years we must emphasize improvement of missiles by the application of technology which
has already been achieved through research. For the longer-range
future, our basic research and development of entirely new methods
and techniques for missiles must be vigorously and imaginatively
pursued. We attach great importance to boldness in our planning for
these future missiles and the initiation and successful carrying through
of fundamental and exploratory work.

The Panel which has reached these conclusions and which continues
to pay close attention to our missile program includes the following:
Dr. George Kistiakowsky (Chairman); Dr. James McRae; and Dr. Herbert York.

<u>The Satellite Programs</u>

We believe that the presently-planned VANGUARD program, consisting of four additional test vehicle firings, followed by six satellite
launching attempts, has only an even chance of success during the year
1958. This is a statement of probability which does not rule out the
possibility that there will be a successful launching of a VANGUARD
satellite during this period. In the event of a successful launching,
further successes will become more probable.

The recent unsuccessful launching of a test VANGUARD system
might have been a failure no matter how thoroughly the vehicle had
been prepared for an all-out test. The complexity of the vehicle and
the limitations of time and quantities which have characterized the
VANGUARD program, greatly increased the probability of this failure.

In view of this technical situation and in view of the uncertainties
surrounding all such tests, we feel that advance publicity should not be
given to test firings.

Fortunately, we have a satellite program as an alternative to VANGUARD.
This is the JUPITER-C program of two launchings which the Department
of Defense has authorized the Army to take through its agencies, ABMA
and JPL.

We believe that <u>each</u> attempt to launch a satellite by the JUPITER-C
group has, in terms of probability, an even chance of success, a sub-
stantially greater probability than that inherent in the VANGUARD program.

Because we have confidence in the JUPITER-C International Geophysical Year satellite program, we urge that it be amply supported and encouraged. In a separate communication to the Secretary of Defense we have made specific recommendations as to ways in which this program can be appropriately expanded.

It is our conclusion that there should be no expansion of the VANGUARD program for International Geophysical Year's needs and that additional resources be made available to the JUPITER-C program in order best to insure the successful launching of satellites. Suggestions, for example, have been made for authorizing six more launching vehicles in the VANGUARD program in order to increase the chances of eventual success in the International Geophysical Year. Considering firing schedules and available manpower, we doubt that such a program would be able to contribute much to our satellite program during the International Geophysical Year.

The Panel which has evaluated our satellite programs includes: Dr. Herbert York (Chairman); Dr. George Kistiakowsky; and Dr. Emanuel Piore.

J. R. Killian, Jr.

The President had many more meetings, of course, but one can tell from the above that he knew exactly what he wanted and his advisors were providing him with information that only convinced him that he was right. He wouldn't combine the military aspects of space with the civilian aspects, as the Soviets had done. In accepting that they had beaten us to space, he knew that our goals were clear and long term. He also knew that we had great military projects in the pipeline, and was very supportive of both the rocket and reconnaissance satellite programs. This was after all, the "Cold War" and he wasn't about to short change the military.

Eisenhower's concern and focus now was to guide the civilian space efforts beyond those of the International Geophysical Year. He knew that the military wanted the whole space program, but his task now moved to convincing the public and the Congress that what was needed was a civilian space program. While he was a military man and had great military men as advisors, he knew that space shouldn't be solely a military competition to dominate space.

The next chapters will explain the interplay between the military, the intelligence community, and the Congress. It is remarkable how "relatively" cooperative all of the diverse organizations were and how rapidly the civilian space program achieved fruition. This truly is a testament to President Eisenhower's leadership.

4

The National Security Council

As an introduction to how the National Security Council (NSC) handled the news of Sputnik, it should be noted that President Eisenhower not only made extensive use of the NSC and created a structured system of integrating policy reviews, but also promoted discussion and debate among the advisors. His NSC had five statutory members: The President, Vice President, Secretaries of State and Defense, and Director of the Office of Defense Mobilization. Other Cabinet members and advisors attended and participated as appropriate to the subject. The agenda included regular briefings by the Director of Center Intelligence. An Operations Coordinating Board brought the subjects to the NSC and implemented its actions. For the meeting about the launch of Sputnik, the key speakers were the Secretary of State John Foster Dulles and his brother Allen Dulles, Director of the Central Intelligence Agency.

When the NSC met on Thursday, October 10, 1957, the day following the President's press conference, it appeared that almost everybody was there including the Cabinet members, their technical assistants, and senior personnel involved with satellites and missiles. The minutes of the TOP SECRET meeting are marked, EYES ONLY but were declassified, with deletions, in 1992. The following describes some comments and quotations.

From CIA Director Allen Dulles: At 1930 hours on October 4 the Soviets had fired their Earth satellite from the Tyuratam range. The orbital path crossed approximately over the range's other end at Klyuchi. Two hours after the successful orbiting of the Earth satellite and after the second circuit of the Earth by the satellite, the Soviets announced their achievement. This delay in the announcement was in line with the previous statements of the Soviet Union that they would not announce an attempt to orbit their satellite until they had been assured that the orbiting had been successful. Moreover, all of the indications available to the intelligence community prior to the actual launching of the satellite pointed to the fact that the Soviets were preparing either to fire an intercontinental ballistic missile or to launch an Earth satellite.

Mr. Dulles then explained that the actual launching of the Earth satellite had not come as a surprise. Indeed, as early as the previous November the intelligence community had estimated that the Soviets would be capable of launching an Earth satellite any time after

© Springer International Publishing Switzerland 2016
M. von Ehrenfried, *The Birth of NASA*, Springer Praxis Books,
DOI 10.1007/978-3-319-28428-6_4

November 1957. Information on the Earth satellite itself was rather sparse, but it was believe to weigh between 165 and 185 lbs.

Mr. Dulles continued by pointing out that the Soviets had joined together their ICBM and Earth satellite programs, which helped to explain the speed of the Soviet launching of its Earth satellite. It was not yet clear whether the satellite was sending out encoded messages. Further launchings of Soviet Earth satellites should be expected during the International Geophysical Year because the Soviets had said they would launch between six and thirteen such satellites.

Mr. Dulles then turned to the world's reaction to the Soviet achievement. He first pointed out that Mr. Khrushchev had moved all his propaganda guns into place. The launching of an Earth satellite was one of a trilogy of propaganda moves; the other two being the announcement of the successful testing of an ICBM and the recent test of a large-scale hydrogen bomb at Novaya Zemlya. Moreover, there had been another Soviet test late the previous night at the same site.

Mr. Dulles responded to a remark made by Mr. Khrushchev that eventually military aircraft would be consigned to museums. Mr. Dulles pointed out that U.S. intelligence hadn't observed as many Soviet heavy bombers on airfields as had been expected. He concluded his remarks by emphasizing that the Soviet Union was making a major propaganda effort which was exerting a very wide and deep impact.

Further discussion addressed the Vanguard Project, foreign government reactions and policy implications, and both the scientific and missile budgets. Mr. Dulles said that at least the Soviets had concentrated more heavily on the guided missiles field than the United States had, ever since 1945. The President, agreeing with Mr. Dulles, said the United States would not decide whether to make an all-out effort in the field of ballistic missiles until after Dr. James R. Killian reported to the National Security Council. He added that of course the Soviets were bound to be ahead of the United States in certain fields and in certain discoveries. The Council continued with detailed discussions of the DOD's missile program.

So what did the CIA know in 1957 and what were they telling the President? What effect did that have on the creation of NASA? Some answers are to be found in declassified documents. Reading them now, these reports seem very simplistic. Most are concerned with the launch of nuclear weapons and the ICBM capabilities of the Soviets. Bear in mind that the stage was set when the Soviets tested their first hydrogen bomb in 1953 and successfully tested an ICBM in 1957, some months ahead of launching Sputnik.

In March 1957 a CIA National Intelligence Estimate (NIE) warned that the "Soviet guided missile program is extensive and enjoys a very high priority; the USSR has the resources and capabilities to develop during this period advanced types of guided missile systems in all categories…"

American U-2s were flying over parts of the Soviet Union as early as 1956 but were getting information on bombers and ships more than on ICBMs. Before the Gary Powers incident in 1960, the U-2s began showing nuclear test sites in Novaya Zemlya and Semipalatinsk and the missile launch facility at Tyuratam.

In December 1957 a TOP SECRET (now declassified) Scientific Intelligence Memo stated, "We believe that the Soviet ICBM and the Soviet Earth satellite vehicles probably utilized the same first and second stage propulsion system. The Soviet ICBM is estimated to have a gross weight of about 300,000 lbs. with a propulsion system consisting of paired nominal 100 metric ton thrust engines or an equivalent single engine in the first stage and a nominal 35 metric ton engine in the second stage. Additionally, although no evidence exists, we believe the Soviets are probably capable of adding a third propulsion stage to this system. The capability of such a staged propulsion system to orbit satellites or propel payloads to the Moon are approximately 5,000 lbs. to orbit or 400 lbs. to the Moon." The report went on to discuss launches of scientific and animal payloads.

This timeframe (late 1957) coincides with the CIA's push for a reconnaissance satellite. As it turned out, the group which the President asked to explore the creation of NASA was the same Scientific Advisory Committee headed by Dr. James R. Killian, Jr., President of MIT. Thus the creation of the CORONA photoreconnaissance system and the seeds of NASA were planted at the same time and by some of the same people!

The President had solved the problem over the debate about the number and capabilities of Soviet missiles targeting the United States and the debate over who should control the civilian space program. In February 1958 he approved the CORONA program, and the next month he accepted the recommendation of Dr. Killian in favor of a civilian space program. The CIA had their spy satellite, the DOD had their missiles, and the Space Act in July 1958 authorized the creation of a new agency to run the civilian space program.

5

The Congress

Members of both houses of Congress were ready to act, but they needed some direction from the President. On November 21, 1957, less than two months after the launch of Sputnik, Eisenhower established the President's Scientific Advisory Committee (PSAC). It was chaired by Dr. James R. Killian, Jr. of the Massachusetts Institute of Technology (MIT). It included 18 distinguished scientists, engineers, and policy makers including the current NACA Chairman, Gen. Doolittle. Dr. Killian had two subcommittees, one on policy headed by Dr. Edward H. Purcell who was a Bell Telephone Executive Vice President, and the other an organization headed by Dr. James B. Fisk of Harvard University. Both subcommittee chairmen were physicists, and they knew what the President wanted.

The Fisk subcommittee report came out in February 1958 and proposed a new agency built around NACA which would emphasize peaceful, civilian-controlled research and development. This was forwarded to the President by the White House Advisory Committee on Government Organization as a formal recommendation and he approved it on March 5. The PSAC produced more rationale and reports that added to the information the President needed in order to make a formal message to Congress, which he did on April 2. He made it clear that outer space was for scientific exploration and not military exploitation. He called for a "National Aeronautical and Space Agency."

It took the Bureau of the Budget a mere 12 days to put some numbers to the proposal, and on April 14 the Administration sent the Bill to the Democratic-controlled Congress, whose special committees promptly began hearings.

In the meantime, all the NACA centers began evaluating what they could offer, and what it might mean to their organizations and staffs. Many of the organizations realized that some of what they were doing was directly or indirectly related to spaceflight. Some of the people had been researchers all their careers and had no program development or operational experience. Some were uncomfortable with the idea of worrying about managing programs, working with contractors, and competing for resources, but others relished these challenges.

© Springer International Publishing Switzerland 2016
M. von Ehrenfried, *The Birth of NASA*, Springer Praxis Books,
DOI 10.1007/978-3-319-28428-6_5

It took the 85th Congress just three months to come up with a Bill that wound up an agency that had been in existence for over 40 years and created a new one; all in a single Bill. On July 29, 1958, with a big smile on his face, the President said,

I have this day, signed H.R.12575 the National Aeronautics and Space Act of 1958.

The enactment of this legislation is an historic step, further equipping the United States for leadership in the space age. I wish to commend the Congress for the promptness with which it has created the organization and provided the authority needed for effective national effort in the fields of aeronautics and space exploration.

The new Act contains one provision that requires comment. Section 205 authorizes cooperation with the other nations and groups of nations in work done pursuant to the Act and in the peaceful application of the results of such work, pursuant to international treaties entered into by the President with the advice and consent of the Senate. I regard this section merely as recognizing that international treaties may be made in this field, and as not precluding, in appropriate cases, less formal arrangements for cooperation. To construe the section otherwise would raise substantial constitutional questions.

The present National Advisory Committee for Aeronautics (NACA), with its large and competent staff and well-equipped laboratories, will provide the nucleus for the NASA. The NACA has an established record of research performance and of cooperation with the Armed Services. The combination of space exploration responsibilities with the NACA aeronautical research functions is a natural evolution.

The enactment of the law establishing the NACA in 1915 proved a decisive step in the advancement of our civil and military aviation. The Aeronautics and Space Act of 1958 should have an even greater impact on our future.

6

The Department of Defense and Other Agencies

The organizations within the Department of Defense (DOD) and their influence upon the STG started long before Sputnik. The Army, Navy and Air Force were independently working on a number of programs that were, in one way or another, related to launch vehicles, although the term used at that time was "missiles." In 1957, there wasn't much emphasis on a "man-rated" missile for putting a "pilot" into orbit.

Charles E. Wilson served as Secretary of Defense from January 20, 1953 to October 8, 1957, leaving just four days after Sputnik. What Secretary Wilson did while in office had a significant impact on all the armed services. He and President Eisenhower were committed to reorganizing the DOD. Secretary Wilson's "New Look" defense concept was a philosophy of maintaining a staunch defense whilst cutting costs and balancing the budget. It was based on the premise that any new world war with the Soviets would be a nuclear war. This put an emphasis on the best means to deliver nuclear weapons; both short and long range. This policy was controversial, to say the least. The Army and Navy felt that the U.S. should prepare for limited war as well as a nuclear retaliatory war. The "New Look" concept had the effect of increasing the Air Force's role and budget at the expense of the Army and Navy's.

In 1956, Secretary Wilson pointed out that the services had eight categories of missiles for various tasks and couldn't agree on the missions. In November 1956 he limited the Army to small aircraft and only surface-to-service missiles whose range did not exceed 200 miles. The Air Force was given responsibility for land-based intermediate range ballistic missiles (IRBM) and the Navy got corresponding ship-based systems.

Four days after Sputnik, Neil H. McElroy took over as Secretary of Defense. It wasn't good timing! His immediate observation was that the United States was ahead of the Soviets in terms of IRBMs, and that a rapid deployment to the United Kingdom and Europe would counter any ICBM threat posed by the Soviets. He ordered production and deployment of the Thor missile developed by the Air Force and the Jupiter missile which had been developed by the Army. He also ordered the accelerated development of the solid-fuel Polaris IRBM for the Navy and the liquid-fueled Atlas and Titan ICBMs for the Air Force. In addition, he authorized the Air Force to begin development of the solid-fueled Minuteman ICBM to be deployed from silos.

© Springer International Publishing Switzerland 2016
M. von Ehrenfried, *The Birth of NASA*, Springer Praxis Books,
DOI 10.1007/978-3-319-28428-6_6

After Sputnik, the democratically controlled Congress wanted to spend more on defense but Eisenhower and McElroy decided to stick to the planned budget and reassign some resources to development, production, and deployment of missiles.

By the time McElroy left office after two years, he had created the Defense Reorganization Act of 1958 and had more clearly defined the roles and responsibilities of the armed services in the development and production of missiles. In addition, the responsibility of the Director of Research and Engineering was elevated to an Assistant Secretary of Defense.

The following sections describe the reorganization of the DOD to deal with the new threat in terms of missiles and orbital mission concepts, and also show the intent of the DOD to grab the missions that the President intended for NASA.

6.1   ADVANCED RESEARCH PROJECTS AGENCY

The Advanced Research Project Agency (ARPA) was created by President Eisenhower on February 2, 1958. As an agency of the DOD, its purpose was to prevent further technological surprises on the scale of Sputnik. The launch clearly showed that the Soviets were ahead of the U.S. in space technology and capability. The President and the DOD didn't want that to happen again. However, this new agency was just an interim measure, pending the establishment of a civilian-controlled space organization. Eisenhower had already asked his Scientific Advisory Committee (PSAC), chaired by James R. Killian, Jr. of MIT, to study space policy and make a recommendation for organizing a national program in space science.

In response to the evident urgency, ARPA investigated how the U.S. might place a man into orbit at the earliest possible date. One plan involved using different Thor and Vanguard stages, and perhaps a new rocket. The Air Force's Air Research and Development Command had a far more ambitious idea; it was seeking to send men to the Moon.

After Sputnik (and before NASA) there was an Air Force program intended to place a man into outer space ahead of the Soviets. The space researchers at Holloman AFB in New Mexico had sent men to extreme altitudes in balloons. They had some good experience with crews and cabins; this was transferrable to spaceflight. As early as February 1958, General Curtis LeMay, the Air Force Chief of Staff, ordered a study of a military manned satellite project. ARPA would oversee the work on what became "Man-in-Space-Soonest." For the next several months the Air Force and its contractors explored the concept.

By June 25, 1958, the Air Force had even selected nine crewmen, many with experience of the Bell/Douglas X-Series aircraft. (Two of these men would later become astronauts; Joseph A. Walker in an X-15 and Neil A. Armstrong, who was the only one of the nine to join NASA.) In addition, meetings were held to resolve technical issues such as the choice of the booster, even development of a new booster; the weight of the space vehicle; and re-entry problems. By July, ARPA, still concerned about the issues and costs, was reluctant to give a go-ahead on the Man-in-Space-Soonest project; especially knowing what Eisenhower wanted.

By the time the Space Act was signed by the President on July 29, 1958, it was apparent to ARPA that the President was committed to a civilian space program. It was also very clear to Max Faget, who represented NACA on an ARPA Man in Space Panel, that ARPA was starting to advise NACA on space matters, even though its traditional role was to provide research and advice to the Air Force and government sponsors of NACA research. Max Faget also knew of the space studies and tests underway at the NACA Langley and Lewis Research Laboratories even before the creation of NASA.

As it turned out, the Man-in-Space-Soonest program was canceled a couple of days after Eisenhower signed the Space Act, transferring the manned space program to the new NASA agency. Now a new National Aeronautics and Space Council would advise the President of the new organization's plans and activities. ARPA focused on other military matters between the services. The Air Force continued with another military space program called Dyna-Soar until this was canceled in December, 1963.

But there was still much work to do after NASA was formed to coordinate interservice and interagency plans and procedures. NASA Administrator T. Keith Glennan and ARPA Director Roy W. Johnson agreed on the bare outline of a joint program. In September the Joint NASA-ARPA Manned Satellite Panel of eight members was established; six from the former NACA and two from ARPA. The panel came up with a two-and-one-half page paper "Objectives and Basic Plan" for a manned satellite by the first week of NASA's existence. This included the objectives, mission, and configuration of the capsule. Hence ARPA assisted NASA from the beginning in defining the basic outline of what would later be called Project Mercury.

6.2   ARMY BALLISTIC MISSILE AGENCY

The Army Ballistic Missile Agency (ABMA) was established on February 1, 1956; well before Sputnik. The Army's focus was on developing an intermediate range ballistic missile (IRBM) at the Redstone Arsenal, where the ABMA was located.

Although the role of Wernher von Braun's team of German missile experts is well known, perhaps what isn't so well known is the number of missiles that were being worked on. In the years after the Berlin Airlift, the Korean War, the Hungarian uprising, and the Suez crisis, the Arsenal was developing a dozen types of missiles for surface-to-surface and surface-to-air roles, using both liquid and solid propellants. Missile technology was rapidly advancing. Even prior to Sputnik, the development of the Redstone missile and an improved Redstone called the Jupiter were well underway. The first launch of a Redstone from Cape Canaveral, Florida was August 20, 1953. Although not capable of launching a satellite into orbit by itself, the Redstone proved capable of firing a man into outer space in 1961.

In 1954, von Braun proposed placing a satellite into orbit using the Redstone that employed clusters of small solid-fuel rockets as its upper stages. This proposal, called Project Orbiter, was rejected by the DOD the following year. The Naval Research Laboratory had already been given the task of managing such a mission as a civilian project as part of the 1957–1958 International Geophysical Year. The Navy intended to use the Vanguard missile built by the Glenn L. Martin Company, which had previously developed the Viking sounding rocket.

On February 1, 1956, General John B. Medaris was assigned to head the new Army Ballistic Missile Agency with instructions to turn the experimental Redstone into an operational weapon and to develop the new longer range Jupiter IRBM. They forged ahead with the three successful Jupiter C launches on the Atlantic Missile Range in late 1956 and mid-1957; all prior to Sputnik. As these were Redstone-related launches to test the re-entry characteristics of the warhead that was intended for the real Jupiter missile, they gave NACA (and the future NASA/STG) people confidence in the capability and reliability of the Redstone as a future manned-rated vehicle.

The Army was "chaffing at the bit" to be given a greater mission responsibility, so when the Vanguard missile failed so spectacularly on December 6, 1957 (two months after Sputnik), the Army was given permission to use the Jupiter C to launch America's first satellite, Explorer 1, created by the Jet Propulsion Laboratory of the California Institute of Technology, and this was achieved on January 31, 1958. This four-stage variant of the Jupiter C was called Juno I in order to make it sound more like a civilian vehicle than a military missile. It was capable of placing a payload into orbit, but that honor was politically reserved for Vanguard.

Two months later in March 1958, ABMA was placed under the new Army Ordnance Missile Command (AOMC) together with the Redstone Arsenal itself, the Jet Propulsion Laboratory, the White Sands Proving Ground in New Mexico, and the Army Rocket and Guided Missile Agency (ARGMA). The ABMA had the Saturn heavy lift launch vehicle in work, but lost control of that vehicle. After NASA was formed in July of that year, the seeds of reorganization were again in the works, and finally, on July 1, 1960, the AOMC space related missions and most of its staff, facilities and equipment were transferred to the NASA George C. Marshall Space Flight Center (MSFC) at the Redstone Arsenal near Huntsville, Alabama, under the directorship of Wernher von Braun.

ABMA/AOMC was subsequently reorganized as the Army Missile Command. It continues its role in advancing both ground and air military missile systems including today's Patriot systems.

6.3   U.S. AIR FORCE

The Air Force had been working on an ICBM in earnest since the Soviets detonated their hydrogen bomb on August 23, 1953, much earlier than the Americans had expected. Trevor Gardner, Assistant to the Secretary of the Air Force, set up the Strategic Missiles Evaluation Group composed of nuclear physicists and missile experts. Informally known as the "Teapot Committee," this was headed by the mathematician Dr. John von Neumann. It also included scientists and engineers from major universities and contractors. They concluded that it would soon be practicable to build smaller, lighter, hydrogen-fusion warheads. These, in turn, would reduce the size of rocket nose cones and propellant loads. They also predicted that the vastly greater yields from such thermonuclear explosions would reduce the need for precise missile accuracy. The military liaison to this group was Bernard Schriever, at that time an Air Force colonel.

Amongst other recommendations, the Committee's report on February 10, 1954 called for a crash program to develop the Atlas missile. It emphasized that a new management organization would be needed which would be free of excessive detailed regulation by government agencies. There was also a parallel study by the Rand Corporation that essentially said the same thing but also estimated that an operational capability could be attained as early as 1958 if enough money and priority were provided. A full-scale prototype of the Atlas A flew on June 11, 1957 but this didn't have the central sustainer engine, and was nothing like the Atlas D that would launch the Mercury spacecraft years later.

Five days after Sputnik, the Air Force Scientific Advisory Board urged the development of a "second generation" of ICBMs that could be used as space boosters and even proposed manned lunar missions by the Air Force, as well as reconnaissance and other satellites. The Secretary of the Air Force appointed a 56 member committee headed by Dr. Edward Teller to recommend a unified space program under Air Force leadership.

A month later, after the Soviets launched their second Sputnik, the Air Force ordered the Air Research and Development Command (ARDC) to prepare an astronautics program. They already had one in work but it was a 15 year plan including the Dyna-Soar program; now it was a 5 year plan. The Air Force assumed that NACA would play a key role as a supplier of research data; as had been the case for four decades. NACA's Hugh Dryden agreed to participate, but advised the Air Force that NACA was working on their own manned capsule designs. Dryden knew NACA would need the Atlas to boost such a capsule into orbit.

It was always planned that the manned spaceflight launches would take place at what was then known as the Cape Canaveral Missile Annex and is now called the Cape Canaveral Air Force Station. It is an installation of the Air Force Space Command's 45th Space Wing that is headquartered at nearby Patrick Air Force Base. NASA's first two manned Redstone launches were from Launch Complex 5 and all of the manned Mercury Atlas launches were from Launch Complex 14.*

Unmanned qualification flights with the Mercury capsule were made in 1960 and 1961. The November 29, 1961 flight of the chimpanzee Enos in orbit would be the last for the STG group at Langley before starting the move to Houston. The first manned orbital flight of Mercury on the now "man-rated" Atlas D was on February 20, 1962.

6.4   U.S. NAVY

You'd think that the connection between the Navy and the STG was primarily the recovery of the Mercury capsules, but the connection goes back to the role the Navy was given after the war to examine scientific applications of the German V-2 missile. This led to the development of a series of sounding rockets called Viking. The Naval Research Laboratory (NRL) was given the responsibility to build advanced liquid-fuel rockets to study the atmosphere and discover how to predict bad weather for the Navy fleet. The Glenn L. Martin Company was given the contract to build the Viking rocket, which was derived from the V-2 but incorporated many improvements.

Twelve Vikings were fired between 1949 and 1955; all but one from the White Sands Missile Range in New Mexico, with the other one from the deck of the USS *Norton Sound* on May 11, 1950. The Viking was deemed to be very successful in both scientific and technological terms.

---

* See the front cover image and its explanation at the end of this volume.

This experience motivated scientists to propose a three-stage rocket which could launch a scientific payload into orbit as part of the International Geophysical Year that would run from mid-1957 through to the end of 1958. Both the Air Force and the Army wanted this mission, but it was decided in 1955 that the NRL's proposal to use a successful scientific rocket, payload and tracking network was the way to go. Project Vanguard was given to the NRL, monitored by the DOD, with funding from the National Science Foundation. The Martin Company would remain as the prime contractor.

Also in 1955, the NRL was investigating how to monitor the satellite using various methods. This led to a proposal to build a network of stations. The Minitrack network of eleven stations was completed in 1957; just in time to track Sputnik. The system went through some changes but within a year became the basis for the NASA Spacecraft Tracking and Data Acquisition Network (STADAN). This would in turn become the Mercury (and later Manned) Space Flight Network (MSFN) that was the responsibility of the Goddard Space Flight Center in Maryland. By 1961, STG flight controllers were manning these and other tracking stations in support of the Mercury flights. See Sections 7.4 and 14.3 on how the Minitrack system expanded into the MSFN.

History vividly records the Vanguard failure of the December 6, 1957 but the satellite which was launched on March 17, 1958 as Vanguard 1 is still in orbit after more than half a century; it has outlived the first two Sputniks and Explorer 1, which have fallen back into the atmosphere and burned up.

But that's just the start of the Navy's story! They would go on to provide Fleet Operational support for the recovery of the astronauts from just about anywhere in the world. See Section 9.4.4 Recovery Operations Branch.

6.5   DEPARTMENT OF STATE

One doesn't immediately think of the Department of State in the context of manned space-flight, but many of the tracking stations were on foreign soil and the U.S. had to secure arrangements with many different governments in order to construct and operate those stations, and that was their bailiwick. And since some of the governments in question were openly anti-American, the diplomatic challenge was serious. On top of that, State Department personnel sometimes had to deal with angry mobs. Indeed, on at least one occasion, a representative took up position in the doorway of a building to prevent a mob from gaining entry.

The Department of State got involved in the space business when the Minitrack system was proposed for the International Geophysical Year. By 1955, there were plans to build stations in six South American locations and there were stations planned down the Atlantic Missile Range on islands owned by the British. By October 1, 1957 the system was operational and three days later was surprisingly given the opportunity to track Sputnik. In November 1957, the eleventh station, Woomera, Australia was added. The network would change over the years, with some stations being dropped and others added.

The role of the State Department expanded for Project Mercury, when the tracking network was extended beyond the Minitrack network to other countries. It was instrumental in securing international cooperation with Third World countries as well as with U.S. allies.

For example, negotiations with Mexico were very difficult and at first seemed likely to be fruitless. In 1959, however, President Eisenhower asked his brother Milton, who was on a trip to Mexico, to make a personal appeal to President Lopez Mateo to open negotiations. Later that summer, there were even talks at the White House with NASA Administrator Keith Glennan, President Mateo, and Ambassador Antonio Carrillo. In January 1960 representatives of NASA's Goddard Space Flight Center went to the U.S. Embassy in Mexico to determine whether Mexican officials were even interested in constructing and operating a tracking station at Guaymas in northwestern Mexico.

Finally, on April 12, 1960 an agreement was signed and fourteen months later the Guaymas tracking station was opened on June 26, 1961. This is just one example of the role of the State Department in dealing with just one country. The Guaymas station went on to support Mercury, Gemini, and Apollo. Another example was the State Department's role in expediting the entry of the Canadian AVRO engineers into the U.S., including transferring their security clearances.

NASA needed the Spanish-owned Grand Canary Islands for a critical tracking station for the confirmation of a spacecraft achieving orbit. In the event of a failure to achieve orbit, the station would be able to assist with the abort situation and the determination of the contingency landing area. The State Department was instrumental in helping to secure this agreement. When NASA approached Madrid on September 10, 1959, William Fraleigh, the First Secretary of the Consul Political Office, was advised the situation was "rather delicate." The Spanish government didn't want the station to have any connection with the U.S. airbase. The agreement that was reached required NASA to tear down the dwellings of migrant farmers and build them new ones off the site.

The case of the tracking station at Kano, Nigeria, is another example of the role of the State Department in achieving an agreement. At the time, Nigeria was a British colony and they had told the local government that the U.S. facility would contribute to the scientific knowledge of the world, but a rumor was circulated that its activities would be related to French atomic bomb experiments. However, NASA Goddard representatives and the U.S. Embassy, with the help of Arnold W. Frutkin, formerly of the National Academy of Sciences and now the NASA Director of International Programs (whom the Emir trusted) convinced the Nigerian government to allow the tracking station to be built. An agreement was reached in the capital city of Lagos on October 19, 1960. The station remained in use until November 18, 1966, when the MSFN was revamped in readiness for Apollo missions.

Zanzibar is a unique story. Located on an island in the Indian Ocean it provided coverage for Project Mercury after the capsule left Kano behind. Zanzibar would be the last station to see the spacecraft before it set out across the vast Indian Ocean. Then part of a British protectorate, it is now part of Zanzania. An agreement was signed with the Sultan on October 14, 1960. With the completion of the Kano and Zanzibar stations in 1961, the MSFN was fully ready to support the orbital Mercury flights.

But it was a time of political instability in Zanzibar, with several pro-independence factions. In July 1963 the State Department warned of imminent potential riots pending the outcome of national elections. NASA Headquarters recommended the station reduce its staff to "caretaker" status by July 3 and develop an "emergency escape plan." They didn't have to evacuate at that time. However, a month after the British granted Zanzibar its independence in December there was a bloody coup. On January 14, 1964 NASA ordered the immediate evacuation of the station. The group assembled at the English Club to await evacuation. Rebels were breaking down doors of neighboring buildings and came to the English Club buildings. The State Department Charge d'Affairs, Fredrick Picard, stood at the front door and confronted the mob. Speaking to them in Swahili, he prevented them from entering. The incident was resolved without tragedy. The U.S. Navy evacuated the entire staff. There was no attempt to remove any station equipment, it was simply abandoned in place.

State Department agreements with British-owned sites such as Bermuda, and those in the British West Indies (Grand Bahamas and Grand Turk) were more easily negotiated. They even negotiated with the remote island of Canton, one of the Phoenix Islands set in the middle of the Pacific Ocean, just south of the equator. Langley Research Center representatives carried out a site survey and then negotiated with the local authorities. The site was constructed the following year and was operational for John Glenn's flight.

So the State Department played a crucial role in supporting NASA for Project Mercury, and continued to do so for subsequent missions. Langley Research Center members of the Tracking System Study Group (TSSG) worked with both the Goddard Space Flight Center and the State Department during the early years of selection and construction of the sites. That group moved from Langley to Goddard once that center was built in 1961. For more stories about NASA's tracking and data network, read the NASA special publication by Sunny Tsiao's entitled *Read You Loud and Clear*.

Part II
Creating the Space Team

7

Creation of the Space Task Group

7.1 NACA – FROM AERONAUTICS TO ASTRONAUTICS

The seeds of the Space Task Group (STG) were planted as early as 1952, when NACA engineers started to focus on flights in the upper atmosphere at altitudes up to 50 miles and at speeds in the range Mach 4 to Mach 10. Almost as an afterthought, it was resolved to devote a modest effort to flights above 50 miles and speeds up to escape velocity.

By 1954, NACA (the Ames, Lewis, and Langley laboratories, including the Wallops Island Station and the High Speed Flight Station), the Air Force, and the Navy were developing joint plans to expand the X-Series of aircraft to the X-15 rocket-powered aircraft. This was certainly a test bed for Project Mercury because the issues of stability and control, and aerodynamic heating, are common to both high speed aircraft and spacecraft. The NACA wind tunnels were studying a whole range of aerodynamic problems for years. The Army and Air Force weapons people were studying similar problems for missiles.

In 1945, Langley's Pilotless Aircraft Research Division (PARD) established a "Station" at Wallops Island on the Atlantic coast of Virginia, from where they launched rockets to evaluate the re-entry heating problems of payloads. Their data related to the re-entry heating of missile warheads as well as capsules. No less than 14 people from this unique group of engineers and other staff became the core of the 36 members of STG in October 1958.

In the mid-to-late 1950s NACA studied various re-entry shapes for a capsule throughout its range of flight characteristics and carried out extensive tests of materials with which to protect the capsule. There were two schools of thought about re-entry heating: One school investigated providing a "heat sink" and the other favored using "ablation" to dissipate heat. Various metals and new metal concepts were studied as possible heat sinks, along with methods of dissipating heat using composites.

© Springer International Publishing Switzerland 2016
M. von Ehrenfried, *The Birth of NASA*, Springer Praxis Books,
DOI 10.1007/978-3-319-28428-6_7

There was considerable "cross fertilization" among the various military, civilian, and aviation industry organizations. Other programs that were in the concept stages had similar aerodynamic and heating issues; for example the Polaris missile. By 1956, four manned aircraft barriers were broken: Mach 1, Mach 2, Mach 3, and an altitude of 100,000 feet. NACA scientists, engineers, and technicians were involved in all of these efforts, and on the basis of their experiences they were all potentially transferrable to a spaceflight project.

Even before the big decision to create NASA, Lewis engineers were commuting to Langley during the summer of 1958 to discuss with their counterparts the concepts and engineering of a manned satellite program. They discussed organizing a "Manned Ballistic Satellite Task Group." This was later shortened to "Space Task Group."

Finally, almost a year after Sputnik and after all the top level meetings from the President on down, the first NASA Administrator T. Keith Glennan declared, "…as of the close of business September 30, 1958, the National Aeronautics and Space Administration has been organized and is prepared to discharge the duties and exercise the powers conferred upon it." During the following week, the Joint NASA-ARPA Manned Satellite Panel laid out a two-and-a-half page report (backed up by a lot of detailed charts, tables, and diagrams) "Objectives and Basic Plan for the Manned Satellite." This paper required joint concurrence by Keith Glennan on behalf of NASA and Roy W. Johnson, Director of ARPA. The Panel also included Dr. Robert R. Gilruth, Max A. Faget, Alfred J. Eggers, Walter C. Williams, George M. Low, Warren J. North, Samuel Batdorf, and Robertson C. Younquist. The work by a lot of scientists and engineers the previous year made a unanimous decision straightforward.

Dr. Gilruth returned to Langley with the decision and reported to his Acting Center Director, Floyd L. Thompson, pointing out, "You know that guy (Glennan) told me to go ahead (with the manned space program) but he didn't tell me how. I haven't got an organization. I don't know how I'm supposed to do this." Thompson replied, "Well, why don't we just create the Space Task Group." And that is how the nucleus of the STG began to take shape in the last weeks of October 1958. They initially met in Langley's Unitary Flow Wind Tunnel, Building 1251.

By November 3, 1958 the STG was formalized in a memorandum by Dr. Gilruth, who had been appointed as its Director. He was the logical choice because he was the Assistant Director of Langley and the former Chief of the Pilotless Aircraft Research Division where much of the research on manned flight was being done. It was at PARD that Max Faget had developed the concept of the manned capsule and that the heat transfer studies were carried out.

7.2   THE CORE TEAM

The first STG strategy meetings in the Unitary Flow Wind Tunnel Building consisted of Dr. Gilruth, Floyd L. Thompson, Charles J. Donlan, and Max Faget. It was here that Dr. Gilruth informed Thompson of his initial "pick of the litter" from among Langley research scientists, engineers, and administrative people. He selected 27 engineers, 12 of whom were his former colleagues in the PARD, plus 9 administrative/clerical people. Hence, counting Gilruth and Donlan, that came to 36. The full list of the first members of the STG is listed along with the abbreviation of their former organization.

Fig. 7.1 Langley's Unitary Flow Wind Tunnel, Building 1251. (Photo Courtesy of the NASA Langley Research Center)

Fig. 7.2 The original Langley Memorial Aeronautical Laboratory was the center section of Building 587. The STG occupied the wing on the right. (Photo courtesy of NASA Langley Research Center)

Fig. 7.3 Entrance to the south side of the building showing the sign for STG Headquarters. (Photo courtesy of NASA Langley Research Center)

Fig. 7.4 The STG sign over the entrance to the building. (Photo courtesy of the NASA Langley Research Center)

Fig. 7.5 Building 60 (now Building 580) used by the astronauts and training personnel. (Photo courtesy of NASA Langley)

Gilruth's informal executive committee (in November and December 1958) comprised:

- Robert R. Gilruth (PARD)
- Charles J. Donlan (OAD)
- Maxime A. Faget (PARD)
- Paul E. Purser (PARD)
- Charles W. Mathews (FRD)
- Charles H. Zimmerman (Stability)

where the codes in parentheses applied to the Langley organization, not STG (see below).

The following people (in alphabetical order) rounded out the initial Langley membership of the Core Team:

- Melvin S. Anderson (Structures)
- William M. Bland (PARD)
- Aleck C. Bond (PARD)
- William J. Boyer (IRD)
- Robert G. Chilton (FRD)
- Edison M. Fields (PARD)
- Jerome B. Hammack (FRD)

- Shirley Hatley (Steno)
- Jack C. Heberlig (PARD)
- Claiborne R. Hicks (PARD)
- Alan B. Kehlet (PARD)
- Ronald Kolenkiewicz (PARD)
- Christopher C. Kraft, Jr. (FRD)
- William T. Lauton (DLD)
- John B. Lee (PARD)
- Norma L. Livesay (Files)
- Nancy Lowe (Steno)
- George F. MacDougall, Jr. (Stability)
- Betsy F. Magin (PARD)
- John P. Mayer (FRD)
- William C. Muhly (Planning)
- Herbert G. Patterson (PARD)
- Harry H. Ricker, Jr. (IRD)
- Frank C. Robert (PARD)
- Joseph Rollins (Files)
- Ronelda F. Sartor (Fiscal)
- Jacquelyn B. Stearn (Steno)
- Paul D. Taylor (FSRD)
- Julia R. Watkins (PARD)
- Shirley Watkins (Files).

The Langley organization codes were:

- DLD – Dynamic Loads Division
- FSRD – Full Scale Research Division
- FRD – Flight Research Division
- IRD – Instrument Research Division
- OAD – Office Associate Director
- PARD – Pilotless Aircraft Research Division
- Fiscal – Fiscal Division
- Files – Office Services Division
- Planning – Technical Services
- Stability – Stability Research Division
- Steno – Office Services Division.

The 10 people from Lewis listed in the following section increased to 46 what is regarded as the original membership of the STG.

7.3 THE LEWIS CONTRIBUTION

At least two years prior to Sputnik, what was then the Lewis Flight Propulsion Laboratory was urging NACA to enter the field of rocket engines because spaceflight was considered to be the logical extension of their work. The Lewis view was that space exploration was imperative for the Nation's survival in the Cold War, which at the time was far from "cold." Unlike Langley and Ames, Lewis wasn't hesitant to get into space research. About this time, several people were commuting from Lewis to Langley/Wallops to work on capsule-related engineering and testing. Amongst their number were John H. Disher, Kenneth Weston and Glynn Lunney.

About the time that President Eisenhower gave his speech to Congress calling for a "National Aeronautical and Space Agency" the Lewis Associate Director, Abe Silverstein moved to NACA Headquarters to assist its Director Hugh Dryden with the formation of a new space agency. After NACA became NASA on October 1, 1958, Silverstein was named as its Director of Space Flight Operations. In particular, this gave him responsibility for the budgets and personnel decisions for what became the program which he decided to name "Mercury." (He would later also name the "Apollo" program.) From this came the Space Task Group at Langley, now supported by people from what had become the NASA Lewis Research Center.

Others from Lewis also moved to NASA Headquarters. George M. Low, Chief of the Special Projects Branch at Lewis, followed Silverstein to Washington as Chief of Manned Space Flight. Warren J. North and Newell D. Sanders both worked at the interface with NASA Headquarters, and John Disher moved there to work advanced manned systems.

The initial group of 10 people from Lewis to join Gilruth's STG at Langley, and their assignments, were:

- John. E. Gilkey – Engineering and Contracts Division
- Milan J. Krasnican – Flight Systems Division
- Andre J. Meyer, Jr. – Engineering and Contracts Division
- Glynn Lunney – Operations Division
- Warren J. North – Headquarters and the Astronauts and Training Group
- Gerald J. Pesman – Flight Systems Division, Life Systems Branch
- G. Merritt Preston – Cape Canaveral for the launch of Big Joe
- Leonard Rabb – Flight Systems Division, Heat Transfer Section
- Scott Simpkinson – Cape Canaveral for the launch of Big Joe
- Kenneth Weston – Flight Systems Division, Heat Transfer Section.

Later, other Lewis engineers were assigned to the STG Preflight Checkout Section at the Cape headed by G. Merritt Preston and Scott H. Simpkinson. Amongst others they included:

- Francis Bechtel
- Joseph Bender
- Dugald O. Black
- Joseph M. Bobik
- Allen L. Bollan
- Jack A. Campbell
- Robert. L. Carlson
- Thomas M. Catalano.

Lewis provided other critical support to the STG. In 1959, Lewis engineers created the Multi-Axis Space Test Inertia Facility (MASTIF) inside the modified Altitude Wind Tunnel. This was used to test the autopilot and attitude control systems intended for the

Big Joe Mercury capsule. It was first tested by Lewis pilot Joe Algranti (who later moved to JSC in Houston), then some other pilots participated before the Mercury astronauts started using it in 1960. The idea was to train pilots to control a tumbling capsule in orbit. Controlling the MASTIF was not simple, and could cause severe vertigo.

Fig. 7.6 The MASTIF at Lewis. (Photo courtesy of NASA)

Fig. 7.7 Testing the posigrade rocket separation of the spacecraft from the launch vehicle. (Photo courtesy of NASA)

STG gave Lewis and Ames the responsibility to verify the reliability and performance of the retro package that contained both posigrade and retrograde rockets. There were concerns about separating the spacecraft from the Redstone or Atlas rockets, as well as potential tumbling after the separation.

There were also concerns about the retrorocket igniters and the rocket performance, because there was no other way for the spacecraft to de-orbit itself should they fail, leaving the astronaut stranded. Using the modified Altitude Wind Tunnel, Lewis managed to verify this system and calibrate the retrorockets so that they would fire through the vehicle's center of gravity and not send it tumbling.

Another responsibility given to Lewis was to qualify the rocket motors for the escape tower. Their performance was tested at Wallops Island on Little Joe launches, as well as in the Lewis modified Altitude Wind Tunnel to a simulated altitude of 100,000 feet.

7.4   THE GODDARD CONTRIBUTION

In November 1958, when Project Mercury was formally initiated, NASA also declared that a new space projects' center would be built. In a memorandum dated January 26, 1959, NASA Administrator T. Keith Glennan stated that "Mr. Robert Gilruth is hereby

Fig. 7.8 Testing the retro package. (Photo courtesy of NASA)

designated Assistant Director, Beltsville Space Center and Director of Project Mercury."
But the future location was just some land owned by the Department of Agriculture's
Agricultural Research Center. People started to transfer to this new organization, which
was actually housed in rented office space in Greenbelt Maryland. The "Beltsville Space
Center" existed in name only, as the facility hadn't yet been built. In reality, the Goddard
Space Flight Center (GSFC) was dedicated on March 16, 1961.

People supporting the military's Minitrack Network intended for ballistic missiles
began to move into this new organization. They came from the Naval Research Laboratory,
where they were developing the Vanguard and Explorer programs. People also came from
the White Sands Missile Range and from NACA Langley. This group's interests were
integrating tracking, data processing, computer operations, and world-wide communications, which was exactly what the Mercury project needed.

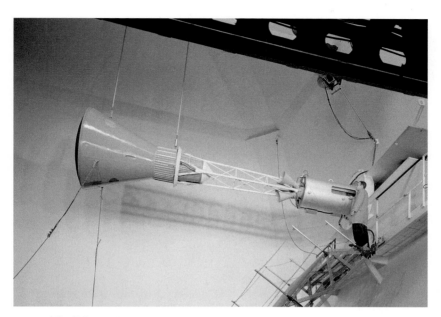

Fig. 7.9 Testing the Mercury escape tower. (Photo courtesy of NASA)

Although the early thinking of the NASA Headquarters management was that the new center would include the STG Mercury effort, there was no room to transfer all those people to offices in Maryland that were required for all the other people appointed. The STG was in full swing by November 3, 1958 with the Langley Research Center providing them with several buildings. As a result the STG reported to NASA Headquarters and never physically moved to Maryland.

By the spring of 1959, four contracts were let to develop the network specifications. Within a year, site surveys were completed and construction had begun at stations around the world. This was still accomplished by the Tracking And Ground Instrumentation Unit (TAGIU) at Langley. It wasn't until April 3, 1961 that Abe Silverstein, Head of the Office of Space Flight Programs at NASA Headquarters, delegated responsibility of the Mercury Network to Harry J. Goett, the first Director of the Goddard Space Flight Center. To implement this major transfer, he was given the services of Ozro M. Covington, Henry Thompson, and George Q. Clark from the White Sands Missile Range.

By June 1959, contracts for the Mercury (later the Manned) Space Flight Network (MSFN) had been let:

- Western Electric – Prime contractor responsible for overall program management, procurement, production, transportation, and installation and testing of equipment; design and implementation of the ground communication subsystem; and training maintenance and operating personnel.
- Bell Telephone Systems – Analysis and development of operation plans and tests; design of command and control displays at both Cape Canaveral and Bermuda; and provide a simulation system for flight controllers and astronauts. (Bell in turn hired Stromberg-Carlson to build and install the flight control displays.)

- Bendix Corporation – Design and fabrication of telemetry and tracking display equipment; systems design, fabrication and integration of all radars not already furnished by the government; and design and fabrication of all Mercury spacecraft communications equipment. (Bendix obtained new radars from RCA and Reeves Instrument Corporation.)
- IBM – Computer programming and operations at both GSFC and Bermuda; and the maintenance and operation of the launch and display subsystem at Cape Canaveral.

On July 1, 1961 NASA officially accepted the new MSFN, just 24 months after awarding the contracts. By the end of Project Mercury, the MSFN had performed beyond the expectations of the designers and was critical to the success of not only Project Mercury but also, as spaceflight evolved, Gemini, Apollo, Skylab, Space Shuttle, and the International Space Station operations. See Section 14.3.

7.5   THE AMES CONTRIBUTION

As early as 1952, the Air Force had urged NACA to begin to investigate a number of problems associated with human spaceflight. Led by now legendary engineer H. Julian "Harvey" Allen, Ames's contribution was in the area of hypersonics. This led to the curved, blunt shape of the Mercury capsule as being the optimum shape for re-entry into the atmosphere. Once his theory was developed, Allen and his colleagues in the High Speed Research Division constructed test facilities which included arc jets, ballistic ranges, and hypersonic wind tunnels, then used them to validate the use of "blunt bodies" for re-entry vehicles.

Alfred J. Eggers was Ames's leading theoretician on capsule design, and a key member of NASA's Research Steering Committee on Manned Space Flight. Eggers worked closely with both Allen and another pioneer in hypersonic spacecraft, Clarence "Sy" Syvertson, who later became Director of Ames. They devised free-flight ballistic ranges in order to create re-entry speeds and validate the damping characteristics that would be required to ensure the capsules would be aerodynamically stable.

Alfred Seiff and Thomas Canning of the Vehicle Environment Division calculated the radius and shape of the capsule's heat shield, and then verified their work in the ballistic ranges.

Eggers designed and developed an Atmosphere Entry Simulator that would propel a Mercury capsule model to 17,000 miles per hour. The test proved that an ablative heat shield would work by vaporizing only 5% of the plastic material.

Larger models of the capsule were also tested in various wind tunnels, with flutter studies of the narrow end of the capsule.

Ames also performed key aerodynamic tests on the launch abort system. Various capsule tests were conducted in the Unitary Wind Tunnels, the Supersonic Wind Tunnel, and the Supersonic Free-Flight Wind Tunnel.

After McDonnell Aircraft was selected to manufacture the capsule, further full-scale flight tests shifted to both Lewis and Langley.

Fig. 7.10 H. Julian "Harvey" Allen. (Photo Courtesy of NASA)

7.6 WALLOPS ISLAND CONTRIBUTION

The Wallops Island "Station" (as it was called then) was established by Langley in 1945. Its contribution to the STG in 1958 was primarily its people from the Pilotless Aircraft Research Division (PARD). It was they, in addition to other Langley personnel, who launched the Little Joe series of vehicles that qualified various aspects of the Mercury missions. In the late 1940s and early 1950s, these people "cut their teeth" on high speed aerodynamic research in order to supplement wind tunnel and laboratory investigations into the issues of such flight that were underway at the Langley Memorial Aeronautical Laboratory (as it was called then) some 80 miles to the south and across the mouth of the Chesapeake Bay.

Seven Little Joe launches were made to qualify Mercury hardware between 1959 and 1961; the STG period. These flights tested the escape and recovery systems, and life support systems. Aborts caused the most violent portions of the flight regime. Re-entry caused the most extreme heating conditions. Both were tested at Wallops. Two flights used Rhesus monkeys known as Sam and Miss Sam. They were named after their "keepers" at the School of Aviation Medicine in San Antonio, Texas. Both animals survived extreme conditions. There were also capsule and rocket failures, and the lessons learned were duly applied to the spacecraft systems.

Fig. 7.11 Alfred J. Eggers, Jr. beside the Atmosphere Entry Simulator used on Mercury models. (Photo Courtesy of NASA)

As related in Section 7.2, many of the former PARD people were assigned to the STG Core Team, including Robert Gilruth who became the Director of the Project Mercury.

Wallops was also used as a Demonstration Site for the Mercury Space Flight Network. Bell Telephone Laboratories conducted subsystems and integrated subsystems tests on the hardware intended for the tracking stations. Revised specifications were later approved by NASA, issued to Western Electric, and distributed to all the sites. These tests served as a

Fig. 7.12 Larger version of the Atmosphere Entry Simulator. (Photo courtesy of NASA)

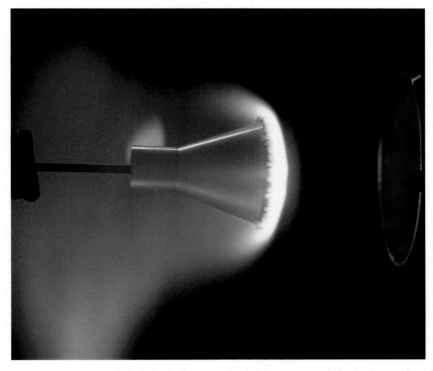

Fig. 7.13 An ablation test of the ¼-inch-diameter plastic Mercury capsule in the Atmosphere Entry Simulator. (Photo courtesy of NASA)

Fig. 7.14 Wallops Island in 1961. (Photo courtesy of NASA)

basis for Goddard's acceptance of the equipment. In addition, STG used the test site as a training facility for flight controllers.

Today, the Wallops Flight Facility (as it is now called) is operated by the Goddard Space Flight Center as a rocket launch site to support science and exploration missions for NASA, NOAA, DOD, and other federal agencies. Its mobile range instrumentation capability can support missions around the world. Since 1945, there have been over 16,000 launches from Wallops Island!

7.7   THE HIGH SPEED FLIGHT STATION'S CONTRIBUTION

This flight test facility has had many names over the years, but is currently the Armstrong Flight Research Center. It was first known as the NACA Muroc Flight Test Unit and during the time of Project Mercury it was known as the NASA High Speed Flight Station (HSFS). It was the home of the X-Series of research aircraft, including the X-15. Much of the research into transonic and supersonic flight was conducted there during the 1950s. The results of this research were part of the Langley Research Center's archives and data bases, and thus available to the STG engineers. Of particular interest was Reaction Control System (RCS) of the Bell X-1B and the X-15, which comprised small thrusters that would fire to stabilize or redirect a vehicle in a near vacuum. This research, development, and testing of hydrogen peroxide thrusters was directly applicable to the Mercury spacecraft.

Fig. 7.15 The Wallops Flight Facility in 2010. (Photo courtesy of NASA)

The first Chief of the High Speed Flight Station under NACA was Walter C. Williams, who had directed a number of flight research programs over the years, including the D-558-II aircraft. He oversaw the NACA portion of the Century Series Program of fighters, and was chairman of the X-15 Flight Test Steering Committee. Much of this research work was directly applicable to Project Mercury, as was Williams's experience of managing researchers, flight test engineers, and test pilots. On August 28, 1959 Williams joined the Mercury team, taking responsibility for overall launch operations. During missions, he served as the Operations Director in the Mercury Control Center. See his biography in Appendix 2.

In the mid-1950s, flight into altitudes with low atmospheric pressure was unknown territory. How would a pilot control his aircraft, what did this mean for control system design, fuel usage, effects of system lag, control effectiveness, and control stick feel? In 1954, the NACA Muroc Flight Test Unit used an Air Force-owned analog computer known as the Goodyear Electronic Differential Analyzer (GEDA) to develop flight simulations to study reaction control systems. Although crude by modern standards, these simulations, along with a mechanical full-motion simulator (called the Iron Cross), started to give the engineers and test pilots good design and operational experience with a completely new way of controlling an aircraft and, ultimately, a spacecraft.

The first aircraft to use the RCS was the Bell X-1B rocket plane. This was an elongated X-1 with higher performance. In July 1957, Neil A. Armstrong and Stanley Butchart were the NACA flight test pilots. This was a year before NASA, and before there were such words as "astronaut" and "Mercury." Armstrong's first flight in a rocket powered craft occurred on August 15, 1957. Despite losing one of the four rocket motors, he achieved

Mach 1.32. In landing on the Rogers Dry Lake the nose wheel collapsed and damaged the aircraft. What many people don't realize is that during his career Armstrong crashed many times and ejected twice. He flew the repaired X-1B a second time on January 16, 1958.

The RCS was also tested aboard the JF-104 research aircraft by Joseph A. Walker on July 31, 1959. Walker would later become the first pilot to fly the X-15. Armstrong also flew the JF-104. The results of the X-1B and JF-104 flights established the design of the RCS system for the X-15 and the Mercury capsule. The thrusters used 90% hydrogen peroxide with pressurized bladders to expel the propellant. The spacecraft had 18 thrusters in three sets of six, giving three different thrust levels.

The RCS development work at the NASA HSFS carried over to the Lunar Landing Research Vehicle (which Armstrong also ejected from), the Lunar Module, and the Manned Maneuvering Units used by spacewalkers. This was one of HSFS's greatest contributions to not only Project Mercury but to spaceflight in general. See Section 13.2.9.

7.8   WHITE SANDS MISSILE RANGE CONTRIBUTION

The Army's White Sands Proving Ground (WSPG) in New Mexico is best known for the first atomic bomb test at the Trinity Site on July 16, 1945 and for the subsequent testing of captured V-2 missiles. Some 341 boxcars of missile materials, including complete V-2s, were shipped to White Sands after the war. These V-2 launches led directly to the development of the Redstone missile, and the men who conducted the tests made an indirect contribution to Project Mercury.

By 1949 the Army had relocated most of the 35 original German engineers and scientists of "Operation Paperclip" to the Redstone Arsenal in Huntsville, Alabama, where the team became the Ordnance Guided Missile Center. Over the years, approximately 1,600 German scientists, engineers, and technicians came to the United States along with thousands of family members. When the Korean War broke out in June 1950 they were assigned to tactical missiles. The team grew significantly during that conflict. The war slowed down the design and development of the Redstone missile, but the first experimental Redstone was launched on August 20, 1953. See Section 6.2.

This section will focus on the men from the White Sands Missile Range; the name given to the range in May 1958, just prior to the creation of NASA and Project Mercury. During those intervening years, the range included the Holloman Air Development Center, which was where the history of research in space biology and biodynamics was written. Graduates of this research included the chimpanzees Ham and Enos. See *Animals in Space: From Research Rockets to the Space Shuttle* by Colin Burgess and Chris Dubbs (details of recommended books are listed in the Reference Section).

White Sands was also a "Minitrack" station for the Vanguard program, and the problems of tracking an orbital object were tested there. The program was managed by the Naval Research Laboratory (NRL), which had a Tracking and Guidance Branch at White Sands.

Some of the people who came from the range to join Project Mercury included:

- Ozro "Ozzie" Covington – Technical Director of the Army Signal Missile Support Agency tracking V-2s. He later relocated to Goddard to develop the STADAN and the MSFN, and was liaison to the STG.

- Henry Thompson – Moved to Goddard to be Covington's deputy.
- Jack Mengel – The NRL representative on the Minitrack system, he moved to Goddard and later became the first Director of the Tracking and Data Systems at Goddard and supported Project Mercury.
- Some of the medical people from Holloman also supported Project Mercury, some of them temporarily and others permanently.

Once Project Mercury got underway, people with missile tracking experience were moved to Goddard to assist in building the Mercury (later Manned) Space Flight Network (MSFN).

7.9   THE ARNOLD ENGINEERING CONTRIBUTION

The STG made good use of the vast Air Force test facility located at Tullahoma, Tennessee. The Arnold Engineering Development Center (AEDC) was named for General Henry "Hap" Arnold, the father of the Air Force. In the 1950s it was known for its Engine Test Facility, its Propulsion Wind Tunnel, and the Gas Dynamics Facility named after Theodore von Kármán. The center was testing jet engines for the B-47 and was used for the Atlas program.

As early as November 1958, even as the STG was in the process of being organized, a scale model of the Mercury capsule, oriented for the re-entry phase, was tested in the 1-foot transonic test tunnel. In 1959, Mercury models were tested at speeds of Mach 8, 16, and 20 to investigate stability, heat transfer, and pressure distribution of various capsule components. By the end of the year, over 70 different Mercury capsule models had been tested using the Arnold and Ames wind tunnels.

In April 1959, two escape tower configurations were tested in the AEDC 16-foot transonic circuit to determine static stability and drag characteristics. By November, they had also tested the solid-fuel rocket motor that was to propel the escape tower and identified the combustion-chamber pressure-time curve. And the following month the AEDC tested the solid-fuel Thiokol retrorockets to evaluate their ignition characteristics. It was also involved in the analysis of the MA-1 failure of July 29, 1960 and contributed test facilities for that investigation.

The AEDC went on to support NASA's Mercury, Gemini, Apollo, and Saturn programs. It is now the most advanced and largest complex of flight simulation test facilities in the world. The complex operates 43 aerodynamic and propulsion wind tunnels, rocket and turbine engine test cells, space environmental chambers, arc heaters, ballistic ranges, and other specialized units.

7.10   MARSHALL SPACE FLIGHT CENTER

On October 21, 1959 President Eisenhower approved the transfer to NASA of all Army space-related activities. This was accomplished effective July 1, 1960 by the creation of the George C. Marshall Space Flight Center (MSFC) on the site of the Army Ballistic Missile Agency (ABMA) in Alabama. It included Wernher von Braun's team, which manufactured the launch vehicles for the Mercury Redstone flights. See Section 6.2. After that series of flights, the center's focus was on the Saturn program for Apollo, including the engines and the vehicle stages. No Saturn ever failed to achieve its mission.

8

The AVRO Canadians

8.1 HISTORY

The AVRO story is best described in the book *Arrows to the Moon* written by Chris Gainor and published in 2001. I shall provide some excerpts along with other information, my interpretation of the events, and where the people worked in the NASA Space Task Group (STG). It is a sad story in a way, illustrating how politics can absolutely destroy an industrial program, impact the lives of thousands of people, and devastate a community and local businesses. Gainor tells that story well. As regards the STG and Project Mercury, the U.S. benefitted greatly from Canada's loss. That can also be said of Gemini, Apollo, Skylab, and the Space Station, since the AVRO Canadians and Brits played a major in those programs as well.

The reader will appreciate the heritage of the company that fostered the great engineers who eventually came to NASA. The A. V. Roe Company was established in 1910 by brothers Alliot Verdon Roe and Humphrey Verdon Roe. Alliot was the aircraft builder and Humphrey was the finance and organizational guy. They built mostly training aircraft for WW-I, although a few saw combat. Financial problems after the war resulted in Crossley Motors buying the majority of the stock. In 1928, Crossley Motors sold the company to Armstrong Siddeley Holdings, Ltd. Alliot Roe resigned and started Saunders-Roe, which then became a subsidiary of Hawker Siddeley in 1935.

During WW-II, AVRO built the famous Manchester, Lancaster, and Lincoln bombers. Some of the more famous Lancasters were built in what was, at the time, the world's largest building, consisting of 1.5 million square feet. Despite its size, the building was disguised to hide it from German planes. The Lancaster is also notable for carrying the largest bomb load of any aircraft during the war, most notably the 22,000 lb. Grand Slam.

After the war, AVRO built the beautiful Vulcan bomber as a nuclear-strike aircraft armed with the Blue Steel missile. It was featured in the 1965 James Bond movie *Thunderball*. Only one restored Vulcan remains. It performed at air shows through to retirement in 2015. AVRO also built a turboprop airliner and four-engine jetliners.

© Springer International Publishing Switzerland 2016
M. von Ehrenfried, *The Birth of NASA*, Springer Praxis Books,
DOI 10.1007/978-3-319-28428-6_8

The Hawker Siddeley Group purchased the former Victory Aircraft firm in Malton, Ontario, Canada and renamed it A. V. Roe Canada Ltd. It employed AVRO as its trading name. During the Cold War period, the Royal Canadian Air Force (RCAF) was worried about Soviet bombers attacking from the north. This led to the design and construction of the CF-100 jet interceptor. AVRO also designed and manufactured the four-engine C-102 Jetliner. As an aside, one of my early supervisors at the STG was C. Frederick Matthews, who worked on this aircraft and later flew in it with Howard Hughes at the controls. See Matthews's biography in Appendix 2.

In 1953 the RCAF issued specifications for the design of a supersonic all-weather fighter to supersede the CF-100. This was designated the CF-105 AVRO Arrow. Several of these design and test engineers later became STG engineers, and even later became quite famous as NASA engineers and managers. Scale models of the CF-105 were launched on top of rockets from the NACA Wallops Island Station by engineers who would later join the STG. AVRO made use of wind tunnels at NACA Langley and Lewis, as well as one in Canada and another at the Cornell Aeronautical Laboratories for aerodynamic research.

In 1957 AVRO gained an IBM 704 computer similar to the one at Langley. See Appendix 3. This was used extensively for stress analysis, aerodynamic stability, as well as for a new CF-105 simulator. Inevitably the cost of the Arrow program started to escalate. There was a new engine, a new missile, a new fire control system, a flight simulator, and associated ground tracking and analysis systems. It was a complex aircraft for its time.

Many in the Liberal Canadian government grew concerned about the increasing costs of the program and it was decided to defer making any program decisions until after the next election, due in 1958. Just a few months prior to the election, the first flight of the CF-105 took place on March 25, posing only minor problems. The aircraft sent telemetry to a control room called the "High Speed Flight Center" where an ex-RAF wing commander talked by radio to the famous test pilot Jan Zurakowski. Later, flight test engineer C. Frederick Matthews likened this to the role of a capsule communicator (CAPCOM) during Project Mercury.

The increasing threat of the Soviet ICBMs made the government question the interceptor; it would be ineffective against such missiles. Mr. Khrushchev claimed the introduction of ICBMs rendered bombers obsolete. Even the Canadian Defense Minister recommended cancelling the Arrow. Cabinet-level meetings of the new Tory government over a period of months achieved a consensus. On February 20, 1959, Prime Minister John G. Diefenbaker informed the House of Commons of the decision to cancel the Arrow development. That day is still considered "Black Friday" in Canadian aviation circles. Approximately 14,000 staff were immediately laid off. It was a tremendous blow not only to the employees but also to the surrounding communities and businesses. The controversy can be seen in a 56 minute YouTube video entitled *CF-105 Arrow Definitive Documentary*.

8.2   THE STG CAPTURES THE TALENT

The cancellation of the Arrow couldn't have been better timed for a new NASA organization in need of aeronautical engineers, flight test engineers, computer engineers, and program managers. At this time there were only about 150 people in the STG. Hundreds more would be needed for Project Mercury.

Consider the sequence:

- March 25, 1958 – First test flight of the CF-105 Arrow.
- March 31, 1958 – John D. Diefenbaker's Tories win the election.
- October 1, 1958 – NASA came into being.
- November 3, 1958 – The STG was formally created.
- December 1958 – The NASA space program was named Project Mercury.
- January 1959 – McDonnell Aircraft was selected to build the spacecraft.
- February 1959 – some 14,000 AVRO engineers were laid off.

AVRO engineers already had a close relationship with NACA Langley, through using their wind tunnels for aerodynamic research. In fact, David D. Ewart was doing wind tunnel tests of an Arrow model at Langley. And Robert Gilruth and Charles Donlan of NASA had a working relationship with Chief Engineer Bob Lindley and Jim Chamberlin, Chief of Technical Design for the Arrow.

At first, Lindley and Chamberlin tried to sell Abe Silverstein of NASA Headquarters on the idea of the Canadian government providing NASA with AVRO engineers. Having Canada share the prestige of developing space travel had a certain appeal to some people. The proposal went around the Canadian Department of Defense and the Canadian Ambassador, earning supporters. But Prime Minister John Deifenbaker rejected it through skepticism over the concept of space travel! Thus the man who killed the CF-105 Arrow, an aircraft more advanced than anything in the U.S., also killed Canada's participation in Project Mercury. In retrospect, he was not a very far sighted individual.

NASA was still interested in acquiring top notch engineers with applicable engineering, flight test, and computer experience. I don't use the term "top notch" lightly. Here are a few examples of their backgrounds up to 1959:

- Peter J. Armitage was a British-born AVRO engineers. He had a master's in aeronautical engineering, had flown with the Royal Air Force, and was trained as a co-pilot and flight engineer. He was a senior flight test engineer on the CF-105 Arrow.
- James A. Chamberlin had mechanical engineering degrees from the University of Toronto and Imperial College of Science and Technology in London. He was chief aerodynamicist on the CF-100 interceptor and the C-102 Jetliner. He was chief of technical design on the CF-105 Arrow.
- C. Frederick Matthews had an aeronautical engineering degree and was also a RCAF pilot during the war. He was a flight test engineer on the C-102 Jetliner and also on the CF-100 twin-engine jet fighter. He also played a role in the redesign of the CF-100 canopy.
- John D. Hodge, a British AVRO engineer, had a degree in engineering from the University of London. He worked on the air loads of the CF-105 and on the flight test program.
- R. Bryan Erb possessed a degree in civil engineering and a master's in fluid mechanics. He conducted aerothermodynamics analysis on the Arrow.

The review team from Langley included Robert Gilruth, Charles Donlan, Charles Mathews, Charles Zimmerman, Paul Purser, and Kemble Johnson. On March 14, 1959 they interviewed about 100 out of 400 who submitted applications. The NASA men soon realized that the AVRO engineers could bring tremendous talent to Project Mercury.

Bob Lindsey and Jim Chamberlin went back to AVRO and explained to their engineers what Mercury was about and the kind of work they might become involved in. NASA was primarily interested in those with flight test and computer experience. Those interested prepared résumés for NASA's consideration.

NASA Administrator T. Keith Glennan approved the hiring of 32 individuals from AVRO, but seven of them declined. Later, more were added. As foreign nationals, they required to be formally processed into the United States, subjected to background investigations and granted necessary security clearances. This process was assigned top priority and cleared in two weeks. The usual time would have been about six months.

8.3   THE AVRO CONTRIBUTION TO THE STG

Initially 25 AVRO people accepted the offers by NASA and the STG, and more came over time. Later, some returned to Canada or England, or moved elsewhere in the U.S. The following is an alphabetical list with just a few remarks about their contributions to the STG. More is available in their biographies in Appendix 2 of this book and, in some cases, also in the NASA JSC Oral Histories. The list doesn't do justice to their unique contributions to spaceflight. Many went on to support Gemini, Apollo, Shuttle, Skylab, Space Shuttle, and the International Space Station, either with NASA or the Canadian Space Agency. Chris Gainers' book gives details through to 2001.

- Bruce Alexander Aikenhead – Worked in the astronaut training group. Returned to Canada in 1962.
- Peter J. Armitage – Worked in the Recovery Operations Branch. See Appendix 2.
- David Brown – Worked in the Structures Branch. Left in 1970.
- Richard R. Carley – Worked in the Flight Dynamics Branch.
- Frank J. Chalmers – Worked in the Flight Control Branch on MCC development, but left after only a few months.
- James A. Chamberlin – Became Chief of the Engineering and Contract Administration Division.
- Thomas V. Chambers – Worked in the Flight Systems Division Dynamics Branch.
- Jack Cohen – Worked in the Mission Analysis Branch developing simulations.
- Stanley H. Cohn – Worked in the Mission Analysis Branch Mathematical Analysis Section. Returned to Canada in 1962.
- Burton G. Cour-Palais – Worked in the Structures Branch.
- Eugene L. Duret – Worked in the Flight System Division Heat Transfer Section and was a remote site flight controller.
- R. Bryan Erb – Worked in the Flight Systems Division Heat Transfer Section. Became Assistant Director of the Canadian Space Station Program. See Appendix 2
- Donna M. Erb – Bryan's wife. She taught school and later went into computer science and worked for Lockheed and MITRE in Houston.
- David D. Ewart – Worked in the Flight Systems Division Flight Dynamics Branch.
- Joseph E. Farbridge – Left after only a few months.
- Norman B. Farmer – Worked in the Flight Systems Division as head of the Electrical Systems Section.

- Dennis E. Fielder – Worked in the Flight Control Branch, Control Central and Flight Safety Section. See Appendix 2.
- Stanley H. Galezowski – Worked in the Flight Systems Division, Dynamics Branch. Left in 1962.
- George Harris Jr. – Did not join STG but worked on NASA's Mercury Space Flight Network.
- John Dennis Hodge – Worked in the Operations Division. Became a Flight Director. See Appendix 2.
- John K. Hughes – Worked in the Flight Control Branch, Control Central and Flight Safety Section.
- Morris V. Jenkins – Worked in the Flight Systems Division Dynamics Branch.
- Robert N. Lindley – He helped to organize the hiring of AVRO engineers by NASA but joined McDonnell Aircraft instead.
- C. Frederick Matthews – Worked for the Flight Control Branch, Control Central and Flight Safety Section training flight controllers. See Appendix 2.
- Owen Eugene Maynard. Worked for the Flight Systems Division, Onboard Systems Branch.
- John K Meson – Did not join the STG. Worked at NASA Headquarters.
- Leonard E. Packman – Worked for the Flight Control Branch, Control Central and Flight Safety Section.
- Tecwyn Roberts – Worked for the Flight Control Branch, Control Central and Flight Safety Section. Was the first Mercury Flight Dynamics Officer (FIDO). Later moved to Goddard.
- Rodney G. Rose – Worked for the Flight Systems Division, Systems Test Branch.
- Leslie G. St. Leger. Did not join the STG. Joined General Dynamics and later joined NASA JSC.
- John N. Shoosmith – Worked for the Operations Division, Mathematical Analysis Branch. Youngest to leave AVRO and last to leave NASA after 36 years.
- Robert E. Vale. Worked for the Engineering and Contract Administration Division, Engineering Branch.
- George A. Watts – Worked for the Flight Systems Division on structural loads.

In summary, the addition of the AVRO engineers to the STG was a brilliant management decision and had a beneficial impact on the entire U.S. space program for many decades and many missions.

9

The STG Organization

9.1 THE DIRECTIVE

Two directives were issued on January 26, 1959. One was from the NASA Administrator T. Keith Glennan appointing Dr. Robert R. Gilruth as the Director of Project Mercury. The other was from Gilruth for all concerned at Langley, setting out the initial organization of the Space Task Group. Gilruth signed his memo as the Project Manager, rather than Director, and it was addressed to 71 people: 65 NASA employees plus 6 military personnel. It took several months for the STG to staff up, consider the tasks ahead, and figure out an initial organization. Gilruth and his staff were located in Building 58.

On August 3, 1959 Gilruth issued a Memorandum to Staff entitled "Subject: Organization of the Space Task Group." This was slightly updated on August 10, with a few additional names. The later memo is included in Appendix 1 of this book. It featured several organizational charts for six staff functions and three main Divisions. These will be discussed below. In preparing for actual flights in 1961 the Life Systems Branch became a Division and there were consolidations in the Engineering and Contracts functions.

9.2 STAFF OFFICES

9.2.1 Liaison

In September 1958, about the time that NASA was formed, NASA Headquarters and the DOD jointly drafted a "Memorandum of Understanding" detailing the specialized assistance which the DOD would provide NASA and the STG. The military liaisons to the STG were Air Force Lt. Col. Keith G. Lindell, Army Lt. Col. Martin L. Raines, and Navy Cdr. Paul L. Havenstein, who were assigned to Gilruth's staff. Each supported a functional Division. Lindell was assigned to the Astronauts and Training Group, Raines supported the Redstone interface, and Havenstein supported the Operations Division in planning. The Langley Research Center also assigned W. Kemble Johnson to Gilruth's staff to coordinate the requirements of the STG with the Center's research facilities and personnel.

© Springer International Publishing Switzerland 2016
M. von Ehrenfried, *The Birth of NASA*, Springer Praxis Books,
DOI 10.1007/978-3-319-28428-6_9

9.2.2   Special Assistants

Paul E. Purser, James A. Chamberlin, and Raymond L. Zavasky were part of Gilruth's staff and were former PARD members. Each received special assignments. Purser was Gilruth's "Go-To" man for any subject area or problem. Chamberlin became Head of the Engineering and Contract Administration Division. As the organization grew, others joined the staff, including Marion R. Franklin Jr., Jack C. Heberlig, and Kenneth S. Kleinknecht, with the latter being assigned as the Project Mercury Manager.

By 1960 it was obvious that a Security Office was required, so Donald D. Blume became the head of this Office. Paul Purser was his liaison to Gilruth. The main concern was the protection of the astronauts and their families. The Security Office dealt with such activities as conducting background investigations of the new people and granting various levels of security clearances. Other security personnel included Ann W. Hill, Joan S. Holden, Theresa M. Peele, and Lloyd D. Yorker. Linda J. Hare was the office receptionist and secretary.

9.2.3   Public Affairs Officer

Lt. Col. John A. "Shorty" Powers came on board in April 1959 to handle the press for the STG. He was the "Voice of Mercury Control" until July 1963. Paul P. Haney then became the Public Affairs Officer. Also in this office was Louis M. Kidd.

9.2.4   Staff Services

Burney H. Goodwin was Head of Personnel. His secretary was Betty S. Knox. Guy H. Boswick, Jr. led the Administrative Services. There were two receptionists: Jo Ann S. Fountain and Ann W. Hill. Norma L. Livesay was in charge of the Files and Library and Margaret B. Burcher ran the Stenographic Pool. Over time these departments grew, with some people moving and others joining. Appendix 1 lists all the people in the 1959 group.

By 1961, the Office of the Director had grown in line with the maturing of the program and the reorganizations that resulted in several other groups moving up to the staff level. The term Staff Services was no longer used. Management Services and Budget functions were separated. Management services now included:

- Administrative Services (Guy Boswick, Jr.)

    Files (General and Classified)
    Mail Desk
    Library
    Teletype Operator
    Stenographic Services
    Report Typing
    Reproduction (Including John and Alphonse Thiel)

- Management Analysis (Roy Magin and Roy Aldridge)

- Personnel Officer (Burney H. Goodwin)

    Classification
    Records and Information

- Security Officer (Donald D. Blume)

    Receptionist (Ann W. Hill).

What had been only a single budget assistant in 1959 had by 1961 become an organization that included:

- Budget and Finance Office/Officer

    Accounting
    Payroll
    Travel Reservations
    Travel Vouchers.

By this time, it was known that the STG would move to the Manned Spacecraft Center in Houston, Texas and that the organization would be reorganized again for following missions.

9.2.5   Technical Services

The coordination of work undertaken by the Langley Research Center, particularly regarding all of its various shops, was coordinated by Jack A. Kinzler, with the help of David L. McCraw and Orrin A. Wobig. They built all of the Mercury models for testing, and provided all the necessary resources and facilities needed to support Project Mercury. The STG initially had people but not facilities, so it relied heavily on the support of the Langley Research Center whose facilities and capabilities were significant. See Section 7.2.

Several technicians were assigned to this organization but actually worked in others. It seems that because they were not scientists or engineers they were not to be permanently assigned to a Division but would instead work out of the Technical Services Office. As the needs of the STG grew, there was an increasing need for technicians. Probably the most well-known of this group was Joe W. Schmitt whose role made him the last man an astronaut saw before launch, because he was the technician who "tucked them in" the capsule, hooked them up, strapped them in, and closed the hatch. In addition, by 1961 there were a lot more. According to Tom Gallagher, here is break down of which technicians supported what areas:

- Space Suits Technicians

    Joe W. Schmitt
    Harry D. Steward

- Space Suit/Altitude Chamber Technicians

    Paul O. Ferguson
    Thomas F. Gallagher
    Alan M. Rochford

Glenn A. Shewmake
Martin Tessler

- Machinists

  La Marr D. Beatty
  William E. Drummond
  Paul A. Folwell II
  Charles C. Nagle
  Charlie E. Rogers Jr.
  Joseph E. Siegfried
  Charles M. Tucker
  James W. Warren

- General Mechanics

  Luther L. Hoover
  Wendell G. Malpass
  Junior N. Mitchell

- Electronics Technicians

  James C. Brady

- Administration

  Westley H. Brenton
  Arthur G. Trader

- Others (assignments unknown)

  James W. Bailey
  Edward A. Carpenter
  Arthur C. Chapman
  Elwood S. Edwards
  Benson B. Gardner
  Francis I. Glynn
  Rodney F. Higgins
  Richard A. Holman
  John L. James, Jr.
  Mark S. Larson
  Miles L. Lockard
  Ralph D. Mann
  Roger Messier
  William S. Pittman.

9.2.6   Astronauts and Training

By late 1959 the role of an astronaut was fairly well defined as involving both flight and non-flight duties. By his very selection, he was expected to contribute to the successful flight of the spacecraft. He was also expected to contribute to the systems design and to the development of the operational procedures. The astronaut training program initially specified the following six areas (more detailed descriptions were added later):

- Programming and monitoring the sequence of vehicle operations during launch, on-orbit, and re-entry
- Systems management involving monitoring and operating the on-board systems
- Vehicle attitude control
- Navigation
- Communications
- Research and evaluation.

It was also known early on that the astronauts would be assigned detailed tasked associated with flight operations. This required close coordination with the spacecraft engineers and STG flight controllers. The astronauts had responsibilities during the countdown and preparation of the vehicle, communications from the ground to the capsule, and recovery operations. All these functions required training.

Few training facilities were available in 1959. Early training was reviewing design drawings and traveling to attend briefings at the various production facilities for the launch vehicles and the spacecraft. The contractors and the STG would provide the astronauts technical lectures on various systems, engineering, and operations. In addition Air Force and Navy facilities would give familiarization lectures for the roles which those services would play. In the latter half of 1959 the astronauts traveled on average one out of every three days.

These trips included:

- U.S. Navy Johnsville centrifuge
- Air Force Flight Test Center
- Air Force Ballistic Missile Division
- Convair/Astronautics
- Zero-G aircraft flights
- B. F. Goodrich pressure suit fittings
- Air Crew Equipment Laboratory
- Customized couch moldings
- McDonnell Aircraft
- Atlantic Missile Range
- Proficiency flying in F-102s.

The Astronauts and Training Group was headed by Col. Keith G. Lindell, who was also the Air Force liaison officer to the STG. This group was large enough to have been made a whole Division but Gilruth wanted the astronauts, physicians, and trainers on his staff. But that was just on paper, because they had their own building and simulators at different locations. Dr. Lt. Col. William K. Douglas was the flight surgeon for the astronauts. Lt. Robert B. Voas was the Training Officer. George C. Guthrie and Raymond G. Zedekar were appointed to the Training Office.

When the astronauts came on-board in the spring of 1959, the initial effort was to bring them up to speed with lectures and familiarization trips. Much of the intended training hardware was not immediately available. As the program progressed in 1960, several comprehensive training manuals were created. McDonnell Aircraft developed an Indoctrination

Manual, a full Mercury Systems Familiarization Manual, and a Capsule Operations Manual which soon became known as the Astronauts' Handbook.

In addition, because no single astronaut would be able to keep track of every development, each closely monitored a specific area and kept his colleagues up to date, as follows:

- Scott Carpenter, navigation and navigation aids
- Gordon Cooper, Redstone booster
- John Glenn, crew/cockpit layout
- Walter Schirra, life support systems
- Alan Shepard, range, tracking and recovery operations
- Donald Slayton, Atlas booster
- Gus Grissom, electromechanical and attitude control systems.

The Training Aids Section in the Flight Control Branch of the Operations Division assisted the astronauts and trained flight controllers. Some of the people within the Operations Analysis Section of the Mission Analysis Branch of the Operations Division took part in the creation of simulations for astronauts and flight controllers. Although former NACA test pilot Warren J. North was in the Headquarters Office of Space Flight Development, he was deeply involved in astronaut training. So did his twin brother Gilbert "Bert" North, who was the McDonnell "in-house astronaut" and also a test pilot engineer.

The STG training duties were approximately as follows:

- Astronauts and Training Group

    Robert Voas concentrated on the coordinating the training and astronaut tasks
    Warren North monitored training for Headquarters
    George C. Guthrie had responsibility for simulation devices and training aids
    Raymond G. Zedekar arranged the lecture series

- Operations Division (combined crew training with flight controller training)

    Harold I. Johnson headed up the Training Aids Section
    Stanley Faber organized the centrifuge training
    Charles C. Olasky ran the crew Mercury Procedures Trainers
    Rodney F. Higgins worked on the Link trainers
    Bruce A. Aikenhead was from AVRO and returned to Canada in 1962
    Arthur E. Franklin was involved in simulations
    Jack Cohen was Head of the Operations Analysis Section
    Harold Miller, simulations
    Arthur A. Hand, simulations
    Glynn Lunney, simulations
    Dick Koos, simulations
    Bob Eddy, simulations.

On March 31, 1960 the Operations Division published the first training document "Plan for Control Center Training Simulations."

9.3   FLIGHT SYSTEMS DIVISION

This group knew the systems and subsystems of the capsule, and the aerodynamic and heating environments in which it would operate. Many of its members came from the PARD and had launched rockets and model aircraft at Wallops Island. Some were former AVRO engineers.

Maxime A. Faget was the Division Chief, Robert O. Piland was the Assistant Chief, and J. Thomas Markley was the Executive Engineer. All had worked closely together before. Faget organized the Division into five Branches and one Computing Group, as follows:

- Systems Test Branch
- Performance Branch
- Life Systems Branch
- On-Board Systems Branch
- Dynamics Branch
- Computing Group.

9.3.1   Systems Test Branch

This Branch was headed by another PARD engineer, William M. Bland, with H. Kurt Strass as his Assistant Head. Jack C. Heberlig was also a former PARD engineer. It included two AVRO engineers: Owen E. Maynard and Rodney G. Rose. This group was involved with testing and evaluating various components of the Mercury system, including the Little Joe components and the capsule systems. As the tests were conducted, the results were fed back into the design. For example, when tests of the landing bag and heat shield indicated interface clamp failures, these clamps were redesigned. The group tested the recovery systems and conducted drop tests with the recovery people. They tested the honeycomb structure for landing, tested scale models in a Langley water tank and full scale models in the Atlantic. They also tested the escape system and later, alternative escape systems.

In addition to those mentioned, the Systems Test Branch also included:

- Lawrence W. Enderson, Jr.
- Edison M. Fields
- Louis R. Fisher
- Jerome B. Hammack
- Robert A. Hermann
- Walter J. Kapryan
- Ronald Kolenkiewicz
- James T. Rose.

9.3.2   Performance Branch

This Branch was headed by Aleck C. Bond, who was another PARD engineer and a member of the STG Core Team. This comprised three sections, Aerodynamics, Loads, and Heat Transfer. It was heavily involved with the Mercury design, test, and evaluation employing models in wind tunnels at Langley and in flights from Wallops Island.

The Aerodynamics Section was headed by Alan B. Kehlet, who was also one of the original members of the STG Core Team and one of the men later honored as Mercury Capsule Inventor. This group was concerned with the stability of the capsule as it went through the various stages of flight. They ran many wind tunnel tests on the vehicle, visited McDonnell in order to discuss their work on stability and control, and were involved in the launches at the Cape by reviewing the telemetry.

In addition to those mentioned, the Aerodynamics Section also included:

- William W. Petynia (Assistant Head)
- Steve W. Brown
- David D. Ewart (former AVRO engineer)
- William H. Hamby
- Dennis F. Hasson
- Bruce G. Jackson (former PARD)
- William C. Moseley, Jr.
- Edward F. Young.

The Loads Section was headed by George A. Watts, an AVRO engineer who had worked with John D. Hodge on the CF-105 inlets and loads analysis. At the STG, Watts was put in charge of structural loads on the Mercury capsule and on the Atlas adapter. The data gathered in real time by Watts during the failure of the adapter on MA-1 provided sufficient information to redesign the adapter. This group assessed the meteorological loads on the Mercury Atlas during both the launch and recovery phases. They also analyzed the landing loads on the capsule in various sea states. They compared their loads analysis with those conducted by McDonnell. The group was involved with all the aspects of computing and analyzing the structural loads on the capsule, in particular during Max-Q, on the escape tower, on parachute deployment, and on landing. They worked with the Heat Transfer people on the heat shield, and when they wanted to change the afterbody shingles and required their assurance that doing so would not change the loading on the capsule.

In addition to Watts, the Loads Section included:

- Joseph E. Farbridge
- May T. Meadows (one of the few women engineers)
- Robert P. Smith
- William Rodgers
- Walter West
- Jim Bergen.

The Heat Transfer Section was headed by Leonard Rabb. This group worried about the heat transfer during re-entry penetrating the capsule's heat shield and afterbody into the cabin. They ran tests in various wind tunnels and examined the work done on missiles that utilized the heat sink approach versus the ablative approach selected for Mercury. They studied beryllium heat sink performance and also the ablative qualities of fiberglass/phenolic resin composites. They took part in the proof-of-concept of an ablative heat shield on the Big Joe test on September 9, 1959. The selection of the heat shield material was one of the major engineering decisions of Project Mercury. The group continued to monitor the performance of the capsule and its heat transfer properties throughout the program.

In addition to Rabb, the Heat Transfer Section included:

- Eugene L. Duret (former AVRO; was also a remote site flight controller)
- R. Bryan Erb (former AVRO)
- Joanna M. Evan
- Archie L. Fitzkee
- Stephen Jacobs
- John S. Llewellyn (also a remote site flight controller)
- Robert O'Neal
- Emily W. Stephens
- Kenneth C. Weston.

9.3.3   Life Systems Branch

This Branch (later a Division) supported the astronauts with experts in aviation medicine. Gilruth arranged for the Army, Air Force, and Navy to send flight surgeons to the STG in order to assist with astronaut training and the manning of the Mercury Control Center in Florida and the remote tracking stations, and also to assist with the recovery of astronauts. Some of these doctors came from the Lovelace Clinic, the Air Force Department of Aviation Medicine and other sites which were involved with aviation medicine.

Some of the physicians/flight surgeons that supported Project Mercury include:

- STG Life Systems Branch

  Dr. Lt. Col. Stanley C. White (Head)
  Dr. Lt. Col. James P. Henry (of pressure suit fame)
  Dr. Capt. William S. Augerson

- Astronauts and Training Group

  Dr. Lt. Col. William K. Douglas (the astronauts' physician)

- Later Mercury and Gemini Aeromedical Monitors

  Dr. Charles A. Berry (MCC)
  Dr. Duane A. Catterson (MCC)
  Dr. Richard Pollard
  Dr. David P. Morris
  Dr. D. Owen Coons (MCC).

When the flight surgeons were manning a console in the MCC their call sign was SURGEON. The physicians mentioned above were full-time, but occasionally a physician was called in on a part-time basis to man a remote site. Most of them were from a variety of military organizations. They were not full-time STG employees. At the remote sites, doctors were called AEROMED. Some would support only one flight, others many flights. Over 100 aeromedical monitors and specialists were deployed on ships and at remote sites for the MA-8 mission, and only slightly fewer for MA-9, which was the final mission in the series.

I couldn't find all of their names, but they included Austin, Bratt, Benson, Beckman, Bishop, Blackburn, Burwell, Davis, Flood, Fox, Graveline, Gull, Hall, Hansen, Hawkins,

Holmstrom, Kawalwiewicz, R. Kelly, F. Kelly, Kratochvil, Lane, Lawson, Luchina, Marchbecks, Moser, Overhold, Pruett, Reed, Rink, Shea, G. Smith, Trummer, Turner, Unger, Ward, and Watertown.

9.3.4   On-Board Systems Branch

This Branch was headed by Harry H. Ricker. There were originally just two sections, Electrical and Mechanical, but in reality this group studied all of the capsule systems and worked with the McDonnell engineers. By 1961, all of the Branch functions were integrated into the new Flight Systems Division, still headed by Max Faget, but with five Branches that had slightly different names which more accurately reflected what people were involved in now that the program was becoming better defined.

The original members of the On-Board Systems Branch included:

- Electrical Systems Section

> Harry H. Ricker, Jr., Acting Head
> Robert E. Bobola
> Norman B. Farmer
> John H. Hohnson
> Milan J. Krasnican
> Harold R. Largent
> Robert E. Munford
> Thomas E. Ohnesorge
> Ralph S. Sawyer
> James E. Towey

- Mechanical Systems Section

> John B. Lee Head
> Philip M. Deans
> James K. Hinson
> Witalij Karakulko
> David L. Winterhalter, Sr.

Later the following people were added to the Mechanical Systems Section:

- Harold Benson
- Walter W. Guy
- Robert H. Rollins, II
- James F. Saunders, Jr.
- Kenneth L. Suit.

9.3.5   Dynamics Branch

This Branch was headed by Robert G. Chilton, a former B-17 pilot with a master's from MIT. It consisted of two sections: Flight Controls and Space Mechanics. The Branch secretary was Jean S. Saucer.

The Flight Controls Section was headed by Richard R. Carley, a former AVRO engineer. They worked on the Mercury flight controls and developed the requirements for the contract proposal, then graded the submitted proposals. When the contract was awarded to McDonnell Aircraft, the group worked with the company and its subcontractor Minneapolis-Honeywell on rate gyros, fly-by-wire control systems and rate control systems, and evaluated various other systems.

In addition to Carley, the Flight Controls Section initially included:

- Thomas V. Chambers
- Stanley Galezowski
- Paul F. Horsman
- Thomas E. Moore
- Fred T. Pearce
- Donald J Jezewski
- Thomas N. Williams.

The Space Mechanics Section was headed by Robert G. Chilton, the Branch chief. It focused on requirements analysis, and the development and evaluation of systems. It also supported the very early Apollo studies, in particular regarding issues of orbital mechanics and trajectories for the re-entry corridor.

In addition to Chilton, the Space Mechanics Section included:

- Robert C. Blanchard
- Harold R. Compton
- Thomas F. Gibson, Jr.
- Jack Funk
- Morris V. Jenkins.

9.3.6 Computing Group

This Group was headed by Katherine S. Stokes who supported all of the Branches in the Flight Systems Division. She had access to the Langley IBM 704 and ran computations for anyone in the Division. Other computer ladies included Patricia D. Link, Mary W. McCloud, and Anne F. Wilson.

9.4 OPERATIONS DIVISION

This Division was headed by Charles "Chuck" W. Mathews. There were two Assistant Chiefs: G. Merritt Preston for Implementation (he soon would head up STG activities at the Cape) and Christopher C. Kraft for Plans and Arrangements (he soon would head up the Mercury Control Center). Chris Critzos was the Executive Engineer (he was the man who hired me). Robert. D. Harrington managed the coordination with the Army Ballistic Missile Division. John D. Hodge (a former AVRO engineer) was Mathews' assistant for operations. He was the Bermuda Flight Director and later a MCC Flight Director. Cdr. Paul L. Havenstein was the liaison officer from the Navy, but was also involved with operations planning. Edith K. Spritzer was the Division secretary.

The Operations Division had four Branches:

- Mission Analysis Branch
- Flight Control Branch
- Launch Operations Branch (later moved to Cape Canaveral)
- Recovery Branch.

9.4.1   Mission Analysis Branch

This Branch was headed up by John P. Mayer. His assistant was Jack Cohen. The front office secretary was Shirley J. Hatley.

The Branch was organized into three sections as follows:

- Trajectory Analysis Section
- Operations Analysis Section
- Mathematical Analysis Section.

Its structure remained essentially the same throughout the STG period.

9.4.1.1   Trajectory Analysis Section

This group was responsible for the trajectory analysis for launch, orbit, and re-entry. Given the capsule and launch vehicle designs, it analyzed various trajectories in terms of the mission rules and constraints. They developed nominal and contingency trajectories for different missions and determined the logic and equations for the computers. They considered the performance of the propulsion systems, guidance accuracies, loads, and heating restrictions. Depending on the abort conditions, various trajectories were computed and assessed in terms of the winds and dispersion effects on the landing area. This group computed many trajectories for specific missions and then performed real-time analysis to provide the MCC with critical data for its "Go/No-Go" decisions. The real-time computers would predict the orbital life-time almost immediately after insertion. The group conducted post-flight analysis and reconstructed the actual flight in order to help to improve the probability of success for subsequent flights.

In late 1959 the Trajectory Analysis Section included:

- John P. Mayer, Acting Head
- Charlie C. Allen
- John A. Behuncik
- James A. Ferrando
- Jack B. Hartung
- Claiborne R. Hicks, Jr.
- Carl R. Huss
- John W. Maynard, Jr.
- John C. O'Loughlin
- Ted H. Skopinski.

9.4.1.2   Operations Analysis Section

This group, informally known as the Simulation Design Section, initially focused on supporting the flight controller training effort and the astronaut training effort. See Section 9.2.6. Although in different organizations, the flight controllers and the astronauts were logically linked for many activities. The common threads were training, training facilities, and simulations.

In 1959 the Operations Analysis Section was primarily involved in defining the requirements for training at the yet-to-be-built MCC and the MSFN. These requirements were sent to Western Electric, the contractor for those facilities. Once the Mercury Procedures Trainers (MPT) were ready in 1960, crude simulations were conducted with mock-up tracking stations but once those facilities were available the simulation effort became more sophisticated. See Chapter 14.

As the simulation capability improved, radar and guidance data was introduced. Over time, the increasing capabilities required more people to create the tapes needed to drive the MPT as well as the MCC displays. At the MCC, even the RCA technicians supported simulations. They would verify various settings to help in diagnosing data problems. Sometimes they were "in" on the simulations and helped to trigger flight controller faults and subsequent actions. Some of the simulation tapes were sent to GSFC to enable it to transmit the simulated data to drive the MCC RETRO and FIDO displays. As the astronauts were presented with ever more complex failures, they strongly voiced their frustrations about what they believed to be "unrealistic" failures, but after some of these situations actually arose in flight they would thank the simulation people for the training. Losing either the MCC or a whole tracking station was thought unrealistic, but both actually occurred. Once a bulldozer severed a power cable and rendered the MCC inoperable for several hours, although fortunately not during a flight.

In due course, many simulations were conducted world-wide and the MSFN and MCC flight controllers began to perform like a professional flight control team. They were challenged many times during Mercury. Even more sophistication was introduced into the simulation process for the more complex Gemini and Apollo. The pace of development in the digital realm was rapid. New capabilities in the Houston MCC included simulation systems, consoles, and displays. An entire new Branch of simulation people was required to design and produce the simulations for new hardware such as the Agena docking target and the Saturn launch vehicle. But that was all after the STG had moved to Houston.

As of August 3, 1959, the Operational Analysis Section included:

- Jack Cohen, Acting Head
- Paul G. Brumberg
- Robert E. Davidson
- Arthur A. Hand
- John H. Lewis, Jr.
- Glynn S. Lunney
- Harold G. Miller.

More people were assigned later, including David A. Beckman.

9.4.1.3   Mathematical Analysis Section

In 1959 this section was headed by Stanley H. Cohn, a former AVRO engineer. At that time, the Section was coming up with standardized constants for mathematical conventions. For example, there were various different constants for measuring the Earth. Later the Earth was redefined as an oblate rotational ellipsoid. Some organizations involved with Project Mercury used different measurements for the radius of the Earth, for example. The Section also found there were many remote sites whose exact locations weren't accurately known, especially islands in the Pacific. It was important to know their locations in order to be able to predict the time of acquisition of a spacecraft as it approached a tracking station.

Cohn visited the Redstone people in Huntsville, Alabama and the Atlas people at Convair in San Diego, California and realized that they used slightly different constants which would give different computer solutions for vehicle trajectories. Finally, the military and the international community developed the World Geodetic System 1960 (WGS 1960). With the experience of many launches over many years, and better methods of measurement, the current model is the Earth Gravitational Model (EGMS 1996; revised 2004). The Global Positioning System would not be practicable without detailed knowledge of the planet's gravitational field.

During his visits to the launch vehicle suppliers, Cohn would compare their calculations to those using the Langley IBM 704. Since the IBM 704 was used by a lot of people at the Center as well as by other STG people, his Section acquired a Bendix G15 computer so they could run many more calculations themselves. See Appendix 3. In 1960, the STG took over the IBM 704 when the Langley Research Center acquired an IBM 7090 similar to those at the Goddard Space Flight Center. Many members of the Mission Analysis Branch and the Mathematical Analysis Section soon became proficient in the FORTRAN programing language. The demand was such that in 1961 a separate Digital Computer Group was established in the Office of the Director so that the entire STG could use that resource.

Eventually, the Langley IBM 7090 became a backup to the Goddard computers and John N. Shoosmith, another AVRO engineer, was the key person for real-time calculations. This team also created plots for the world maps and trajectories. This entire computer effort became more formalized with procedures for how programs were "de-bugged" and how to support a mission for pre-flight, flight operations, and post-flight analysis.

The original Mathematical Analysis Section included the following, but in 1961 some were transferred to the new Digital Computer Group (DCG) and other people moved in:

- Stanley H. Cohn, moved to the DCG
- Jerome N. Engel, moved to the Trajectory Analysis Section
- John N. Shoosmith, moved to the DCG
- Mary S. Burton
- Nancy K. Carter
- Shirley A. Hunt (later Hunt Hinson)
- Elizabeth P. Johnson, moved to the DCG
- Pattie S. Leatherman
- Catherine T. Osgood.

In 1961, as actually Mercury flights began, more people moved into this group including:

- Paul G. Brumberg
- Lynwood C. Dunseith
- Pauline O. Leonard
- John H. Lewis Jr.
- Athena T. Markos
- John Maynard, Jr.
- Emil R. Schiesser.

9.4.2 Flight Control Branch

This Branch was headed by Gerry W. Brewer. Nancy C. Lowe was the Branch secretary until the astronauts came onboard, whereupon she transferred to that organization. Joan B. Maynard took her place. The Branch was located in Building 104 and carried out all the planning for operating the MCC and the remote sites, including the training of flight controllers. It originally comprised the Control Central and Flight Safety Section and the Training Aids Section. However, in 1961 it reorganized into the Flight Control Facilities Section and the Flight Control Operations Section in order to better reflect what people were doing; the Training Aids Section was moved to the Spacecraft Operations Branch and the Flight Simulations Section. The organization of the STG was beginning to stabilize as it prepared for actual flights and as the control center and remote site facilities became operational.

9.4.2.1 Control Central and Flight Safety Section

In 1959 this group focused on getting ready to build and man a new control center, manning the remote tracking stations, learning the capsule systems and how to operate in the new spaceflight environment. It was headed up by Gerald W. Brewer. His assistant was Howard C. Kyle. It did not have any of the original STG engineers, but had three of the AVRO engineers and some of the people from the Langley Research Center.

Those that focused on the new Mercury Control Center were Howard C. Kyle, C. Frederick Matthews, and Tecwyn Roberts. Kyle became an MCC CAPCOM on the early flights and then worked on the requirements for the tracking stations and their performance. Matthews became a backup Flight Director and also focused on flight controller training. Roberts focused on launch, orbit, and re-entry flight dynamics and became the first FIDO. The following year Gene Kranz joined this effort and focused on operational procedures, countdowns, and coordination with the remote sites. He became the first PROCEDURES officer in the MCC. A year later I joined this group. By that time, it was called the Flight Operations Section and I was assigned to Kranz to work on mission rules and the communications interface to the flight controllers at the remote sites.

Arnold D. Aldrich and Dennis E. Fielder focused on the operational aspects of the remote sites. Aldrich was initially a remote site CAPCOM flight controller at the remote sites and later the SYSTEMS flight controller in the MCC. Fielder initially focused upon the flight controller requirements for the remote sites and interfaced with the Langley Instrument Research Division which was working with Western Electric to build the tracking stations, and he traveled to some sites to work on the communications interface with the MCC. He was later involved in training flight controllers for the remote sites, and worked with the simulation people.

By 1961 some people in the Control Central and Flight Safety Section had moved on to other organizations. The staffing of the two newly named sections of the Flight Control Branch were as follows:

- Flight Control Operations

    Gerald W. Brewer, Acting Head
    Howard C. Kyle, Assistant Head
    Arnold D. Aldrich
    Robert E. Ernull
    Eugene F. Kranz
    John T. Koslosky
    C. Frederick Matthews
    Richard F. Schultheiss
    Manfred von Ehrenfried

- Flight Control Facilities

    John H. Dabbs
    Dennis E. Fielder
    David T. Myles
    Leonard E. Packham
    Tecwyn Roberts
    William L. Davidson (came late 1961).

9.4.2.2   Training Aids Section

This group was headed up by Harold I. Johnson who came from the Langley Flight Research Division, as did his colleague Stanley Faber. It supported the Astronaut Training Group with simulators and other training equipment. As such, it coordinated with George C. Guthrie and Raymond G. Zedakar of the Astronaut Training Group. Initially, Faber was involved with the training of the astronauts utilizing the Navy's centrifuge at the Aviation Medical Acceleration Laboratory in Johnsville, Pennsylvania. Back then it wasn't known how much "g" load a pilot could take. Many runs were made for each astronaut in their specially constructed couches, to determine the optimum position to facilitate breathing, to operate controls, and to stave off unconsciousness.

When the Mercury Procedures Trainers became available in 1960, Charles C. Olasky, Jr. took the lead in developing and conducting simulations with this new trainer. He worked with Stanley Faber and the Operations Analysis Section which, as explained in the previous section, was also involved in simulations. Now there were three different organizations involved with training astronauts and flight controllers, and either obtaining or operating special training facilities.

In reorganizing the Flight Control Branch in 1961, the Operations Division transferred the support for training the astronauts to a Flight Simulation Section in the Spacecraft Operations Branch. There was still some mission and trajectory simulation work done by the Operations Analysis Section related to driving the FIDO and RETRO displays in the MCC.

In late 1959 the Training Aids Section included:

- Harold I. Johnson, Head
- Bruce A. Aikenhead
- Stanley Faber
- Arthur E. Franklin
- Rodney F. Higgins
- Richard A. Hoover (1960)
- Alfred J. Meintel, Jr.
- Charles C. Olasky, Jr.

9.4.3 Launch Operations Branch

Initially Charles Mathews headed up this Branch. As early as May 1, 1959, B. Porter Brown was sent to Cape Canaveral to coordinate with the Air Force Missile Test Center, which operated the Atlantic Missile Range. Brown also coordinated with the Army Missile Firing Laboratory, which was a contingent of the Army Ballistic Missile Agency. This effort required more people, and an AMR Project Office was set up with Brown, Philip R. Maloney and Elmer H. Buller to work the interface with the organizations at the Cape. There was always conflict between the military and civilian use of the Cape facilities.

In January 1960, G. Merritt Preston moved to the Cape to head up all the launch operations, including pre-flight checkout of a capsule. Scott Simpkinson would coordinate with McDonnell on the checkout there and continue the final pre-launch checkout in Hanger S at the Cape. He was the capsule operations manager. This team grew to about 35 people and it was always an intense activity to get a capsule ready by the planned launch date.

9.4.4 Recovery Operations Branch

Robert F. Thompson headed up this Branch, which was involved in qualifying the capsule for landing on land as well as at sea. Drop tests were carried out to evaluate the landing bag and to verify the capsule would float properly. The various recovery aids were also tested, such as the beacons, dye markers, flashing lights, shark repellents, and radio aids, as well as the SOFAR (Sound Fixing And Ranging) bombs. It was also involved with testing the drogue parachute, main parachute, and reserve parachute.

Branch personnel were involved with the design of recovery systems, manning of Navy's recovery ships, and coordinating with the mission planners concerning the planned recovery positions. Once in orbit, Thompson would coordinate with the Navy on planned and possible contingency capsule landing points.

The Recovery Operations Branch originally included the following people:

- Robert F. Thompson, Head
- Peter Armitage
- John B. Graham
- Eziaslav N. Harrin
- Enoch M. Jones
- Carl J. Kovitz
- Charles I. Tynan, Jr.
- Julia R. Watkins
- Milton L. Windler.

Later, the following people were added:

- Gilbert M. Freedman
- Harold E. Granger
- Jerry Hammack
- William C. Hayes
- Leon B. Hodge
- Walter C. Hoggard, Jr.
- John C. Stonesifer.

9.5   ENGINEERING AND CONTRACTS DIVISION

There is one major function that the STG had to perform which the Langley Research Center didn't; namely they had to manage a major space program with many subcontractors. Because the engineering of the Mercury capsule was still in flux, this Division needed very experienced engineers as well as contract administrators. James A. Chamberlin was the Chief of Technical Design for the AVRO CF-105, which was arguably more complex than the Mercury capsule. Andre J. Meyer, Jr., who was Chamberlin's Assistant, was one of the original Lewis engineers who worked with Max Faget on the heat shield. Norman F. Smith, the Executive Engineer, had significant experience with the Little Joe at Wallops Island. The Division staff was rounded out by John C. French, an experienced systems engineer. He was the STG's chief of reliability and flight safety and worked on overall systems reliability, including the capsule and launch vehicle. Acquilla D. Saunders was the front office secretary.

In 1959, this Division had the following four groups:

- Field Representatives at McDonnell Aircraft
- Capsule Coordination Office
- Contracts and Scheduling Branch
- Engineering Branch.

It became apparent in late 1959 that the STG was in desperate need of 718 people, and that this Division needed 200 more positions for technical and administrative support. By 1961, it also was clear that the job was too big for one Division, so Chamberlin broke it up and gave the procurement and supply function as well as some of the contract administration and purchasing to the Office of the Director. Most of the engineering functions that didn't involve projects and contracts were given to Max Faget's Flight Systems Division. Chamberlin also reorganized the remaining functions into the Engineering Division consisting of the Project Engineering Branch and the Contract Engineering Branch.

9.5.1 Project Engineering Branch

In 1961 this new Branch focused on coordinating with the launch vehicle manufactures and the issues that came up after each flight and/or the next flight. The Special Projects people worked on issues related to non-launch vehicle problems, such as parachutes, landing bags, and various problems of the day. Some of these people also worked on the Mercury Scout project to place a satellite into orbit in order to test the MSFN but the launch on November 1, 1961 failed.

Some people, such as Rodney Rose and Donald Arabian worked with the Atlas people only briefly and later moved to either the Flight Systems or Operations Divisions.

- Atlas Project

    Lewis R. Fisher
    William T. Lauten Jr.
    Albert J. Saecker
    Harry C. Shoaf

- Redstone Project

    Jerome B. Hammack
    Joan P. Samonski

- Special Projects

    Joe W. Dodson
    Donald T. Gregory.

By August 1961 the two Mercury Redstone flights had been made and the team was working toward the orbital flights. The STG organization was still making adjustments but the Divisions were fairly well defined, with many people transferring from Langley and the others being new hires. By this time, the following additional people were in the Project Engineering Branch:

- Donald D. Arabian
- Leonard J. Boler
- Evelyn B. Fitzgerald
- David C. Grana
- William R. Humphrey
- Carol L. Johnson
- Walter J. Kapryan
- Carroll D. Lytle
- Archibald E. Morse, Jr.

- Edward H. Olling
- Rodney G. Rose
- Albert J. Saecker.

9.5.2   Contract Engineering Branch

This reorganized group retained George F. MacDougall, Jr., as its Head and Joseph V. Piland as Assistant Head. Margaret Marshall was the Branch secretary. They were located in Building 104. In addition to these three, the following people were part of the new organization. Those with an asterisk were part of the original 1959 Contracts and Scheduling Branch:

- Contract Section

    Richard F. Baillie*
    Jack Barnard*
    James A. Bennett
    James E. Bost
    Bryant L. Johnson
    Francis s. Karick
    Paul H. Kloetzer
    Diane F. Sawyer
    John B. Goslee

- Scheduling Section

    John A. Rann
    Nicholas Jevas
    William C. Muhly*
    Carol C. Reed
    Lester A. Stewart*
    Paul M. Sturtevant*
    Kenneth J. Vogel
    Ralph L. Westphal

- Files Unit

    Ray S. Woodman
    Kathryn E. Linn
    Janette H. Beck
    Earnestine H. Wolfer

- Transportation

    John R. Bailey*.

10

Representatives and Contractors

10.1 MILITARY

From the very start, Robert Gilruth knew the importance to the STG of the various branches of the military. It relied on the Army for the Redstone launch vehicle, the Air Force for the Atlas, and the Navy for its Vanguard experience and the Johnsville centrifuge for astronaut training. Also the astronauts were all military, and required aeromedical and physiological support. Even though their pressure suits were made by B.F. Goodrich, they were of a Navy Mark IV design.

In Gilruth's memos dated August 3 and 10, 1959 (See Appendix 1) to the STG organization, the following military personnel were assigned:

- Col. Keith G. Lindell, USAF – Staff Assistant to the Director and Head of the Astronauts and Training Group
- Lt. Col. Martin L. Raines, USA – Staff Assistant to the Director and the Liaison Officer for the Army Ordnance Missile Command; the former Army Ballistic Missile Agency and the Redstone Arsenal
- Cdr. Paul L. Havenstein, USN – Staff Assistant to the Director, Assistant to the Operations Division for Operational Planning
- Lt. Col. John A. Powers, USAF – Public Affairs Officer. Served as mission commentator in the MCC for all six manned Mercury flights
- Lt. Col. William K. Douglas, USAF – Flight Surgeon, Astronauts and Training Group
- Lt. Robert B. Voas, USN – Training Officer for the astronauts
- Lt. Col. Stanley C. White, USAF – Head of the Life Systems Branch
- Lt. Col. James P. Henry, USAF – Specialist for Biological Flight
- Capt. William S. Augerson, USA – Life Systems Branch.

For the most part, these men were on temporary assignment to NASA and maintained their rank, but they usually wore civilian clothes. During the STG period through 1961, Dr. Stanley White manned the SURGEON console in the Mercury Control Center. For later flights, it was manned by other flight surgeons including Lt. Col. Dr. William K. Douglas, Dr. Charles Berry, and Dr. Duane Catterson.

© Springer International Publishing Switzerland 2016
M. von Ehrenfried, *The Birth of NASA*, Springer Praxis Books,
DOI 10.1007/978-3-319-28428-6_10

The most visible and well known of the military supporting Project Mercury was the Public Affairs Officer, Lt. Col. John A. Powers, better known as John "Shorty" Powers, who was seen on TV many times and was known as the "Voice of Mercury Control." He sat in the top row of the Mercury Control Center (MCC) next to the NASA Mission Director Walter Williams. Also on that row were two other military officers; one an Admiral who represented the Navy forces supporting recovery, and the other a General representing the Atlantic Missile Range support. Both were always in uniform. In the event that the military services were required to provide support, these men would assist the Mission Director or Flight Director in the coordination of military assets.

10.2   STG CONTRACTORS

10.2.1   McDonnell Aircraft

The interface between the STG and McDonnell was sufficiently important for there to be NASA representatives at the factory to oversee the manufacture of the Mercury capsules. Those on duty there were headed by Wilbur H. Gray, and they included William J. Nesbitt and Louise E. Kase. On occasion, McDonnell personnel would visit Langley for coordination and training of flight controllers on the Mercury spacecraft systems. They also provided a Capsule Flight Operations Manual that was used by the astronauts and flight controllers. This led to the Flight Controllers Handbook. Two company representatives were detailed to the Flight Control Branch. They were Ed Nieman and Dana Boatman. McDonnell had contracted out the Mercury Procedures Trainer to the Link Trainer Company. One trainer was at Langley in Building 643 and the other was at the Cape in the Engineering Support Building (Telemetry No. 3). Riley McCafferty would assist the simulation people with incorporation of the latest Mercury capsule data.

10.2.2   Philco

The largest contractor team on-site at the STG was the Philco team of 16 engineers/technicians who provided the Operations Division with communications and technical support. They had previously manned tracking stations for Discoverer, Vanguard, and Explorer satellites as well as various classified missions. They had been to Vandenburg in California, Kodiak in Alaska, and Kwajelein in the Pacific and were knowledgeable about construction and checkout of facilities. They were assigned to the Flight Control Branch, and during Mercury missions many of them would be assigned to the remote tracking stations as SYSTEMS flight controllers. They brought some instant maturity and experience to the relatively younger NASA flight control team.

10.3   MERCURY CONTROL CENTER CONTRACTORS

The MCC was designed by the U.S. Army Corps of Engineers and Burns and Roe, Inc., and built by Carlson-Ewell, but they had no representatives in the MCC. The Cape Canaveral Air Force contractor for all facilities was the Guided Missile Range Division of Pan American (PanAm) Airways.

Fig. 10.1 The Philco team at the south entrance to Langley STG Headquarters in Building 58. Front row (l-r): Jim Tomberlin, Wilbur Hubert, Ted White, Jim Strickland, Dan Hunter, Sy Rumbaugh, Lou DeLuca, Lloyd White. Back row (l-r): Harold Stenfors, John Gorman, Larry Wafford, Al Barker, Harry Hopp, Dick Cross, Marv Rosenbluth, Dick Rembert. (Photo courtesy of NASA)

10.3.1   Radio Corporation of America (RCA)

RCA had the PanAm technical support subcontract for the MCC, as well as other facilities in the Atlantic Missile Range. They handled all of the technical support functions. From an operational perspective, the MCC flight control team interfaced closely with the RCA communications, data and display people. This included communications technicians for voice, teletype, and television. There were radar technicians present to support the range tracking functions. The Operations and Procedures Officer, designated PROCEDURES, coordinated the voice and teletype support with the RCA team of Andy Anderson and J. Eshelman. The Support Control Coordinator, known as SUPPORT, was John Hatcher, and he was the primary interface with the team for resolving any technical support problems. Due to his knowledge of data flow interface from the various Cape facilities into the MCC, he also played a training role to some of the flight controllers.

10.3.2   Western Electric

The Goddard Space Flight Center selected Western Electric as the prime contractor and overall program manager for the construction of the Mercury (later Manned) Space Flight Network that included the MCC. One of their responsibilities was the training of maintenance and operating personnel. Their representative to NASA was Paul Johnson, who wrote a manual on the MCC structure, functions and plans for checking out the facility. He wasn't permanently assigned to the MCC or STG, since he also worked on the remote sites. Johnson coordinated with the RCA team and initially worked with Operations and Procedures Officer, Gene Kranz, and introduced him to the new control center operations in November 1960 in time to conduct the launch of MR-1 as the first operational use of the MCC.

10.3.3   Bell Telephone/Stromberg-Carlson

Bell Telephone was one of the Goddard Space Flight Center MSFN major contractors involved with the design of the command and control displays in the flight control centers at the Cape and on the island of Bermuda. Stromberg-Carlson was their subcontractor for building and installing the displays for the flight controllers. The company was also involved in providing a simulations system. John Hibbert was their representative. Dick Koch was the voice systems expert.

10.3.4   IBM

As early as 1956, the Office of Naval Research contracted with IBM to install and operate on a 24-hour basis an IBM 704 analog computer (at that time known as an "electronic calculator") in downtown Washington, DC, to process satellite data. As it turned out, Sputnik provided the test and checkout of the Minitrack system created for Vanguard. This world-wide system was later transferred to NASA, and the Goddard Space Flight Center received not only the hardware but also hundreds of Naval Research Laboratory people.

Although the STG made use of the IBM 704 computer, it was initially owned by the Langley Research Center. When Langley acquired an IBM 7090, the old 704 was transferred to the STG. Similarly, the IBM computers used by the Mercury Control Center were owned by the Goddard Space Flight Center. During a mission, IBM personnel Ira Sachs and Al Layton would run data flow tests from the MCC to the Goddard computers to verify the quality of the data lines. Both the RETRO and FIDO flight controllers relied on these lines for their plot board displays.

10.3.5   Redstone/Atlas

The Mercury Redstone launch vehicle was built by Chrysler, but it was represented at the STG by Lt. Col. Martin L. Raines, who was the Army Ordinance Missile Command liaison officer. Robert D. Harrington was on Charles Mathews' staff in the STG as a Ballistic Missile Division coordinator. For launches, most of the Redstone team was in the blockhouse that was part of the pad complex. For MR-1 only, one representative was in the MCC; Dr. Joachim "Jack" Kuettner. Typically, the Flight Director would interface with the (Redstone or Atlas) Launch Vehicle Test Conductor via voice communications; they wouldn't have had any representatives in the MCC.

Although the Mercury Atlas launch vehicle was built by Convair, it was represented at the STG by a USAF representative. There was a great deal of coordination between the Air Force, Convair, and NASA on the Abort Sensing and Implementation System which was new for the man-rated Atlas.

11

The Need for More People

It was clear from the outset that the STG was going to need many more people than the original Core Team. According to a memorandum from Robert Gilruth dated January 26, 1959 – barely four months after the creation of NASA and three months after the creation of the STG – there were only 71 people in the entire organization; including six recently assigned military officers. Ten of the 71 were secretaries and file clerks on the administrative side and several more ladies were in computer support.

The January 26, 1959 memorandum by NASA Administrator Glennan clearly made Gilruth responsible for Project Mercury. It also told the Director of the Langley Research Center (LRC) to provide the STG with such administrative and supporting services as it might need, and told Gilruth to feel free to obtain these, which he rapidly did. By the end of 1959, the STG staff was already up to around 400. The memorandum didn't address technical support but it was evident that the STG would require to draw upon its host's areas of expertise and unique facilities, both of which were considerable.

11.1 LANGLEY SUPPORT TO PROJECT MERCURY

In addition to transferring the initial Core Team of people to the STG, Langley Acting Center Director Floyd L. Thompson worked closely with Gilruth to support Project Mercury with his people and capabilities. There were many people at Langley that were traditional researchers, technicians, and shop fabricators who remained at the center but were interested in supporting the project.

In 1958 the LRC employed 1,151 professionals and 2,145 non-professionals for a total of 3,296 people. Many of them were involved in traditional aeronautical research and testing of military and civilian aircraft, as well as working on NASA research aircraft such as the X-15 rocket plane. But now they were also required to support the STG and Project Mercury. Max Faget's Flight Systems Division needed support from the Instrument Research Division and the Applied Materials and Physics Division. Jack Kinzler, head of

© Springer International Publishing Switzerland 2016
M. von Ehrenfried, *The Birth of NASA*, Springer Praxis Books,
DOI 10.1007/978-3-319-28428-6_11

Technical Services on Gilruth's STG staff, interfaced with the Langley shop to fabricate items such as models for wind tunnel tests and launching from Wallops Island. The following are general summaries of the support provided.

11.1.1   Little Joe and Big Joe Support

In 1959 there were four flights with Little Joe out of Wallops Island Station and one Big Joe out of Cape Canaveral, followed by three more flights in 1960, one of which was a beach abort test. This support was provided by the former PARD personnel who remained with the LRC, led by Joseph A. Shortal. They included engineering, fabrication, testing, and flight operations people. For example:

- Use of various wind tunnels to study pressure distributions, determine loads and wake surveys and parachute tests
- Launch and recovery support, including helicopters and boats
- Stability tests in water with and without the heat shield and airbags, and effects of escape rockets on stability
- Structural testing of vehicle components, including recovery hooks, parachute lines, and bolts
- Vibration, flutter and noise tests
- Shop support including fabrication of vehicle and capsule components
- Pyrotechnics testing at the hypersonic physics test area
- Heat transfer tests and temperature distribution tests of capsule panels
- Breaking-rocket techniques
- Post-flight analysis.

11.1.2   The Tracking Systems Study Group

Almost immediately after the creation of NASA, Langley's Assistant Director Hartley A. Soule pulled together a team primarily from the Instrument Research Division. This Tracking System Study Group (TSSG) defined the network support requirements for Project Mercury. It initially included:

Hartley A. Soule
Edmund C. Buckley
James J. Donegan
Ray W. Hooker
George B. Graves
Frances B. Smith
Paul H. Vavra
H. William Wood.

Howard C. Kyle was initially part of the IRD, but transferred to the STG as the interface between the Flight Control Branch which required the network and the TSSG (and later the TAGIU) which was to design it. As this group developed the requirements, they, and others, realized that tracking and communicating with the Mercury capsule in orbit was a bigger job than had been envisaged. It was not the same as tracking a Vanguard or Explorer satellite. In only a few months, Charles Mathews, head of the Operations Division, recommended to Abe Silverstein at NASA Headquarters that this responsibility be assigned to the TSSG. Thus, on February 16, 1959 the TSSG was officially named the Tracking And Ground Instrumentation Unit (TAGIU). The team grew to 35 people and was responsible for preparing the contracts to build the world-wide network. In the spring of 1959, the four major contracts for what would become the Mercury (later Manned) Space Flight Network were awarded. The team stayed at Langley until the Goddard Space Flight Center was built in 1961, then they transfer there. See Section 14.3.

11.1.3   Other LRC Support to Project Mercury

In addition to supporting the Wallops Island Little Joe launches and the Cape launches of Big Joe, Langley supported other Mercury activities, including:

- Support for the Mercury tracking station scaled mockup, and supervising the Western Electric contractors
- Theoretical trajectory and error studies
- IBM 704 support
- Wind tunnel parachute tests
- Capsule instrumentation for MA-1 and MA-3
- Fatigue testing of vehicle components
- Flight model tests
- Three-axis hand controllers (also known as pilot's sidearm controllers), control center simulators
- Astronaut egress training.

During the STG's first year, Langley provided the support of 325 professionals, the cost of which was borne by the LRC budget. During the next year, it was still providing a significant amount of support to Project Mercury. With authority to rapidly hire, the STG was up to about 600 personnel by the end of 1960, taking some of the load off its host. By its end in November 1961 the total had grown to about 750 people.

11.2   STG HIRING

Although by August 1959 the organization of the STG had started to settle down, it was still growing in numbers. There was a Staff Services Office under the Office of the Director headed by Burney H. Goodwin, whose title was Personnel Assistant. It included Guy W. Boswick, Jr., and two receptionists, Jo Ann S. Fountain and Ann W. Hill. Gilruth's Special Assistant, Paul E. Purser was very much involved in hiring and worked with the STG Division Chiefs who made known their personnel needs. There was a whole Personnel Division in the LRC, as you would expect for a national laboratory. Some of their methods and procedures carried over to the new organization but with much less formality and less of the usual government red tape. While the STG was to do the hiring, each organization defined the kind of people it wanted, and some of the assistants to those Division Chiefs were active in the hiring process. There were some who knew people outside of NASA whom they believed would fit in. As a result, personal contacts were made outside of the normal personnel hiring practices.

The astronauts came onboard in April 1959, and with them came the hiring of many military flight surgeons, aeromedical people, and trainers. The requirement to protect the astronauts and their families necessitated hiring security personnel or transferring them from the LRC. A new Life Systems Branch was staffed up by August. Over the summer of 1959 the AVRO engineers began reporting in to the STG, and the growth of the organization required more administrative support including secretaries, payroll, travel, and stenographers (yes, there were such people in those days).

On August 3, 1959 Gilruth sent a staffing memorandum to the STG that included 322 people and several organization charts. But the plan was to have 488 authorized positions by the end of the year. See Appendix 1. At about this time, Max Faget was saying he was greatly understaffed and James Chamberlin said he urgently needed 200 more people. Hiring of more people was an urgent STG priority and they moved out smartly.

There were several colleges and universities in Virginia (home of the LRC) and neighboring states that were prime sources of new graduates possessing science, technology, engineering and mathematics degrees. Nowadays we label these STEM (Science, Technology, Engineering and Mathematics) academic fields, but that term wasn't in use then. The NACA and LRC typically looked for aeronautical and mechanical engineers as well as physicists and mathematicians. The relatively local schools that provided people to NASA and the STG during the 1950s and 1960s included:

- University of Virginia
- University of Richmond (my alma mater)
- College of William & Mary
- Old Dominion University
- Randolph-Macon College
- Virginia Military Institute
- Virginia Polytechnic Institute
- Mary Washington College
- Washington and Lee University
- North Carolina State.

There wasn't a "Monster.com" in those days; hiring was done in the old fashion way. It was personal. You talked with a friend that you thought might be interested in the program. Project Mercury was really exciting and in the news. The local colleges knew what was going on at the Langley Research Center because it traditionally took their graduates. Someone told his friend about the program and encouraged him to go see "so and so" at NASA for an interview. If the NASA guy liked what he saw he would go to his supervisor and report he had found a good fit for some task or other. The supervisor either carried out his own interview or told the personnel office to make a hire. The personnel person would telephone the guy and either set an interview or simply invite him to report to the STG Staff Services Office at Langley on a specific date and time.

In my case, I had just left the Air Force side of the field and saw a sign that said NASA, so I walked in the door. I didn't know anyone and my interview was with the Executive

Engineer of the Operations Division, Chris Critzos. See Appendix 7 for my hiring experience.

The same thing would happen if the prospect was at another company, typically in aviation. There was relatively little formality but there was the usual detailed government form to fill out, and in most cases a security clearance would be required later. A lot of people were hired from universities and aviation companies by the STG between 1959 and 1961. Some military people were temporarily assigned, but some actually hired on. When the STG ended in late 1961 about 750 people transferred either to the new Manned Spacecraft Center, to the Goddard Space Flight Center, or to NASA Headquarters.

11.2.1 Procurement and Contracts

If there was one area in which the Langley people had little training and experience, it was in the procurement of large systems and managing contracts. It was one thing for a Langley engineer to manage the construction of a wind tunnel or to purchase special equipment, and quite another to manage a major task which had never been attempted before, such as the procurement of a space capsule. Robert Gilruth recognized this was going to be a challenge and directed the Engineering and Contracts Division headed by James A. Chamberlin to manage this effort. See Section 9.5.

In August 1959, this Division had around 50 people. In January 1960, it was effectively split into two, with Chamberlin leading the Engineering Division and George F. MacDougall leading the contracting effort with Joseph Piland, who interfaced with McDonnell Aircraft's contracting officer. Eventually, McDonnell Aircraft would draw upon 50 prime contractors and in excess of 5,000 subcontractors. The hiring process continued.

11.2.2 Flight Systems

On January 26, 1959, only 21 people made up the original staff of the Flight Systems Division headed by Max Faget. Eight of the 21 were members of the STG Core Team. Several came over with the PARD group and the others came from either the LRC or were hired afresh. Four of the group were women; either secretaries or computer types. For the most part, they were engineers whom Faget knew or had worked with. The Life Systems Branch headed by Dr. Lt. Col. Stanley C. White hired aeromedical and life support people with applicable flight experience. That group alone added 16 people to the STG total. By August the Flight Systems Division had grown to 98 people in five Branches and a Computing Group. See Section 9.3.

11.2.3 Operations

To support the areas of pre-launch, launch, on-orbit, and recovery, the Operations Division had to grow rapidly. On January 26, 1959, three months after the creation of the STG, this Division had only 16 people. By August 3, 1959 it had grown to about 130; 150 if you count contractors. Many of those came from previous Langley organizations like the PARD. About 40 came from the Lewis Research Center in order to support the capsule checkout at the Cape, about 10 came from AVRO, and 16 were Philco employees. There were additional transfers in from other LRC organizations and new hires. See Section 9.4.

11.2.4   Engineering

After the breakup of the Engineering and Contracts Division, the Engineering Division was left with only around 24 people. It had field representatives at McDonnell, a Capsule Coordination Committee, and an Engineering Branch headed by Caldwell C. Johnson. In order to manage the incessant changes to the capsule and all the flight hardware, this group had to continue hiring as well.

In summary, the STG expanded from the original 36 from Langley and the 10 from Lewis in October 1958 to about 750 by its end in 1961. It was a period of rapid hiring in order to support a very active space program. But if you add up all the support from the other NASA agencies, the DOD range and recovery support, the prime and subprime contractors, and the military up to the end of Project Mercury in 1963, approximately 2 million people were involved. Yet it all started with the small STG team of space pioneers at Langley Field, Virginia.

12

The End of the Space Task Group

When the STG was formed in late 1958 it was administratively under the Goddard Space Flight Center which didn't yet have a facility and so it was temporarily based at Langley. The Langley people thought that eventually they would move to the new Goddard site at Beltsville/Greenbelt, Maryland. However, it was soon realized that another new center would be required for Project Mercury and later manned space programs. The STG was extremely busy with carrying out its flight schedules and didn't need the distraction of having to pick a site and make the move. The STG personnel were informed they would ultimately move to some yet to be disclosed location. When the decision finally came in 1961, families had to cope with it; some actually cried, some were eager, some were resigned, some refused to move… but about 750 did.

12.1 THE DECISION

In August 1961 John F. Parsons, Associate Director of the Ames Research Center (ARC) was tasked by NASA Headquarters to head a survey team to recommend the permanent site for the new center for manned space missions. The team, which included the STG's Martin A. Byrnes, came up with the following selection criteria:

- Available facilities for advanced scientific study
- Power facilities and utilities
- Water supply
- Temperate climate
- Adequate housing for center personnel
- At least 1,000 acres of land for the installation
- Industrial facilities available
- Transportation facilities including water transportation for shipping cumbersome space facilities by barge
- A first class, all-weather jet service airport
- Local cultural and recreational assets.

© Springer International Publishing Switzerland 2016
M. von Ehrenfried, *The Birth of NASA*, Springer Praxis Books,
DOI 10.1007/978-3-319-28428-6_12

Twenty sites seemed to meet these criteria:

- Tampa, Florida
- Jacksonville, Florida
- New Orleans, Louisiana
- Baton Rouge, Louisiana
- Shreveport, Louisiana
- Houston, Texas
- Beaumont, Texas
- Corpus Christi, Texas
- Victoria, Texas
- St. Louis, Missouri
- Los Angeles, California
- Berkeley, California
- San Diego, California
- Richmond, California
- Moffett Field, California
- San Francisco, California
- Bogalusa, Louisiana
- Liberty, Texas
- Harlingen, Texas
- Boston, Massachusetts

On September 19, 1961, James E. Webb, who had succeeded T. Keith Glennan as NASA Administrator on February 14 of that year, announced that the new Manned Spacecraft Center (MSC) would be established on a 1,000 acre tract near Houston to be transferred to NASA by Rice University. It now occupies 1,620 acres. At first, it was called the Manned Space Flight Laboratory. There were many who thought the decision was purely political; imagine that! Consider that all of the following were Texans: Vice President Johnson was chairman of the National Aeronautics and Space Council, Samuel T. Rayburn was Speaker of the House of Representatives, Representative Albert Thomas chaired the House Appropriations Committee, and Olin W. Teague not only served on the House Committee on Science and Astronautics but was also in charge of the Subcommittee on Manned Space Flight. The Government denied any improper influence! By October 13, 1961, NASA had established the building requirements for as many as 3,151 personnel.

The center had to provide:

- Project Management
- Flight Operations and MCC
- Life Systems Laboratory
- Technical Services
- Technical Shop
- Structures Laboratory
- Central Data Processing

- Research and Development
- Equipment Evaluation Laboratory
- Support Offices
- Warehouses and offices
- Project Test Laboratory
- Auditorium
- Cafeteria.

Some of the STG people from beautiful Virginia couldn't believe they were being asked to move to Texas. Some decided to remain behind at the Langley Research Center. Some went to Goddard or to Headquarters in Washington, DC. NASA photographer "Pat" Patneski took the accompanying photograph of the proposed site in Texas. He said he spent some time to get the cows to all look at the camera. Between this image and people's perception of it being straight out of the "wild west," there was a lot of concern about the move.

Fig. 12.1 The building site for the Manned Spacecraft Center. (Photo courtesy of NASA/ Andrew "Pat" Patneski)

Fig. 12.2 Just another view of where we're moving! (Photo courtesy of NASA)

12.2   THE MOVE

Shortly after Webb's announcement, Gilruth and members of the STG staff went to Houston to scout out some temporary quarters. Weeks later, Martin A. Byrnes and a team of procurement and personnel people opened an office in the Houston Gulfgate Shopping Center.

Back at the STG, a special relocation center was established in the Public Affairs Office to provide information on the office locations where people might be assigned, as well as essential information on schools, churches, etc. There was no mass exodus; individuals timed their move to accommodate the mission schedules and their particular support roles. Some flight controllers were spread out over the globe and deployed to the Mercury Control Center for the MA-5 flight of the chimpanzee Enos in November 1961. Some did not move until after John Glenn's MA-6 flight in February 1962. By that time, the NASA Mercury personnel total had peaked at 850. Of course the number of contractors across the country and the military support personnel involved in recovery operations numbered in the thousands. As the transition to Houston progressed, the number of Mercury people decreased and the number of people working on Gemini and Apollo increased.

Within a month after Glenn's historic first American orbital flight there was a grand showing of appreciation for the NASA STG team. On March 17, 1962 there was a 25 mile motorcade of 40 open convertibles through the cities of Newport News and Hampton in Virginia to the Darling Memorial Stadium in Hampton. The astronauts and their wives waved to the cheering crowds, as did the NASA STG managers, including Robert Gilruth, Floyd Thompson, and other senior LRC and STG people. The crowds waved their good-byes and cheered. At the Stadium, Public Affairs Officer John A. "Shorty" Powers introduced the astronauts. The Mayor of Newport News and Governor Albertus S. Harrison delivered speeches. The public that once called the employees of Langley "those NACA Nuts" now called them "NASA Wizards."

By July 1962, the move was essentially complete. The STG was no more. It had become the Manned Spacecraft Center, albeit scattered across many diverse locations, including:

- Gulfgate Shopping Center
- Stahl-Myers Building
- Houston Petroleum Center
- Ellington Air Force Base
- Phil Rich Fan Building
- Farnsworth-Chambers
- First Pasadena State Bank
- Ben Gordon Loan Building
- East End State Bank Building
- Franklin Development Complex
- Roberts Carpets Building
- Lane-Wells Building
- Canada Dry Bottling Plant
- Minneapolis-Honeywell Building
- TV Studio on Cullen Street
- Veteran's Administration Building.

When the construction work that began in April 1962 approached completion in 1964, many people thought it was absolutely beautiful. It was a campus with ponds complete with ducks and newly planted trees. The sparklingly new Mission Control Center was backup to the old Cape Mercury Control Center for Gemini 3 and then prime for all subsequent manned missions. The STG people who had pioneered spaceflight with Project Mercury were moving on to programs that would reach even greater heights – indeed all the way to the surface of the Moon.

Fig. 12.3 The tall building in the center of the photo without windows is the new Mission Control Center in 1964. (Photo courtesy of NASA)

Fig. 12.4 The completed Manned Spacecraft Center in 1965. (Photo courtesy of NASA/Dick Holt)

13

Some Key Project Mercury Decisions and Lessons Learned

13.1 MANAGEMENT

13.1.1 Decisions from "On High"

For a program as original and monumental as Project Mercury, very little is certain at the start. Nevertheless, management must have some clear direction from the top. President Eisenhower was very specific that he wanted a civilian space program in spite of the rather obvious military implications of Sputnik and the military's pushing hard to manage the manned space program. His directive solved at least two problems immediately. It made it clear that NACA would get the space program instead of either the Army or Air Force. And by directing the Department of Defense to support the civilian space program Eisenhower relieved NACA of the need to come up with independent resources and personnel for launching and recovering flights.

Eisenhower's directive and ongoing influence quickly put things into a clear perspective for all of the interested organizations. NACA already had the ideas, facilities, scientists, engineers, and reputation to carry out excellent work on budget. It also made it clear that the Army would stick to tactical missiles and the Air Force would stick to ICBMs, but provide NASA with man-rated launch vehicles. The Navy would carry out the recovery job. As it turned out, there was a great deal of cooperation between the civilian and military organizations, with everyone taking great pride in their roles in Project Mercury.

Another clear lesson was that there are always people at the top who, if not actually negative, are vociferous in their criticism. After Robert Gilruth of NACA briefed the President's Scientific Advisory Committee, some people were not enthusiastic. One member told Gilruth, "Your plan will provide the most expensive funeral a man has ever seen." However, NACA's Hugh Dryden was very pleased with the presentation. President Eisenhower signed the National Space Act on July 29, 1958. Gilruth's briefing to Congress on August 1 was delivered to a filled hearing room. NASA was established on October 1.

© Springer International Publishing Switzerland 2016
M. von Ehrenfried, *The Birth of NASA*, Springer Praxis Books,
DOI 10.1007/978-3-319-28428-6_13

So within one year of the shock of Sputnik, the U.S. created a national space program. Could such a program be achieved in today's political and high-debt environment? What does this say for a manned Mars program?

Certainly, one of the greatest programmatic decisions made by any President is the one that President John F. Kennedy made on May 25, 1961. With only one manned suborbital flight of experience in space, how could anyone make any bold statements about the future of American spaceflight, let alone the one which Kennedy made? Many NASA managers and staff reacted with incredulity to his famous speech, "I believe this Nation should commit itself to achieving the goal, before this decade is out, of landing a man on the Moon and returning him safely to the Earth." Sometime during the three weeks that followed Alan Shepard's flight, there had been a "sea change" in the public's perception of spaceflight. Members of Congress sensed the change and believed the American people were ready to support an expanded and ambitious long-term program. While the speech had an immediate and obvious impact on the STG, the positive and far reaching impact on America over the remainder of the decade will never be bettered by any President for decades to come.

A good example and lesson learned regarding negative political influences from the top is the decision by Canadian Prime Minister John G. Diefenbaker to cancel the CF-105 Arrow aircraft. This was a program started under Diefenbaker's political adversaries and, some would say once in power he seized the opportunity to cancel it at the first sign of trouble. Others tried to justify his decision in terms of the developing Soviet ICBM threat. The lesson here is that if a properly positioned politician with enough power, influence, and disregard for an ongoing program wants to cancel that program, then he certainly can.

13.1.2   Programmatic Management Lessons

Another management lesson was characterized by Christopher C. Kraft as the "Not Invented Here" syndrome. When a group of different organizations initially come together to work on a project, there can be a sense of distrust of the other person or organization's knowledge and/or capability. An excellent example of this was the meeting of minds needed in order to man-rate the Atlas booster. It involved the Ballistic Missile Division of the Air Force, its contractor the Space Technology Laboratories (STL; in 1960 the Aerospace Corporation), the manufacturer Convair/Astronautics, and finally the user, namely the NASA STG. The problem immediately became one of shared responsibility, who was responsible for what, and who would pay for the proposed changes. The situation was further complicated by the fact that there were some very experienced German rocket experts in these organizations. Imagine the conference room where the STG explained to the vested interests what it wanted to do with "their" Atlas. What, it was wondered, could "researchers" at a laboratory on a peninsula in Virginia possibly know about large military missiles? When, in discussing how they intended to man-rate the Atlas, the STG people questioned specific failure modes that would endanger the astronaut, the meetings often became contentious. The solution was to put a full-time NASA/STG representative in the STL offices to work out and coordinate the engineering modifications. Over time, a degree of mutual respect was gained and, after many technical and managerial meetings the desired changes were indeed made to the Atlas.

Another issue that wasn't so clear at first was why were there no women astronauts. It arose when female test pilot Jacqueline Cochran suggested the "Mercury Thirteen" female astronaut program to match the "Mercury Seven" group of male astronauts. But Eisenhower had declared early on that the astronauts must be recruited from the ranks of military test pilots. At that time the military test pilot schools didn't accept females, so this negated further discussion of female astronauts. This is an example of how an early programmatic decision can unwittingly lead to a later decision.

13.1.3   Inter-Agency Coordination

Abe Silverstein at Headquarters and Robert Gilruth at the STG soon realized the difficulties of working with so many external organizations, in particular those in the Department of Defense (DOD). The Army, Air Force, and Navy already had their own space-related programs and each very much wanted to claim the manned space program for itself as well. But now they were in a support role to NASA with Project Mercury. The Army had the Redstone, the Air Force had the Atlas, and the Navy had the recovery forces. And it did not help NASA that these organizations were spread out all across the country.

The DOD representative for military support to Project Mercury was Major General Donald N. Yates, Commander of the Air Force Missile Test Center at Cape Canaveral, and his tasking included pre-launch and launch support, Navy search and recovery operations, Army tracking and communications facilities, and joint service bioastronautics resources. His counterpart in NASA was Walter C. Williams from the High Speed Flight Station, who had been involved in the X-Series of aircraft including the X-15.

Here is a list of many of the organizations that had a supporting role for Project Mercury:

- Weather Bureau of the Department of Commerce
- Aviation Medical Acceleration Laboratory
- El Centro Naval Parachute Test Facility
- Wright Air Development Center
- Aero Medical Field Laboratory
- Air Force Survival School
- Aerospace Medical Division
- China Lake Naval Ordnance Test Station
- Navy School of Aviation Medicine and the Pensacola Naval Air Station
- Eglin Air Force Base
- Air Force Chart and Information Center
- Public Health Service
- Cherry Point Marine Corps Air Station
- Military Transport Service
- Pacific Missile Range
- Point Mugu Naval Air Station
- Corpus Christi Naval Air Station
- Navy Daingerfield Test Facility
- Navy Aircrew Equipment Laboratory
- Air Force Flight Test Center

- Army Audit Agency and Audit Office
- District Coast Guard
- Walter Reed Army Medical Center
- Navy Comptroller
- Navy Bureau of Ships
- Naval Research Laboratory
- Navy Bureau of Weapons and Plant
- Marine Corps Air Facility
- Air Rescue Service
- Air Force Communications Service
- Air Weather Service
- Air Force Tactical Air Command
- Air Force Surgeon General's Office
- U.S. Army, Fort Eustis
- U.S. Army, Europe
- U.S. Army European Command, Paris
- U.S. Army Research and Engineering Laboratory.

Each of these many supporting organizations had their clearly defined role and a single point of contact; especially with the military elements. There were Army, Air Force, and Navy liaison officers with the Office of the Director at the STG.

The lesson for all programs involving many diverse but supportive organizational elements was to have both formal and informal lines of communication. At the highest levels the formal lines needed people of stature who had respect for one another's experience and achievements. Whilst these individual attributes might be very different, they had to engender respect for the opinions and positions of others. High-level management of major organizations often attracts leaders possessing dominant personality traits, but they are able to appreciate the positions of others. After agreeing on the best course of action for the program they must provide guidance and direction to the supporting elements within their organizations. People at a more informal level may, or may not, have direct line authority but may still be able to influence the decision process. This level of decision making can often be subtle but effective.

13.1.4   Intra-Agency Coordination

Coordination between the new NASA field centers was fairly straightforward because the new people at Headquarters included people who, in many cases, came from those centers and had worked together in the NACA days. There were others from the military and from academia. In some cases, there were decades of association and close working relationships between the new high-level managers, with each aware of the reputations of their counterparts and their strengths and weaknesses. Consider the following Headquarters managers:

- T. Keith Glennan – the first NASA Administrator was a Republican political appointee on the advice of James R. Killian to President Eisenhower, being hired for his administrative leadership. He became President of the Case Institute of

Technology in Ohio in 1947 and was given extended leave of absence in order to take up the NASA appointment. He was also a member of the Atomic Energy Commission.

- Hugh Dryden – the first Deputy Administrator of NASA. He had been Director of NACA since 1949. He was chosen for his scientific and technical background and knowledge of NACA facilities, programs, people, and skills. He was personally picked by Glennan and provided a continuity of management as NACA evolved into NASA.
- Richard Horner – the first Associate Administrator (to act as General Manager). He was the former Assistant Secretary of the Air Force. He was personally picked by Glennan but was able to commit for only one year and was superseded by Robert C. Seamans.
- John A. Johnson – the first General Council. Formerly the Air Force General Council. He was personally picked by Glennan.
- Abe Silverstein – the Director of Space Flight Development had experience in managing aeronautical research as the Associate Director of the Lewis Flight Propulsion Laboratory of NACA. He worked closely with Dryden and Glennan in the early preparations for the new NASA organization.
- George M. Low – as Chief of Special Projects at Lewis he participated in the early work in 1958 to establish NASA; he was then personally picked by Silverstein as Chief of Manned Space Flight.
- Warren J. North – as Assistant Chief of the Aerodynamics Branch at Lewis he was picked by Silverstein to head up Space Flight Programs. As a former test pilot, well known to the X-Series pilots, he worked with the astronauts on training.

This first NASA group of people met in the Dolley Madison House at 1520 H St. NW, near the White House, many times during the period October 1, 1958 to October 20, 1961. Actually, as NASA sorted itself out, it was spread out over five locations within the District of Columbia and nearby Silver Spring, Maryland. Abe Silverstein, Newell Sanders, John W. Crowley, George Low, and many others, settled in the nearby Scientific & Technical Building at 1512 H St. NW. By June 1959, there were over 100 people in the Dolley Madison House and at least 150 in the Scientific & Technical Building working on the various programs and projects which had been initiated by NASA.

In late 1958 the focus of attention became the new STG, temporarily based at the recently named NASA Langley Research Center in Virginia. Effective intra-agency coordination was almost immediate. Dryden and Gilruth were old friends. Silverstein, Low and North were all from Lewis and had worked with Langley on many projects. Although Ames in California was not very visible at Headquarters it was well known at Langley and Lewis, and was very much involved in the Mercury aerodynamics heating issues. As yet the Goddard Space Flight Center existed in name only, but many people from Langley and White Sands eventually moved there.

Of course, intra-agency coordination was not "peachy keen" all the time. There were heated discussion between reliability and quality control people at Headquarters and those at the STG and at McDonnell Aircraft. These issues were resolved, but in the process a

Fig. 13.1 NASA's first Headquarters from October 1958 until October 1961. (Photo courtesy of NASA)

number of people resigned and moved on. There was also some animosity between some at Langley and the new Director at Goddard, Harry Goett, who was known to be a difficult person to work with. All in all however, the various NASA organizations worked well together to pull off Project Mercury.

13.2 ENGINEERING

13.2.1 The Capsule's shape

Sometimes science, engineering, management, and operations are all interrelated without that being obvious from the start. A good example is the shape of the Mercury capsule. The Army had problems with their warheads not surviving the re-entry heating. In the 1950s Dr. Harvey Julian Allen from NACA's Ames Aeronautical Laboratory had studied the shapes of re-entry vehicles at high Mach numbers and realized that a pointed projectile shape would encounter a greater degree of aerodynamic heating than a more rounded shape. After studying a variety of shapes, Dr. Allen published his "blunt body" theory in 1953. This was the science part of the problem.

Back at Langley, Max Faget and others at the Pilotless Aircraft Research Division (PARD) were studying ways to operate an aircraft above Mach 3; that being the maximum speed of the X-1 series. They were already involved with the X-15, reaching speeds of up to Mach 6. They wanted to examine the next phase, which was called "Round Three." They had the benefit of German work during WW-II, they had Dr. Allen's work, and they had their own Langley work. Then in October 1957 the Soviet Union launched Sputnik. It was time to forget about pushing aircraft to higher Mach numbers; the task was to think about using rockets to get spacecraft to those velocities. Faget, Benjamin J. Garland, and James J. Buglia gave a paper "Preliminary Studies of Manned Satellites-Wingless Configuration: Nonlifting" at a conference at Ames in March 1958, describing the engineering aspects of the basic capsule design.

Competing with two Langley designs was an Ames design proposed by Alfred J. Eggers Jr., which was called the "half backed potato" because it was a cone with a spherical nose that was cut in half with just the lower half serving as the capsule. This was essentially a "lifting body," and it would absorb a lot more heat over a longer period of time than something that followed a purely ballistic trajectory. Unfortunately, calculations showed that this design would be far too heavy to be lifted by the Atlas, which was the most capable booster in the U.S. inventory at the time. As a result, the decision was clearly in favor of Langley. (The capsule wouldn't be given the name Mercury until November 26, 1958.) Hence the management part of the decision was straightforward because the Atlas was the only available rocket to boost a capsule weighing one ton (later one and a half tons) into orbit. Once the shape of the capsule was decided on, the next engineering decisions were how big to make it to fit on the Atlas and to accommodate the pilot and equipment, and how to protect it from the heat of both launch and re-entry. The chosen heat shield would go on to play an operational role in John Glenn's flight. The capsule shape wasn't dictated by bureaucrats or military scientists and engineers. It was defined by NACA scientists and engineers with very little input from above.

13.2.2   The Capsule's Heat Shield

One of the most significant decisions to be made for the Project Mercury capsule concerned the material for the heat shield. Having defined its shape, deciding what it should be made of was a complex problem of physics, engineering, manufacturing, and testing. Years before NASA, the Army and the Air Force were testing both ICBMs and IRBMs. These flights were ballistic, not orbital. They had their ideas for ballistic missile nose cones, and so too did their subcontractor General Electric. Langley's PARD was launching rockets at Wallops Island, as well as aircraft models. The science of thermodynamics was applicable to both aircraft and rockets/missiles. A great deal was being learned in the mid-1950s. Wind tunnels were being developed to study the thermal heating on various objects at very high Mach numbers.

There were basically two approaches to nose cones and heat shields; the "heat sink" method and the "ablative" method, and all of this research was highly classified. The heat sink method was well understood because it used metallic heat sinks. Metallurgy was well understood. The manner in which a metal absorbs and radiates heat was well defined. There was a lot of data on metals. The aerospace industry had tremendous experience with metals. There was not so much data available on materials that "melt, vaporize and dissipate" using the process of ablation. In ablation, the heat necessary to change the state of a material from solid first to liquid and then to a gas is vastly greater than the heat absorbed by that material in raising its temperature. If such a material can carry heat away from the capsule, rather than absorbing it, then so much the better for the occupant.

When NASA was formed in October 1958, STG's Gilruth and Faget were leaning toward a metal heat sink. But when the contracts were let for the Mercury capsule, the proposals from industry focused on a beryllium metal or alloy heat shield. When McDonnell won the contract they were told to design the capsule so that it could have either a beryllium or an ablative heat shield.

In the meantime, STG's Andre J. Meyer, Jr., one of the original Lewis engineers, now at the Engineering and Contract Administration Division, worked with the McDonnell subcontractors; Brush Beryllium Company for the metal heat shield, and General Electric and Cincinnati Testing and Research Laboratory for the ablative shield. The technology available to produce either type was very primitive. Meyer was familiar with the use of laminated plastics in aircraft structures. What complicated matters was that there were only two suppliers of beryllium, neither of which had produced beryllium of acceptable purity. Meyer collected all of the data on both types, then met with STG engineers from the Performance Branch of the Flight Systems Division and from the Systems Test Branch. They were working on the Big Joe project that was to launch a "boiler plate" capsule on an Atlas in order to test the heat shield. They preferred the ablative type made from a fiberglass-phenolic material. On an earlier drop test, a metal heat shield was deployed in order to simulate getting rid of the heat and this adopted a "falling leaf" mode which resulted in it colliding with the capsule. The idea of jettisoning a hot heat shield after re-entry was rejected. This swung the decision in favor of using an ablative heat shield, because that wouldn't need to be jettisoned.

By mid-summer 1959 the decision was made to employ the recently proven Atlas to drive a highly instrumented "boiler plate" capsule to re-entry speeds to test an ablative heat shield. The first such heat shield was delivered to the NASA STG pre-flight checkout team at the Cape on June 22, 1959. The team comprised 45 people, including many of the original Lewis engineers, led by STG's Scott H. Simpkinson of the Launch Operations Branch.

The entire capsule operation was managed by Charles W. Mathews, Chief of the Operations Division. In attendance were Aleck C. Bond, who had managed the effort for nearly a year, B. Porter Brown, the STG liaison with the Atlantic Missile Range, and the Air Force/Convair team. Brown was also the Mercury Atlas Test Coordinator and the main interface with Melvin Gough, the Headquarters representative.

The Atlas with the capsule and its test heat shield, called Big Joe in comparison to the Little Joe series on Wallops Island, lifted off at 3:19 AM on September 9, 1959. While the 13 minute flight didn't go quite as planned because the Atlas failed to stage, it did achieve a speed of nearly 15,000 MPH and a heat pulse on the capsule that was shorter but more severe than planned, and sufficient to prove the value of the ablation approach. The capsule was retrieve by the destroyer *Strong* and shipped back to the Cape. It was midnight by the time the capsule got back to Hanger S and was opened. Gilruth, Faget, and the whole launch and checkout team were eager to see the heat shield. They were delighted. The instrumentation also confirmed what they had hoped, and what they observed. When they opened the capsule there was a note inside from the whole crew of 53 people under Charles Mathews that read:

> This note comes to you after being transported into space during the successful flight of the "Big Joe" capsule, the first full-scale flight operation associated with Project Mercury. The people who have worked on this project hereby send you greetings and congratulations.

As a result of this test the conical section of the capsule's afterbody was thickened and the cylindrical section which used René 41 nickel alloy shingles was replaced by thick beryllium shingles. The heat shields on the Mercury Redstone flights of Alan Shepard and Gus Grissom used beryllium, but all orbital Mercury, Gemini, and Apollo spacecraft employed the ablative type of heat shield, even for the more severe heating imposed by a re-entry returning from the Moon.

13.2.3   The Capsule's Couch

For decades the military had been researching ways to mitigate acceleration loads on pilots. It was even studied by German aeromedical researchers during WW-II and after the war many of those scientists worked for our government on the same problems, but on aircraft attaining even higher speeds. The work of Lt. Col. John P. Stapp and Eli L. Beeding of the Aeromedical Field Laboratory at Holloman AFB in New Mexico is legendary, and represents the limits of human tolerance for deceleration. The centrifuge studies at the Navy's Aviation Medical Acceleration Laboratory at Johnsville, Pennsylvania and at the Aeromedical Laboratory at Wright-Patterson AFB in Ohio could simulate gradual build-ups of "g" forces and modify the position of a pilot. After experimenting on many subjects, E. R. Ballinger concluded that 8 "g" was the maximum safe acceleration for spaceflight.

The concern for spaceflight was related to both the possible accelerations during launch and decelerations during re-entry. Now the constraints were somewhat different. The forces applied over a longer period and could be quite high depending upon the abnormal conditions, such as during a launch abort or during a steeper re-entry. During the mid-1950s, Max Faget was Head of the Performance and Aerodynamics Branch at NACA Langley. He was working on a space capsule design that would eventually become the Mercury capsule. Even before NACA became NASA, he and his colleagues William M. Bland and Jack C. Heberlig considered the whole "g" envelope for the pilot for both normal and possible abort launch profiles as well as a normal and abnormal re-entry. They also considered the size and weight of any devices that might mitigate the forces upon the astronaut. In addition to having to fit into the capsule, such devices had to be lightweight because weight was a critical factor. Faget was fully aware of the research that had been done in this area. The lifting capability and the launch profiles of the Redstone and Atlas missiles were already well defined for normal cases. It was the launch aborts and steeper angles of re-entry that posed the highest forces.

In April 1958, just weeks after Eisenhower called for the creation of a civilian space agency and Faget published his report on "Preliminary Studies of Manned Satellites," he and NACA test pilot Robert A. Champine began to make rough sketches of a "couch" for the capsule. Champine was probably more qualified to be an astronaut than the others because he had not only flown the X-1 but also both the Douglas Skystreak and Skyrocket.

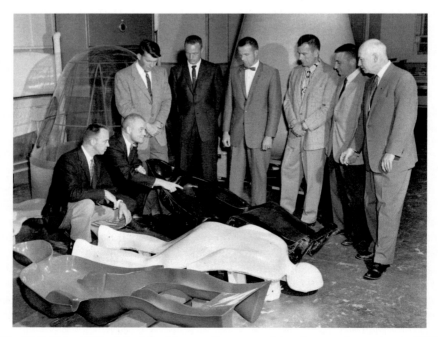

Fig. 13.2 Robert Gilruth and the Mercury Seven examining the capsule couches in 1959. (Photo courtesy of NASA Langley Research Center)

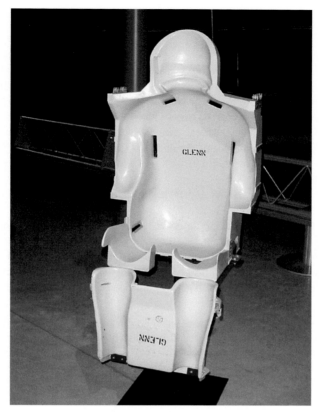

Fig. 13.3 John Glenn's couch manufactured by McDonnell Aircraft.

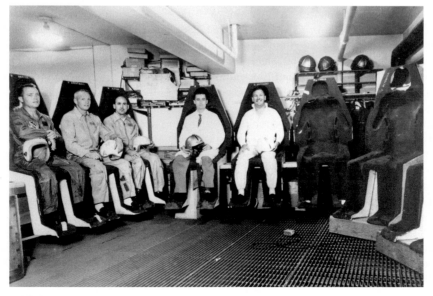

Fig. 13.4 Pilots testing the couches at the Johnsville centrifuge in 1960. Notice Neil Armstrong, second from left. Others are Navy pilots and aeromedical personnel.

Although he was considered the NASA Test Pilot Astronaut, he proved too tall to qualify for the program. Working with Langley's West Area Model Shop (famous for making just about anything), a prototype couch made of fiberglass was contoured for Champine's body in May. The next task was to test it at the Navy's Johnsville centrifuge; the largest ever built. On July 29, Champine, lying on his couch, was installed in the centrifuge and accelerated to 12 "g" with no adverse effects. The following day, Navy Lt. Carter Collins rode it to 20 "g". After a full evaluation of the test runs, the West Area Model Shop was given the go-ahead to fabricate couches for all seven of the astronauts when they came onboard the following year.

On June 12, 1962, Faget, Bland and Heberlig were granted a patent for the "Survival Couch."

The lessons learned from this engineering work apply to all current and future flights in space. While in the 1950s much of the physiological/aeromedical work was pioneered by others, it took some basic engineering to provide a practical solution to protecting a crew from extreme forces. The solution was relatively simple in retrospect. The one-and-a-half page patent explanation laid out the parameters. What was claimed as new and patentable was:

> A couch comprising a sheet of lightweight, rigid material having individualized recesses formed therein conforming to the lateral and posterior contour of a specific preselected living occupant, and a posterior layer of crushable cellular honeycomb shock absorbing material applied to said sheet, said occupant, when positioned in said couch recesses, being substantially protected thereby from physical injury during application through said couch of omnidirectional high accelerative G forces upon said occupant.

13.2.4   The Capsule's Hatch

The side hatch of the capsule was a point of discussion as soon as the astronauts were introduced to the design in 1959. The original design for the capsules which were flown by the chimpanzee Ham and Alan Shepard had a latch system which weighed 69 lbs. and was too heavy for orbital flight. The new hatch design for Gus Grissom's *Liberty Bell 7* weighed only 23 lbs. and still had 70 titanium bolts which were fastened on ingress, but it also had a quick-release mechanism to facilitate the astronaut's exit after landing and, if necessary, to enable the Navy recovery divers to access the capsule to assist him. The new design consisted of a continuous double explosive train to ensure that all of the bolts were severed upon activation. It would also propel the hatch out about 25 feet. The divers were cautioned to stand clear. In order to activate the mechanism, the astronaut had to remove a cover and pin, then press a plunger with a good force. The hatch was as strong as the capsule and included a special heat resistant window that was made by the Corning Glass Works to give the pilot greater visibility than was provided by the periscope.

The premature detonation of the hatch of *Liberty Bell 7* and Grissom's near drowning is well documented. The media attacks were unwarranted. Grissom was really affected by the criticism. I think the proof that Grissom didn't hit the detonator was given by John Glenn, who received a scuffed knuckle when he hit the detonator on *Friendship 7* and similarly by Wally Schirra upon his blowing the hatch on *Sigma 7*. This bruising of the hand was a result of the kick back of the detonator. Grissom had shown his hands to the flight surgeons during the recovery physical and his hand wasn't bruised; as it would have been if he had intentionally hit the plunger in order to blow the hatch.

The incident changed the mission rules and some procedures. The astronaut was not to pull the hatch pin until he was ready to hit the plunger to blow the hatch and egress. He would secure his suit inlet valve and inflate the neck air dam prior to egress. The external hatch release handle was better secured (it only had one screw and was a possible cause of the premature detonation) and the Navy divers would recover the astronaut first and his capsule second.

The ironic and sad ending to the hatch story is that the Apollo 1 crew couldn't open the hatch when a fire broke out in their spacecraft during a routine pad test on January 27, 1967, resulting in the deaths of Gus Grissom, Roger Chaffee and Ed White. This was because the hatch had been designed to be opened inward, which became impossible as the internal pressure built up. It was not able to be opened by the pad crew until after the pressure had burst the wall of the craft. The hatch was then redesigned to enable it to be opened rapidly, even against internal pressure. As a personal note, I was on the GUIDO console in the Mission Control Center in Houston for the test and heard the entire accident. There is not a spacecraft design to this day that doesn't consider all the ramifications of crew safety in many different hatch egress situations.

Liberty Bell 7 was lifted from the floor of the Atlantic in 1999 by Oceaneering International, Inc., with the team led by Curt Newport being funded by the Discovery Channel. However, the beryllium heat shield and the hatch were not recovered. Total restoration was accomplished by the Kansas Cosmosphere and Space Center in Hutchinson, Kansas.

Fig. 13.5 *Liberty Bell 7* recovered on July 20, 1999 by Curt Newport during the Oceaneering Expedition. (Photo courtesy of Oceaneering International, Inc.)

Fig. 13.6 *Liberty Bell 7* awaiting restoration. (Photo courtesy of The Kansas Cosmosphere and Space Center)

Fig. 13.7 *Liberty Bell 7* being restored by Greg "Buck" Buckingham at the Kansas Cosmosphere and Space Center.

13.2.5 Little Joe

Little Joe was the designated name of the solid propellant launch vehicle that was designed by NASA to test the Mercury capsule in certain critical conditions that it would encounter during the launch phase. This vehicle was designed to attain these objectives in a relatively simple and inexpensive way. The Little Joe was about one-fifth the cost of a Redstone rocket and had much lower support costs. It could propel a Mercury capsule to Mach 6, so most of the critical launch abort conditions could be tested. STG management chose it not only to save a lot of money but also to limit the impact of such testing on the Redstone development and flight schedule. Many of the capsule development test objectives involving launch escape could be studied and solved by the Little Joe. In addition, these flight tests could be conducted at the Wallops Island Station that had the facilities and personnel

Fig. 13.8 The fully restored *Liberty Bell 7*. (Photo courtesy of the Kansas Cosmosphere and Space Center)

for flight test and launch operations; after all, they had been launching rockets since the 1940s. See Section 7.6.

Almost as soon as the STG was established in November 1958, an invitation for bids on the airframe for the Little Joe was sent to twelve companies. North American Aviation (NAA) won the contract a month later, on December 29. They had been awarded the contract to build three X-15s in November 1955, and the major manufacturing work was over now and the project was in the early operational stages. The Little Joe was a fin stabilized vehicle consisting of a cluster of solid-fuel rocket motors – four main motors plus four auxiliary motors with some variations. They were all built by NAA's subcontractor Thiokol; the four larger (greater specific impulse) main motors were either the XM-33E2 (Castor) or the XM-33E4 (Pollux). The four auxiliary motors were of the XM-19E1-C12 (Recruit) type. Depending upon the launch requirements, it was possible to assemble four configurations. Theoretically the Little Joe could send a two ton payload on a ballistic path to over 100 miles, simulating an Atlas launch. The flight trajectories were predicted utilizing the Langley/STG IBM 704. The Little Joe pioneered the use of rocket clusters for operational launch vehicles, as follows:

- LJ-1 was planned for launch on August 21, 1959 but faulty wiring triggered the abort sequencer and both the capsule and escape tower plunged into the ocean.
- LJ-6 was launched on September 9, 1959 less than a month after Big Joe. The 5 minute flight was successful in qualifying the launch escape system.
- LJ-1A was intended to repeat the plan for LJ-1 and was launched in only two months on November 4, 1959. All went well until the abort motor fired, but it didn't

build up thrust fast enough to achieve the intended objective of an abort at maximum dynamic pressure (Max-Q).

- LJ-2 was a successful high altitude abort carrying the Rhesus monkey Sam.
- LJ-1B launched on January 21, 1960 was a repeat of LJ-1A and was a successful Max-Q abort carrying female Rhesus monkey Miss Sam.
- LJ-5 launched on November 8, 1960 was to be a Max-Q abort with a production Mercury capsule fitted with an escape tower, but 16 seconds after liftoff the escape tower and tower jettison rockets fired prematurely without separating and the entire booster, spacecraft, and tower fell into the ocean.

Note that all of the above flights took place prior to the first Mercury Redstone MR-1 flight.

- LJ-5A launched on March 18, 1961 was intended to repeat the plan for LJ-5, but again the failure sequence was repeated. However, this time a ground command was sent to separate the capsule and escape tower so that the flight qualified production Mercury capsule could be saved and used again the next month on LJ-5B.
- LJ-5B was launched on April 28, 1961 (a week before Alan Shepard's MR-3) and this time the Max-Q abort test was successful, with the capsule being recovered once again.

The Little Joe paved the way for the Mercury Redstone flights and gave the NASA and STG managers and astronauts confidence that the capsule and escape tower would function properly through the launch phase.

13.2.6   Big Joe

Another excellent example of NASA's intra-agency decision-making and the ability of its field centers to cooperate closely on Project Mercury was the December 1958 decision to delegate to the Lewis Research Center the task of designing the electronic instrumentation and automatic stabilization system for the Big Joe capsule. Robert Gilruth appointed Aleck Bond, the former head of the Structural Dynamics Section at Langley as the STG Project Engineer for Big Joe. The STG manager at Lewis was G. Merritt Preston. Lewis also fabricated the capsule's lower section, which contained a pressurized volume with the electronics and two nitrogen tanks for the retro rockets. Lewis assembled the entire capsule, including the ablative heat shield which was provided by General Electric, the Langley afterbody and recovery canister, and the Lewis electronics. A capsule lower section was tested in the MASTIF apparatus in order to verify the performance of the autopilot control system that would be critical for maneuvering the capsule subsequent to its separation from the Atlas. See Section 7.3.

Another NASA decision that affected not only this mission but many later ones was that to assign the Lewis team as the first capsule pre-launch test and checkout team. On June 9, Scott Simpkinson's team of 45 people followed the Big Joe capsule to Cape Canaveral, where they spent months in Hanger S preparing the complete capsule for launch. This marked the start of NASA capsule/spacecraft engineers having a full-time presence at the Cape; it would continue for decades.

The purpose of the first Mercury Atlas launch on September 9, 1959 was to test the ablative heat shield and the heating on the afterbody of the "boiler plate" capsule. See Section 13.2.2. All these materials had been tested on the ground, and to an extent on some of the Little Joe flights. This flight would also test the ability of the capsule to control its attitude in space, the dynamics of re-entry, and the recovery operation.

Prior to launch, the Atlas was subjected to two flight readiness firings. During the first one a faulty timer caused a shutdown before the engines fired. There was no damage to the Atlas. The second firing was normal and the shutdown occurred on time after 19 seconds.

The launch was normal up to booster engine cutoff (BECO), when the Atlas failed to jettison its booster engines. The central sustainer engine performed normally, but was carrying the extra weight of the booster engines and ran out of propellant 14 seconds early, after a flight lasting 13 minutes. Although the test conditions weren't exactly as planned, the capsule re-entry tests were considered to be satisfactory, as was the recovery operation. The STG engineers considered it to have verified the design of the spacecraft, because it showed that the automatic attitude control system and the heat shield and thermal protection system worked, and that the dynamics during re-entry were stable.

The engineers and management agreed that this Big Joe flight had produced sufficient data, and so a second planned test was canceled; this helped the overall schedule. This was both an engineering and a management decision. The second Atlas was released for another program.

Merritt Preston became Director of Launch Operations for Mercury (and later Gemini) and then Manager of the Johnson Space Center's Florida Operations. Scott Simpkinson went on to manage testing programs for Gemini and Apollo, and flight safety on the Shuttle. Many of the original team remained with the space program; some in Florida, some in Houston, but others returned to the Lewis Research Center where they had families and homes.

13.2.7   Atlas "Belly Band" AKA "Horse Collar"

The launch of MA-1 took place on July 29, 1960 in bad weather. It apparently had a structural failure at 58 seconds into its flight, about the time of maximum dynamic pressure (Max-Q) and at an altitude of about 30,000 feet. This first launch of the Mercury Atlas since the Big Joe test carried a production Mercury capsule rather than a "boiler plate." Among other things, the aim was to prove the booster-capsule combination for launch and re-entry flights.

The bad weather meant there was no photography of the vehicle breakup but many lessons were learned, including management's handling of the diagnosis of the launch failure, a new mission rule against launching without photographic coverage, and the engineering solution described here. There was a lot of initial finger-pointing. One side argued that it was capsule's fault because a fairing on top of the capsule broke loose and punctured the Atlas liquid oxygen tank. Another side said the Atlas was to blame because its design relied upon pressurization for structural support and its skin, only 0.1-inch thick, wasn't strong enough to support the capsule through Max-Q.

Owen Maynard (one of the AVRO engineers) worked on the systems engineering of Project Mercury and on this mission participated in the recovery effort, in which he made a 30 foot free-dive to recover a missing component of the capsule. His post-flight calculations showed the skin of the Atlas just below the spacecraft would have buckled as a result of the combined drag and bending loads at the Max-Q point, thereby exceeding the tensile stress in the skin yielded by its internal pressure.

It took engineering teams on both coasts almost six months to resolve the problem, and thus clear the way for MA-2. Meanwhile there was tremendous bad press and political pressure. The 12-member Rhode-Worthman Committee met during the last week of December 1960 and the first week of January 1961 at Convair/Astronautics in San Diego and at the Air Force Ballistic Missile Division in Los Angeles. Paul E. Purser and Robert E. Vale represented the STG. There was disagreement between what the STG and Convair wanted to do, and what Aerospace/STL wanted to do. Some wanted to wait until a "thick skinned" Atlas could be made available even though that would have slipped the whole program.

One day prior to the Inauguration of President Kennedy on January 20, 1961 the Rhode-Worthman Committee issued their report with the recommended fix to the MA-2 Atlas. This involved adding an 8-inch wide stainless steel band to the interface between the adapter of the spacecraft and the "thin skinned" Atlas to distribute the load of Max-Q over the Atlas airframe. Some referred to this as a "Belly Band" but the reluctant Aerospace/STL representatives of the Air Force called it a "Horse Collar." The fix was verified using wind tunnel tests at Ames and at Tullahoma. NASA accepted the responsibility of the fix. There was also a requirement for more accelerometers and strain gauges, plus good photographic coverage of the launch. When MA-2 was launched on February 21, 1961 it was successful.

Paul Purser informed Richard Rhodes, "The joint team effort required for these decisions admittedly has not always been easy but we believe it has worked. Resolutions of conflicts of technical judgement have been achieved by mutual discussion and education rather than by manager edicts."

13.2.8   Man-Rating the Launch Vehicles

It is all about risk versus gain (and cost)! The STG management had watched missiles blow up. I watched (with binoculars) an Atlas explode. It must have taken what seemed like 30 minutes for all the little pieces of metal to fall to the ground like pinwheels fluttering down from a tree, but then pushed upward by the rising hot air from the fireball beneath. The main engines and heavy metal fell quickly, with the little pieces coming down much later. So now you want to put a man on top of a missile that was really intended as an ICBM! Just how does one get the confidence to do that? How does the pilot get the confidence to fly it?

NASA management knew that if any one of the first astronauts were to be killed, that could end the nascent space program. The launch vehicles had to be man-rated and be reliable enough to make the decision to undertake a manned flight. The question was how to quantify, justify, or rationalize the decision. Experience had shown that when things did go awry, they often did so faster than a pilot would be able to react. This began to impact the design of the capsule as well as the launch vehicle. What would be the maximum "g" load the astronaut would experience in the worst case of a launch abort? Would he be sufficiently conscious to be able to carry out the appropriate actions? Would he suffer any permanent damage? Man-rating the Mercury launch vehicles was complicated, in part because two very different rockets were to be employed: the Redstone and the Atlas.

Even before the creation of NASA, the Army had proposed to launch a manned capsule in a ballistic suborbital mission called Project Adam. The pilot would be simply a passenger. During a launch abort the capsule would be tossed into a tank of water alongside the launch tower. There wasn't any serious thought given to man-rating the Redstone in 1957. ARPA and the DOD didn't approve Project Adam. Even the Navy had a project for a manned mission with a to-be-defined booster. These were just studies, not funded projects. As early as 1957 the Air Force and Boeing had proposed the Dyna-Soar program but the studies did little to address the man-rating of any proposed launch vehicle and the project was canceled early on.

During 1958, both before and after NASA was created, the aerospace missile industry began to perform studies about using their military missiles for civilian space projects. The following companies were involved:

- Convair/Astronautics (A Division of General Dynamics)
- Chrysler Corporation
- Avco Manufacturing Corporation
- General Electric Company/Burroughs
- Rocketdyne Division of North American Aviation.

They were already supporting either the Army's Redstone or the Air Force's Atlas for DOD missions. Then on October 1, 1958 NASA was created. It would be NASA that would make the decision as to how and when an astronaut could safely ride a missile or safely escape from it in the event of something going amiss. The STG wasn't organized until the end of 1958 and it had very few people. Their involvement in the man-rating decision did not occur until many months later. Their initial focus was to staff up for the design of the capsule and to get ready to compete a contract.

The Redstone was to be the first booster to launch a man. The task of man-rating the vehicle fell to German scientist Dr. Joachim P. Kuettner, a member of Wernher von Braun's team at the Army Ballistic Missile Agency at Huntsville. Who better to study and characterize the problem? He was an experienced glider pilot, a Messerschmitt test pilot, and had actually flown a V-1. In early 1959 he began a series of conferences between the ABMA team and Jerome B. Hammack who was the STG Mercury Redstone Project Manager and Charles "Chuck" Mathews, the STG Operations Division Manager.

The ABMA Redstone people proposed a purely automatic system, regarding the astronaut as merely a passenger who would be unable to make a timely decision in the case of an exploding vehicle. The NASA STG view was that the astronaut was a test pilot and should always be "in the loop."

It was decided that the Mercury Redstone launch vehicle would incorporate both manual and automatic abort sensing capabilities. Either the astronaut or the flight controllers in the Mercury Control Center would be able to trigger an abort. The automatic system would be designed to be activated by any of the following conditions:

Fig. 13.9 Gus Grissom inspecting his Redstone vehicle at Marshall Space Flight Center. (Photo courtesy of NASA)

- Pitch, yaw, or roll angle deviating too far from the programmed flight profile
- Pitch or yaw angle changing too rapidly
- Pressure in the engine's combustion chamber falling below a critical level
- Loss of electrical power for the flight control system
- Loss of general electrical power which could indicate a catastrophic failure.

If the Range Safety Officer were to decide to send a destruct command when the vehicle was heading toward a populated area, then the automatic system would delay three seconds to allow time for the escape tower to fire and carry the astronaut a safe distance. Other changes were also made to the Redstone to increase its reliability.

Things were a little different with man-rating the Atlas missile. There was a whole new cast of characters. The Air Force Ballistic Missile Division (BMD) was the government owner of the Atlas missile and their systems engineering contractor was the Space Technology Laboratories (STL). The manufacturer was Convair/Astronautics (CV/A). Bernard A. Hohmann was the lead engineer for the BMD. Like Kuettner, he also was a former German engineer and test pilot. He had worked at Peenemünde, and was project engineer on the rocket powered Messerschmitt 163 "Komet" interceptor. The BMD Mercury Project Officer was Lt. Col. Robert H. Brundin.

The STL effort was led by Edward B. Doll. STL had significant systems engineering, missile development, and business management experience. The abort sensing team was

Fig. 13.10 Gordon Cooper's Atlas being off-loaded at the Cape. (Photo courtesy of NASA)

headed up by Convair's Philip E. Culbertson, who many years later worked for NASA. At this point in early 1959, the STG was not very involved with the initial STL/Air Force/Convair discussions. With only about 200 people at this time, the STG was more focused on the design of the capsule and staffing up.

The Atlas group was well aware of the Redstone work but this was the Atlas, the only vehicle able to launch a 3,000 lb. Mercury spacecraft into orbit. Convair had built many variants of the Atlas, with the first flying in 1957, but these were all designed to transport a nuclear warhead on a ballistic trajectory. The Atlas that would be man-rated was the Atlas D LV-3B. This benefitted from the previous Atlas development and the lessons learned from a great many failures.

In April 1959 Christopher C. Kraft and his STG colleagues attended the Abort Sensing and Implementation System (ASIS) oversight meeting in San Diego. Kraft listened to the briefings and it became clear that the Atlas team would talk about how the system *worked* but not about how it *failed*. Kraft made sure the Atlas team considered the failure modes, and how the ASIS would deal with them. This prompted a degree of friction between the STG team and the Atlas team. Kraft assigned Robert Harrington as the STG representative at STL in Los Angles with instructions to stay on top of the ASIS development. Much of the discussion focused upon the reliability and cost of what would be an STG-only requirement with low production runs. The ASIS wasn't required for an ICBM, and the Atlas team didn't welcome the additional cost and work load. However, the ASIS was tested on several ICBM launches before being installed on MA-1 on July 29, 1960. Some of the changes included:

- Changing the aerodynamics of the Atlas to compensate for the Mercury capsule and its escape tower
- Loss of power would trigger an abort
- The electromechanical autopilot was replaced with a solid-state model
- Extra sensors were added to monitor combustion levels
- The vehicle would be held down on the pad to ensure smooth thrust prior to launch
- The guidance antenna was modified to reduce interference
- Later vehicles would be made with a thicker skin
- An upgraded propulsion system was added for the MA-7, -8 and -9 flights
- The Rocketdyne engines were realigned for MA-9.

Many lessons were learned about how to man-rate a launch vehicle for manned flights. The engineering decisions to accommodate the STG requirement to man-rate the Atlas were driven by the concern for the safety of the pilot. After seeing many Atlas failures, these concerns were highly justified. The number of organizations and people involved in the decisions added to the complexity of the decision process. The existing Air Force/Convair/STL design was driven by military requirements. The new STG requirements were not needed for an ICBM, and not only increased the cost but also slowed down the manufacturing of other Atlas missiles.

After the decision was taken to man-rate the Atlas military missile for Project Mercury, the various engineering changes seemed to be advantageous to both military and civilian missions. Certainly the modifications made to remedy the combustion problems that had led to previous failures fell into that category, as did changes to the autopilot and the alignment of the engine. And changes to improve the reliability of the vehicle added to the overall safety of the mission and the pilot. Changes made to the structure of the Atlas were primarily due to its thin skin. Its original design requirements hadn't envisaged it having to lift a payload as heavy as a Mercury capsule and its escape tower. The changes made to give the pilot a chance to escape were rather minor in comparison to those that improved the vehicle's inherent weaknesses.

In the late 1950s reliability engineering was thought by many to be a "black art," but slowly mathematicians, scientists, and engineers were starting to apply some rigor, albeit reluctantly by some. John C. French, Technical Assistant to James A. Chamberlin became the lead engineer in the STG for these studies. Redundancy was the most obvious solution to increasing reliability. Throughout Mercury, there were single point failures which caused concern and consequences. The lesson that was carried over to future programs with very complex hardware and software was to "design in" reliability and then to conduct rigorous component, subsystem, and system testing in order to increase reliability and the probability of mission success.

The lesson for the managers of very complicated systems is to recruit the brightest and most experienced people you can find. While the BMD Atlas team had that "in spades," they had to listen to the customer that, despite having no missile experience, knew a little something about how a flight operations team would handle failures.

13.2.9   Reaction Control System

In accordance with the Project Mercury Objectives and the supporting Guideline which stated, "Existing technology and off-the-shelf equipment should be used wherever practical," the STG and McDonnell Aircraft capsule designers adapted the work of the NACA High Speed Flight Station's work on hydrogen peroxide thrusters.

First, a little background. The concept for using hydrogen peroxide goes back to the German program, which used this chemical in the turbo pump of the V-2. Hellmuth Walter was the first to use it as early as 1937 to assist the Heinkel Kadett to take off. After WW-II, then Director of Aeronautical Research for NACA, Hugh L. Dryden (later to be NASA's Deputy Administrator) and Theodore von Kármán visited German laboratories in order to obtain as many aeronautical documents as possible. Walter's report dated in 1943 was translated after the war as *Report on Rocket Power Plants Based on T-Substance* (the latter being shorthand for highly concentrated hydrogen peroxide) and published as NACA TM #1170 with limited distribution. By 1946, the German rocket team was in New Mexico.

Also by 1946, the X-1 program was going on at the NACA Muroc Flight Test Unit. Chuck Yeager became famous on October 14, 1947 for "breaking the sound barrier." The X-1A, X-2, X-3, D-558-I, and D-558-II soon followed, all of which reached altitudes where attitude control became a problem. In 1954 the Air Force acquired a Goodyear Electronic Differential Analyzer (GEDA) and this was made available to NACA to develop flight simulations of the X-2 and to study the effectiveness of the RCS. The driver for this research was the recently approved X-15 program.

Fig. 13.11 NACA research pilot Stanley Butchart flying the Iron Cross, circa 1957. (Photo courtesy of the NASA Armstrong Research Center)

In 1955, Neil Armstrong transferred from NACA Lewis to Muroc. In 1956 North American Aviation, just chosen as the X-15 airframe contractor, proposed a hydrogen peroxide system for attitude control in the upper atmosphere where the air was too thin to use aerodynamic control surfaces. Using the GEDA and data from the X-1B flights, engineers created a simplified flight simulator to enable pilots to investigate controllability using different thrusters. The next phase was to build a mechanical ground simulator nicknamed the Iron Cross.

This simulator used nitrogen thrusters because it was primarily intended to explore piloting responses and possible RCS controls. Despite the Iron Cross's appearance, it revealed control problems that were too subtle for the analog simulator to manifest; in particular the importance of aligning the thrust axes of the reaction controls. The results from this type of simulator were factored into the X-1B, which was fitted with hydrogen peroxide thrusters. On August 15, 1957, Neil Armstrong had his first checkout flight on the X-1B. After release from the launch aircraft, the number 2 rocket chamber failed to ignite. Despite this he reached Mach 1.32. On coming in to land on the Rodger's Dry Lake,

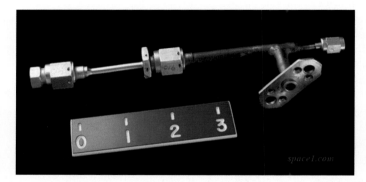

Fig. 13.12 Mercury 1 lb. thruster. (Photo courtesy of Historic Space Systems)

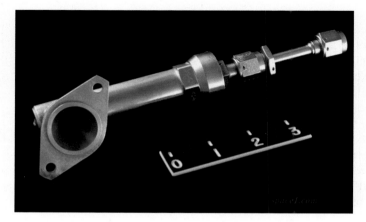

Fig. 13.13 Mercury 24 lb. thruster. (Photo courtesy of Historic Space Systems)

Fig. 13.14 Thrusters installed on the Mercury capsule. (Photo Courtesy of Historic Space Systems)

the aircraft skipped, sheared off its nose wheel and damaged the fuselage and liquid oxygen tank.

A pre-production F-104 Starfighter aircraft was delivered to NACA and modified in order to conduct further RCS flight tests using the hydrogen peroxide thrusters. By this time NASA had been created and Muroc was the Flight Research Center. Hugh Dryden was still in charge from NASA Headquarters and Walt Williams was engaged in the planning of research requirements; indeed, Williams became chairman of the X-15 Flight Test Steering Committee.

Project pilot Joseph Walker made the first flight in the newly designated JF-104 on July 31, 1959. Neil Armstrong also flew it on September 25, and said it was a "surprising ride." All the JF-104 flight test data and experience was factored into the X-15 airframe by North American Aviation and into the Mercury capsule by McDonnell Aircraft.

The relationship between the data collected from the JF-104 RCS flights and the final design of the X-15 RCS is clear, because each had eight yaw and pitch thrusters and two roll thrusters. Like the X-15, the thrusters of the Mercury capsule used 90% hydrogen peroxide delivered by pressurized bladders. The capsule had 18 thrusters in three sets of six, with different maximum thrust levels, namely 6 lb. and 24 lb. "high torque" thrusters and 1 lb. "low torque" thrusters for fine attitude control. This propulsion technology was transferred to the both the Lunar Landing Research Vehicle (LLRV) and Lunar Landing Training Vehicle (LLTV), which used hydrogen peroxide thrusters.

Armstrong has said of making the final approach in his lunar landing using the RCS, "I felt comfortable. I felt at home. I felt like I was flying something I was used to, and it was doing the things that it ought to be doing." Here he was crediting his experience in these free-flying lunar landing trainers.

The decision by the STG management to fully utilize the experience and lessons learned from the High Speed Flight Station's work with hydrogen peroxide RCS thrusters was an example of efficient cooperation between elements of NASA. The technology was incrementally improved and transferred to the various aerospace contractors that manufactured Mercury for possible use on other applications.

Many thanks to Christian Gelzer and Curtis Peebles of the Armstrong Flight Research Center for input to this section.

13.3   OPERATIONS

The discussions, decisions, and operational lessons learned from the Mercury flights are grouped by launch vehicle and are not in flight order. First the Redstone missions are listed, and then the Atlas missions.

13.3.1   MR-1/MR-1A Pad Shutdown and Relaunch

Although engineers can design a system and believe they know how it works, they may not have considered all of the ways in which it might fail and how the astronaut or flight controllers might work around the failure. In the early days of Mercury, there would be several examples of system failures that required operational work-arounds. There would also be failures where no one could do anything; they simply occurred. In addition to knowing what to do in case of a failure, there is also the element of time. The astronaut or flight controllers may have time to fix or work around a failure, or perhaps not. Often, if a decision isn't made by a certain time then something worse will happen.

In the early unmanned flights, the Mission Control Center had very little control. But in one example, MR-1, the blockhouse had to make a decision which no one had ever considered and the team learned some valuable lessons. The Redstone was a product of Wernher von Braun's German rocket scientists in Huntsville. Some of his men were in the blockhouse managing the launch preparations and countdown. There was also one highly experienced German manning the BOOSTER console in the MCC; Dr. Joachim Kuettner.

On November 21, 1960 when the countdown reached $T = 0$, the Redstone lifted off and rose about 4 inches, shutdown, and miraculously settled back onto its tiny cradle. The escape tower fired instantly, leaving the capsule in place. As the smoke cleared, the capsule, believing it was in the recovery mode, deployed its parachutes and scattered small strips of tin foil meant to aid the recovery radar. The wind was blowing the parachutes around the live and fueled Redstone. The blockhouse was as stunned as the MCC. Flight Director Kraft called the BOOSTER flight controller, one row of consoles in front of and below his own positon, several times to find out what had happened but Kuettner was so excited that he started to converse in German with his colleagues in the blockhouse. Kraft was so furious that Kuettner wasn't answering calls that he stepped down to the BOOSTER console and pulled out the communications plug. Words were exchanged!

Then all kinds of ideas came from the blockhouse and elsewhere suggesting what to do with the live rocket sitting on the pad at risk either of being blown down by the parachute or even of exploding on its own. The MCC was powerless to do anything about what they saw on the TV. Fortunately, it was unmanned. The blockhouse's ideas ranged from having someone go out and pull the umbilical of the fully fueled rocket, to having a "cherry picker" maneuver up alongside the capsule in order for someone to cut the parachute away so that it wouldn't topple the rocket, to firing bullets into the fuel and oxidizer tanks. Yes, that was seriously considered! Finally, the launch test conductor came up with the idea of just waiting until the batteries depleted; then the relays and valves would open and depressurize the Redstone. The blockhouse watched all night and the next morning, then the rocket was safed and, after some repairs, later reused.

There were several lessons learned by both the launch team in the blockhouse and the MCC. Kraft declared that the first rule of flight control would be, "If you don't know what to do, don't do anything."

Also, when the Flight Director calls you, you'd better answer, and do so in English! Another lesson was based on the fact that the people in the newly established MCC were not a team. The booster engineers were from the blockhouse and a separate organization. The systems engineers were from a different branch of the STG than the flight operations people. The CAPCOMs were from the Astronaut Office, and the life support controller was from yet another organization. In addition, some STG training people were present to conduct simulations. The support people in the control center were all contractors. Gene Kranz had been hired the previous month. He was manning the PROCEDURES console and this was his first mission! On returning to Langley he realized that what he must do was to develop and prepare a flight control team. But before that could be achieved, he had to return to the Cape to attempt to launch MR-1A. That flight test on December 19, 1960 was successful, and qualified the spacecraft for the upcoming launch of the chimpanzee Ham. See Section 13.2 for some of the engineering lessons learned.

13.3.2   MR-2 Ham

The objective of this flight on January 31, 1961 was to alleviate the concerns expressed by the aeromedical people who, if they'd had their way would've launched many more chimpanzees. But it had been decided that if the three year old highly trained chimpanzee Ham survived the accelerations of launch and atmospheric re-entry, and the brief period of weightlessness of this suborbital mission, then they would clear an astronaut to fly. The launch went into an extended hold due both to weather and to some technical problems. Flight Director Kraft had to make a decision. He polled the Flight Surgeon (SURGEON) regarding Ham's status, and the Recovery people. He and Operations Director Walt Williams made the decision to pick up the countdown. The launch went well for the first two minutes, then FIDO saw that the rocket's trajectory was higher than planned. Other than that though, it looked acceptable. At 2 minutes 17 seconds the ASIS triggered an abort by firing the escape tower, subjecting Ham to a harsh but brief 15 "g" spike! Kraft polled the SURGEON again, who said that Ham had recovered well despite being electrically shocked through his feet a few times when he slipped behind in his assigned tasks.

After a flight lasting 17 minutes the capsule came down hard in rough seas. The rim of the lowered heat shield snapped back violently and breached the titanium bulkhead in two places, enabling sea water to seep in. A cabin relief valve had also jammed open, allowing even more water to penetrate. It took some time for a ship to arrive and hoist the capsule aboard. When it was opened, there was an estimated 800 lbs. of water inside. Fortunately it had not infiltrated Ham's container. These problems were fixed for subsequent flights.

The Mercury part of the mission was successful, but the Redstone had suffered some fuel performance and vibration problems which had to be fixed. Kraft was pleased with how the MCC and the capsule had performed. Dr. von Braun wasn't pleased with the Redstone, and wanted another "Booster Development" (BD) flight to fix the problems and verify the rocket prior to attempting a manned flight.

13.3.3   MR-BD Booster Development

The over-acceleration and vibration problems encountered by MR-2 were diagnosed and fixed and the "extra" test flight occurred on March 24, 1961 carrying a "boiler plate" capsule and an inert escape tower. Its success qualified the Redstone to carry an astronaut.

13.3.4   MR-3 Shepard

I have a personal connection to this mission, as I interviewed with the STG on May 4, 1961, the day before the flight and sent in my application on that day. As I was teaching at the time, I was told I was hired and to report back as soon as I could. Many years later, I was in the astronauts' gym jogging with Al Shepard, who was preparing for the Apollo 14 mission. After our workout, we had a sauna together and I told him that I was hired by NASA the day prior to his flight and that it had started a great career for me as well as a life full of memorable experiences. I wished him well on his lunar flight.

This MR-3 mission is well documented and doesn't need much in the way of explanation or commentary. From an operational perspective, the flight control team in the MCC worked well and everyone was quite pleased with their individual performances as well as that of the whole team. This aspect of the mission is best read in Christopher Kraft's memoir "*Flight*." The most incredible thing about this text-book flight was the impact it had on President Kennedy. A mere three weeks later, Kennedy urged the Nation to commit itself to landing a man on the Moon by the end of the decade. See Section 13.1.1 for the rationale for this bold call and responses to it.

13.3.5   MR-4 Grissom

From a flight operations point of view, the MR-4 flight on July 21, 1961 was nominal both in terms of the astronaut and the MCC team. However, from a recovery operations standpoint it was nearly a disaster. In engineering terms it demonstrated that adding complexity can cause a reduction in safety and reliability. See Section 13.2.5 for a discussion about the new explosive hatch.

The launch phase was almost a duplication of its predecessor, with the Redstone performing as advertised. Grissom performed his tasks as planned. He had the advantage of a new window for observations and a new rate damping control mode that allowed finer attitude control. The jeopardy occurred after splashdown. Many lessons were learned from the recovery operation for the designers, crew, and recovery people. While some accused Grissom of accidentally blowing the explosive hatch and causing the loss of the spacecraft, this was proven false. The post-flight medical examination showed John Glenn had a bruised hand, and Walter Schirra, who stayed in his capsule until it had been hoisted onto the deck of the recovery ship before he hit the manual trigger, also had, as expected, an injured hand from the plunger's kickback. When Grissom was given his post-recovery medical examination aboard ship, he did not possess a kickback injury. Case closed! This flight paved the way for a manned orbital flight.

Six years later, the new Apollo spacecraft would not have an explosive hatch for fear that it might fire inadvertently, as it did on MR-4. When the Apollo 1 fire occurred, the crew couldn't escape because the inward opening hatch was sealed by the pressure inside the cabin. Grissom, Roger Chaffee and Ed White all died. For this and other reasons, extensive modifications were made to the Apollo spacecraft.

Grissom's *Liberty Bell 7* capsule was recovered 38 years later by Oceaneering International Inc., from the Atlantic at a depth of 15,000 feet. The operation was sponsored by the Discovery Channel and led by Curt Newport. The hatch was not recovered.

13.3.6   MA-1 Launch Failure

This was to be the first flight of a Mercury capsule on an Atlas. The objectives included testing the structural integrity of the capsule and its heat protection from an abort. The launch was on July 29, 1960; about 9 months before Shepard's flight. The weather at the Cape was rainy and overcast but the Air Force didn't care because the Atlas was designed for launch in almost any weather during wartime. They pressed Walt Williams, the Mercury Operations Director, for a decision. He gave the go-ahead. Some of the STG people voiced concern because they wanted photographic coverage of the ascent. The MCC was just being built, so Gilruth, Mathews, and Chris Kraft were just observers at the Range Safety Officer's center and Williams was with the launch team in the blockhouse. This was an Air Force/Convair/STL launch operation.

The launch looked normal up to 58 seconds, about the time of Max-Q, when the Atlas broke up and the ASIS system kicked in. The capsule separated and continued to provide telemetry all the way to impact. The Atlas broke apart and fell into the Atlantic. There was a scramble to find out what happened. In his JSC Oral History Owen Maynard relates the story about free diving in the ocean to recover vital debris. The Air Force launch team was eager to blame the capsule for the loss of the Atlas. But Bob Gilruth and Jim Chamberlin, with their significant backgrounds in structures, seized upon the problem. The next month a joint meeting of the parties discussed the failure and the best remedy; there were many ruffled feathers. The issue was that the skin of the Atlas was too thin to take the aerodynamic loads during Max-Q when the capsule was mated to the Atlas in the manner that it had been. General Bernard Schriever of the Air Force wouldn't budge on their position and Bob Gilruth, confident in the STG analysis and with the support of NASA Headquarters, recommended that the decision be made by the Secretary of the Air Force, Dudley Sharp. A few weeks later the Secretary agreed, but with the stipulation that because the proposed fix was not approved by the Air Force the responsibility would be borne by NASA. This was a classic example of having sufficient confidence in your engineering analysis and judgement that if necessary you take the decision to the top. The result was the MA-2 "Belly Band." See Section 13.2.2 for a thorough discussion of the decision process and the resulting engineering fix.

13.3.7   MA-2 Belly Band

Only three months after Ham's flight and seven months after MA-1, the modified Atlas with its "Belly Band" AKA "Horse Collar," lifted off on February 21, 1961. Everyone in the MCC and the blockhouse breathed a sigh of relief as the Atlas breezed through Max-Q. It was a perfect flight and recovery, with all of mission objectives being accomplished. The lesson learned was that the STG engineers knew why the MA-1 Atlas had failed, and they fought the Air Force to overcome the inability of the vehicle's thin skin to support the weight of the capsule at Max-Q. So when you know you are right, hold your ground even when you are blamed for the failure.

13.3.8   MA-3 Mechanical Man

This flight was to test the Mercury capsule and both the control centers at the Cape and on the island of Bermuda in an abort just prior to orbital insertion. The flight would also test the new "thick skinned" Atlas fitted with an improved telemetry system and the ASIS. Installed in the capsule was a "mechanical man" which could simulate the load on the environmental control system by consuming oxygen, emitting carbon dioxide, and generating heat like an astronaut.

But in a surprise move on April 12, 1961 Yuri Gagarin made an orbital flight on Vostok 1. The Russians had done it again! They were the first to put a man in space and, in orbit no less. NASA management quickly redesigned the MA-3 mission to perform a single orbit. This was a relatively easy flight preparation effort. The MSFN had recently become operational, so flight controllers were sent out around the world for the first time in order to support our first attempt at orbital flight.

As soon as the STG decided on an orbital mission, the simulation team began to develop the materials necessary to exercise the MCC and the MSFN. This required developing tapes to be sent out with the deploying teams that would be played for their simulated orbital passes. This was very new, and required an element of synchronization on the part of the remote site flight controllers and the support technicians there. It also required the MCC simulation supervisor, known as SimSup, to plan how to exercise not only the MCC people but the remote site flight controllers with radar, command, telemetry, and voice links for their orbital passes. Whenever the simulation would trigger a fault or situation, the flight controllers and the person who was playing the role of astronaut for that pass had to respond appropriately. Prior to the launch, the simulations and the responses of the flight controllers left much be desired. But it was the first use of the MSFN and the first time that some flight controllers were deployed. In some cases, it could take a week for a team to reach a remote site such as Zanzibar or to get a tracking ship to the desired location.

When MA-3 lifted off on April 25, 1961, the guidance system of the Atlas failed to start the roll and pitch program that would place the missile onto the proper trajectory, so it just climbed straight up. The Range Safety Officer watched it to see if it would correct, then sent the destruct command at 43 seconds. The escape system worked perfectly, and the capsule was recovered to be refurbished for use with MA-4. The disappointed remote teams returned home, because the next attempt at an orbital mission wasn't due for many months. The MCC and Bermuda teams remained in place and trained for Shepard's MR-3 flight.

13.3.9 MA-4 First Orbital Flight

With Shepard and Grissom's flights behind them, the STG flight operations teams were looking forward to getting a real orbital flight under their belts. This would test both the MSFN and the DOD recovery forces in an essential step toward verifying the total system prior to attempting a manned orbital flight. The Atlas was modified for the propellant sloshing problem, the autopilot, and faulty transistors. It was the second flight of the "thick skinned" Atlas. The capsule was the refurbished one from MA-3.

As the STG prepared, Gherman Titov was launched aboard Vostok 2 on August 8, 1961 on a mission that really took the wind out of NASA's sails because it lasted an entire day and ran for 17 orbits. NASA wouldn't accomplish this feat for two more years with Gordon Cooper's flight of MA-9. This was a time when some aeromedical people were calling for launching fifty more chimpanzees. However Titov's flight put to rest a number of medical concerns about prolonged weightlessness inducing "space sickness."

MA-4 was launched on September 19, 1961. There were several anomalies in the trajectory but the Atlas proved that it could put the capsule into an acceptable orbit. Initial computations after insertion indicated that it could last seven orbits but the mission was planned for just one. The Mercury capsule suffered a few anomalies, with high fuel usage caused by some electrical and mechanical connections as well as thruster failures. Had there been an astronaut on-board instead of the "mechanical man," he could have taken the requisite actions. The heat protection system performed as planned, with only a few anomalies. The MSFN and the remote site flight controllers got a good workout.

There was to be a test of the MSFN by the first Mercury Scout mission but on November 1, 1961 this launch failed. Nevertheless, future such flights were canceled because the MSFN had supported MA-4, and the MA-5 mission would occur before it would be possible to schedule a second Scout launch.

The STG management was quite happy with the results and came to accept the fact that the Atlas, with all of its previous failures, could safely put a man in orbit and the Mercury capsule could bring him back safely. Operationally, the MCC flight control team was pleased with the performance of the teams. In just two months, they would be tested again with the launch of a chimpanzee.

13.3.10   MA-5 Enos

The MA-5 flight on November 29, 1961 was the dress rehearsal for the first manned Mercury Atlas orbital flight. The doctors still wanted to see how a chimpanzee reacted to the prolonged weightlessness of an orbital flight. This even though the Russians had two orbital flights under their belts; one lasting a full day. But MA-5 would be an excellent checkout of the world-wide tracking stations and the ability of the flight controllers to handle any in-flight emergency in a timely manner.

The pre-launch phase suffered problems, but the launch was perfect. The "Horse Collar" on the Atlas worked. The capsule separated and automatically adopted the appropriate attitude. The chimpanzee Enos performed his duties properly. The mission was planned for three orbits. Then the remote sites started to observe roll disturbances and abnormal fuel usage. The cabin and suit temperatures started to increase. The surgeons became concerned. All the tracking stations were alerted and monitored the situation when the spacecraft came over their sites. The temperatures seemed to stabilize, but the thruster problem and fuel usage continued. If an astronaut had been onboard he could have switched off the malfunctioning automatic system. And unknown to the surgeons, Enos's test apparatus was malfunctioning and delivering electrical shocks to his feet even though he was correctly performing his assigned tasks. He was so upset that he ripped out his urinary catheter and was beginning to bleed. The weightlessness of the space environment was the

Fig. 13.15 Arnie Aldrich demonstrating how he threw the retrofire switch to bring Enos home. (Photo courtesy of Arnie Aldrich)

least of his problems. This was not known on the ground. The big concern was having enough fuel to control the re-entry.

Flight Director Chris Kraft warned the remote sites of the problems and alerted them to the possibility of re-entering on the next orbit, ahead of schedule. As the spacecraft came up on the Point Arguello site in California to cross the U.S. for the second time, CAPCOM Arnie Aldrich reported the errant roll maneuvers were still occurring and that the fuel situation was low. Kraft asked his MCC SYSTEMS flight controller for a recommendation that seemed slow in coming. Aldrich already had the correct retrofire times loaded into his console. Kraft made the decision, ordering, "California, Flight. Retrofire on my mark." Aldrich was ready. With only seconds left, Kraft called, "...five, four, three, two, one, mark!" The re-entry sequence was perfect. Enos was coming home!

Several lessons were learned from this flight. It was the responsibility of the flight controllers to get the astronaut home; albeit he was only a chimpanzee. The teamwork around the world was excellent in getting the critical information to the MCC. As a lot of little problems mount up, the decision area can become rather gray. How many small problems does it take to prompt a "No-Go?" This had some lessons for the mission rules process. Whilst there was some hesitation in the MCC systems area of decision making, in this case it had been better to make a conservative decision and end the flight early than to wait for the perfect decision and maybe lose everything along the way. Time was of the essence if the capsule and Enos were to be recovered in a prime recovery area. This was the first time that Kraft made a life-and-death decision on his own as the Flight Director. Arnie Aldrich has the distinction of being the only remote site flight controller to have terminated a flight. This was also a time when flight controller performance was critical and would be evaluated following the flight for subsequent assignments. There were also engineering lessons to be learned. See Section 13.2.

13.3.11   MA-6 Glenn

The first American into orbit was John Glenn, who was launched aboard MA-6 on February 20, 1962. It had initially been scheduled for January 27 and the STG flight control teams had begun deployment to the remote sites on January 12. After many scrubs, the launch finally took place and everything was normal for about one orbit. When Glenn switched the control system back to automatic the capsule yawed to the left. He tried to control it manually but it would swing again. He was advised to switch off the offending thrusters. Then the MCC SYSTEMS controller got a Segment 51 signal; the now famous "Landing Bag Deploy" signal. Because it was not a normal occurrence in orbit, this event had never been simulated. The remote sites were told to verify the signal. The local McDonnell personnel and NASA systems people were asked about the signal. Capsule schematics were consulted. What did this mean if it wasn't an erroneous signal? All the remote sites were to check their telemetry patching. Did this mean that the heat shield had come loose? What then?

While the systems engineers were looking into the problem, the Flight Director was exploring the operational ramifications. What do we need to do to verify the event or get Glenn to assist in determining the status of the situation? Glenn was asked if he was hearing any noises as he was maneuvering. The answer was "negative." It was suggested to leave the retro package on during re-entry. But was that safe? What if the retrorockets had remaining fuel or if they didn't all fire? Would they explode when the re-entry heating reached them? Would the retro pack burn off and then the heat shield would come off?

Mission Director Walt Williams got together with Max Faget the Mercury designer and John Yardley, the McDonnell chief engineer, in a side room and discussed the situation. The proposal was to retain the retrorockets in place after they fired. The reasoning was that if the heat shield was loose, the retro pack might hold it in position during re-entry, at least until the pack burned away. The Flight Director was also concerned about the aerodynamics of the capsule with the pack still on.

Coming up on the end of the third orbit, Flight made the decision to retain the retro pack in place. The CAPCOM in California was Wally Schirra, who had deployed as part of the flight control team to communicate with Glenn in such a situation. He relayed the decision to Glenn, who wanted to know why. He knew from previous questions that the ground was working on a problem but he didn't know the extent of the issue and lacked the benefit of all the discussions that had been going on for the past hour or so. The necessary changes to the re-entry checklist were read up to him. The re-entry was very fiery and noisy until the straps which held the retro pack in place burned through. There was silence in MCC during the communications blackout. Glenn landed 40 miles long but only five miles from the destroyer, *Noa*, stationed down range.

There were several lessons in every category from this first manned orbital mission. From an operational perspective there needed to be more redundant telemetry to facilitate verification of critical events. If the three switches involved with the deployment of the landing bag had each provided a telemetry signal, then it would have been clearer that a reading of only one of them meant it was a faulty signal. In addition, flight controllers must have sufficient data to provide the Flight Director with the input he requires to make the final decisions. Flight Director Kraft made a note to have another rule: "Never depart from the norm unless it is absolutely required. Once you do, you enter a regime where events are unpredictable. Make a change on the fly and it might bite you, and bite you hard." He also knew that flight operational decisions would have to be made by the Flight Director with input from his team of flight controllers. Management may have an input, but the final decision must be the Flight Director's. Management input might be valuable, but it might not be timely and it might not benefit from all of the data the whole flight control team can bring to bear on the problem.

Another key lesson was the need to have more depth to the systems knowledge immediately available in the MCC. The Mercury spacecraft was limited in the telemetry that it could provide. Subsequent spacecraft would require to provide more data to the ground. The rushed calls to the McDonnell systems engineers and the NASA engineers to get their input on a situation that had not been simulated or discussed, led to later missions having "hot lines" to the manufacturers of equipment and support staff rooms filled with various experts.

13.3.12   MA-7 Carpenter

The first manned Mercury mission had shown that the spacecraft (the word "capsule" was now discontinued) worked well and that the astronaut could be more than a passenger, therefore the next flight was to allow the astronaut to do more science experiments. These experiments were devised by the Ad Hoc Committee on Scientific Tasks and Training for Man-in-Space, headed by both Headquarters and Goddard scientists after obtaining input from scientists from various other organizations. The committee came up with five different experiments for astronaut Scott Carpenter to carry out and he was given special preflight training for these tasks as well as the appropriate equipment. They were:

- The release of a 30-inch multi-colored balloon that was to be unreeled on a 100-foot line. Carpenter was to comment on the relative visibility of the various colored panels and take photographs. Strain gauges would measure the atmospheric drag on the balloon. It was to be jettisoned prior to re-entry.
- Observing the behavior of liquid in a weightless state inside a closed glass bottle. This had implications for the design of tanks and pumps for liquids of future spacecraft.
- Use of a special light meter to determine the visibility of million-candle-power flares that were lit on the ground near the Woomera tracking station as the spacecraft flew over. Poor weather and cloud cover prevented Carpenter from seeing the flares.
- Making weather observations and taking photographs.
- Making observations and taking photographs of the "air glow" layer.

Although there were engineering and scientific lessons learned on this flight, the experiments had operational ramifications. The Atlas and the Mercury spacecraft both performed well, as did the launch countdown and the world-wide network. But Carpenter used more fuel than planned while conducting these experiments. Flight controllers advised him of his fuel state as he passed over the tracking stations. Starting at the beginning of the last orbit he compensated by not using the thrusters and instead coasting, letting his attitude drift. Up to this point, Flight Director Kraft was pleased with the mission except for the high fuel usage.

Then, on the final orbit, an unplanned series of observations of the "fire flies" that had been reported by Glenn and which Carpenter renamed "frost flies" pre-occupied him. They proved to be particles of frost from the capsule that drifted close by and shone in sunlight. This distraction put him behind in the pre-retro checklist. Then he found that the automatic stabilization control system wasn't holding the re-entry attitude. He switched to fly-by-wire but forgot to turn off the manual system, causing both systems to consume fuel. Alan Shepard was the CAPCOM at the Port Arguello station in California. He reminded Carpenter to reset the switch positions. These problems changed the retro fire sequence. As it turned out, Carpenter was not in the optimum attitude at retrofire. As a result the spacecraft overshot the target by 250 miles. Radar tracking, the Goddard computers, and the MCC all agreed that the landing was far beyond the recovery ships.

This error caused problems for the recovery forces. Air Force SA-16 recovery aircraft began flying toward the splashdown area. A Navy P2V detected the beacon of the spacecraft from 50 miles away and several Air Force SC-54 with frogmen flew towards it. A decision in the MCC directed the Navy to pick up Carpenter, not the Air Force. This was done by an HSS-2 helicopter from the carrier *Intrepid*, which found the astronaut patiently floating in his life raft tethered to the capsule. The delay in recovery was thoroughly discussed post-flight, as was Carpenter's re-entry performance.

Several operational lessons were learned from this mission. In future the experimental work load on the astronaut must be carefully planned. Changes were being made right up to the last minute. The heavy use of fuel was somewhat related to the astronaut's mismanagement of the multiple control systems, as well as work load. The general observation was that whenever an astronaut falls behind in the flight plan and checklists, especially a critical one such as the pre-retro fire checklist, he becomes rushed and is more likely to forget critical switch settings and actions.

Seeing the high fuel usage the ground repeatedly asked Carpenter to conserve fuel, which he began to do during the third orbit. They did their job in monitoring the spacecraft and astronaut, and communicated their concern and provided direction. This seemed to solve the fuel problem, but then Carpenter's fixation with the "fire flies" consumed valuable fuel and time, which led to his re-entry error. This in turn led to the recovery problem, which could have been much worse. This situation was exacerbated by the fact that Carpenter was so far from the ships that there was no communications between him and the recovery forces and the MCC. Furthermore, there was no radio in the life raft for him to call or receive voice from the aircraft or ships.

The overall recovery problem raised the flight management problem of who was in charge at that time. Should the Air Force have recovered the astronaut, or the Navy? Flight Director Kraft concurred with Recovery Coordinator Robert Thompson, who made the decision to let the Navy pick him up, as was the normal plan. However, the Air Force challenged that decision during the mission and post-flight, and this matter was even discussed in Congress. The changes made for MA-8 included providing the Recovery room in the MCC with communications to the tracking sites and recovery forces, and installing a long cable to run from the spacecraft to the life raft to enable the astronaut to talk to the recovery forces. Such a capability would have greatly changed the MA-7 situation by making the MCC aware of Carpenter's status. And from now on, other parties wouldn't be able to challenge operational decisions made by the MCC.

13.3.13   MA-8 Schirra

Planning for MA-8 started right after John Glenn's flight. It was to be an intermediate step to an extended duration flight which at the time was planned for 18 orbits. Extensive work was done on the spacecraft in order to extend its power and life support capabilities. A lot of analysis of MA-6 and MA-7 gave the engineers the data that they needed to make the desired changes. A look at the recovery planning for extended duration flights led to the conclusion that a mission rule for contingency recovery of the astronaut would be violated. He required to be recovered within 18 hours after landing. As a six orbit mission would meet that requirement, the original plan for seven orbits was reduced to six.

This flight was intended to be more about engineering than science, so that was minimized. Astronaut Wally Schirra was to re-evaluate the issues Carpenter experienced with determining yaw; one of the reasons he was long on re-entry. Schirra was to use both the periscope and the window in daylight and in darkness, and conserve his fuel by minimizing his use of the RCS thrusters. He practiced these maneuvers in Mercury Procedures Trainers at Langley and at the Cape.

Changes were also made to the Atlas, which was going through checkout at the same time, including a flight readiness static firing. New baffled fuel injectors were installed to eliminate combustion instability at engine start.

The MSFN, having been designed for three orbits, was also modified for the six orbit flight. Five Air Force C-130s were added for voice relays. The recovery forces were now huge; about 17,000 men on 19 ships, including 100 aeromedical monitors.

Pre-flight preparations and the checkout of the MCC and MSFN went well with only minor glitches. Schirra lifted off in *Sigma 7* on October 3, 1962. There were several minor problems during the launch phase. In orbit, Schirra worked through his engineering tasks to evaluate the spacecraft systems. He managed minor suit cooling problems himself, with the ground making suggestions that he basically ignored because he felt better than implied by the readings on the ground. He made extensive control evaluations, especially of yaw at night employing celestial references. On the fourth orbit Schirra adopted drifting flight and used the time for photography and some experimental observations. Coming up on Point Arguello in California he talked with John Glenn over a "live" nationwide broadcast.

The remainder of the flight passed as planned, with the spacecraft functioning normally and Schirra taking pictures and performing his systems checks. The ground continued to advise him of his "Go" status, including his fuel usage. By the sixth orbit he was preparing the cabin for re-entry, stowing his gear and going through the pre-retro sequence checklist. On passing over the Pacific Ocean command ship, Alan Shepard readied Schirra for the re-entry maneuver. Schirra commented that he was in retro attitude using the automatic system. Shepard counted him down and re-entry began. He followed the plan; shifting to fly-by-wire and then the rate stabilization control system. The descent was perfect, the landing was perfect, and the recovery was perfect. Unlike his predecessors, he remained in the capsule until after it had been hoisted aboard ship, then he "blew" the hatch. This "text book flight" greatly simplified the planning for the "day long" MA-9 mission.

Besides all the engineering and operations data obtained, the main lesson learned from the MA-8 flight was that a mission can be "executed as planned." Lessons learned from previous flights were carried forward. The management lessons from the MA-7 recovery were resolved before a vast armada of Navy and Air Force aircraft were deployed. Having learned from the mistakes of his immediate predecessor, Schirra knew of the problem, trained for it, thoroughly explored it, and found the solution. That's flight operations at its best!

13.3.14   MA-9 Cooper

There were some who were ready to end Project Mercury and push on with Gemini, but a "day long" mission was intended to bridge the knowledge and experience gap between Mercury and Gemini. By this time there were only about 500 NASA personnel working on Mercury; the rest had moved on to Gemini and Apollo.

Meanwhile the Atlas was having some difficulties, with the Air Force ICBM version having suffered two failures. And when Cooper's launch vehicle was rolled out it failed the inspection. The Mercury spacecraft was growing in weight with all the added fuel and life support needed for the extended duration. In addition, to exploit the additional time in orbit this was to be more of a scientific mission involving a lot of photographic equipment. It was to fly over most of the Earth between latitudes 33 degrees north and south, so it would require even more recovery and tracking forces than the six-orbit flight. In fact, MA-9 required 28 ships, 171 aircraft, and about 18,000 servicemen. The aeromedical teams were reduced slightly, reflecting the flight surgeons' increasing confidence of the astronauts' reaction to prolonged weightlessness and the ability of the spacecraft's environmental control systems.

On the first attempt to launch *Faith 7* there was a delay caused by the gantry's diesel engine failing to move the gantry. Fixing that imposed a two hour delay. By then the Bermuda station was no longer fully operational, so the launch was scrubbed. The next day, May 15, 1963, the weather was good and all systems were "Go." It was a perfect launch; the Atlas did its job and Cooper was good for at least 20 orbits.

Cooper carried out the flight plan, conducted the various experiments, and slept on occasion. His acute observations of features on the ground were amazing to some. His management of the fuel supply was excellent. On the 19th orbit, the 0.05 "g" light came on. This was intended only to illuminate at the start of re-entry. Cooper suspected this to be a faulty signal because he could feel no change in weightlessness. The California site confirmed they didn't have the indication. On the next pass, he lost the attitude readings. Then on the 21st orbit, he lost the 250 volt main inverter, which meant he also lost the automatic stabilization and control system. The levels of carbon dioxide were also rising, both in his suit and in the cabin. After checking the spacecraft systems the MCC and the MSFN revised the pre-retrofire checklist.

Coming up on the final orbit and over the *Coastal Sentry Quebec* tracking ship stationed off the Japanese coast, Cooper reported that he was in retro attitude and using the manual control system. John Glenn gave him the 10 second countdown, and Cooper fired off the retro rockets manually. The re-entry was perfect and Cooper landed within a couple of miles of the primary recovery carrier *Kearsarge*, located south of Midway Island. He stayed in the spacecraft, was towed to the ship and hoisted aboard.

Like Schirra, Cooper's personal performance was perfect but unlike his colleague he had to overcome systems faults that further demonstrated that the pilot was an essential component of the spacecraft and therefore critical to mission success. The operational lessons learned on this flight were few in number, but nevertheless significant. The flight control team, now scattered around the globe, had performed as planned. So had the MSFN. The teams analyzed the faulty indications and the astronaut's physical status, kept him informed about how his systems were performing, supplied him up-to-date changes to the checklist, confirmed his retrofire attitude, and counted him down. All of the preparation, training, and the experience gained from earlier flights was evident by the teamwork demonstrated by the flight controllers and tracking station people.

Project Mercury was at an end, having served its purpose. It proved the ability of America to put together a spaceflight system from scratch and achieve all of its objectives in the full light of day; as opposed to the Soviets who did not announce their plans, didn't admit their failures, and reported their successes only after they had been achieved.

Mercury proved man's ability to work in space, to take necessary action to save the mission, and to survive the rigors of "g" loads, weightlessness, and all of the other conditions which had concerned the medical community. It was time to move on to Gemini and Apollo.

13.4 SCIENTIFIC

Some readers may be astonished that none of the Project Mercury mission objectives or their supporting guidelines mentioned science. As part of their training, the astronauts were given basic knowledge of the space sciences in order to enable them to function better as observers. They were also given brief courses in mechanics, aerodynamics, guidance and control, space navigation, space physics, and basic physiology. It was apparent that the 15 minute suborbital flights of Shepard and Grissom couldn't provide an opportunity for much science other than a quick look out the window. But longer duration flights would offer periods of time where the astronaut could conduct some simple experiments.

The major constraints imposed on an experiment were the weight and volume requirements, the power it would consume, and the fuel that would be used by any maneuvering. The overall weight of experiments grew from 11 lbs. on MA-6 to 62 lbs. on MA-9; a physical indication of the move to conduct ever more experiments with each orbital flight. Finding a place to stow the equipment became an issue. After the aeromedical people had gained sufficient information to satisfy them that an astronaut could not only survive in the space environment but also perform as an integral part of the system, they became comfortable with increasing his work load. As the program evolved, the experiments fell into three broad categories: biomedical, physical sciences, and engineering.

Early on in Project Mercury, there was no formal means of requesting an experiment or how they should be handled. Suggestions would be submitted by the STG or the various NASA field centers to the training group, requesting that the astronaut make this observation or photograph something or other. Shortly after John Glenn's flight, the new Manned Spacecraft Center and the new Mercury Project Office established the Mercury Scientific Experiment Panel (MSEP). This group, which represented all the Divisions, would perform the following:

- Evaluate proposed in-flight experiments
- Suggest the order of priority of acceptable experiments
- Seek out and foster the generation of experiments.

The MSEP worked closely with the NASA Headquarters Office of Space Sciences and the Goddard Space Flight Center. By MA-9, it had become the In-Flight Experiments Panel (IFEP) and its remit had been extended to address the forthcoming Gemini and Apollo programs. The process of adding and approving experiments for a spaceflight then became much more formal.

13.4.1 MA-6 Science

Since this was Project Mercury's first orbital mission, the planners didn't give John Glenn much to do in the way of space science, but he did plan to undertake a lot of weather and astronomical observations and to take photographs. However, he had a thruster problem on the first orbit and spent more time than planned in controlling his spacecraft manually. Then he had to prepare for re-entry on the third orbit. The aeromedical and physiological experiments are covered in a later section.

The four principal space science observations were:

- The observation of the luminous particles that Glenn called "fire flies"
- A luminous band of light above the night-time horizon called "air glow"
- The flattened appearance of the Sun at sunset
- Ultraviolet photography of stars in the Orion region.

To conduct these observations Glenn had cameras, filters, film canisters, binoculars, and a photometer. He snapped a lot of pictures for the Weather Bureau, but much of his ground track was masked by cloud. Nevertheless, it was hoped that some of the observations would be useful to those designing optical systems for Nimbus and Tiros weather satellites. Many of these basic observations were planned for later flights. He was surprised at his ability to discern small items on the ground and at sea. He assessed his night vision as well. He was also surprised at how well he adapted to weightlessness, as demonstrated by the ease with which he handled the cameras, film, and filters.

From a post-flight perspective, Glenn's comments concerning decision-making and lessons learned related more to the operational aspects of the Segment 51 deployment issue than about conducting space science experiments. With its first orbital mission accomplished, NASA was eager to assign more scientific experiments to the next flight.

13.4.2   MA-7 Science

There were several scientific experiments on Scott Carpenter's flight. They were sponsored by several organizations and integrated into the flight plan. One key constraint was weight. It is an example of both inter-agency coordination and intra-agency coordination, as well as integration into the manufacturing of the capsule. Four of the NASA centers were involved, one university, and the Weather Bureau. They also had engineering, systems and operational implications. The program included the following:

- Balloon Experiment – This was sponsored by the Langley Research Center. It built a 30-inch diameter balloon to measure air drag at orbital altitude and the visibility of colors in space. The drag would be measured using strain gauges on the balloon's 100-foot tether. When Carpenter deployed the balloon it failed to inflate properly but he did observe the "confetti" that had been folded into the packed balloon. Since it hadn't fully inflated, the drag measurements weren't of any use. He was able to see only two of the colors. It was not possible to jettison the balloon because it became entangled with the capsule during maneuvers; it burnt up on re-entry with no effect on the vehicle. While this was a rather rudimentary experiment by modern standards, it was simple and inexpensive and despite having failed, it provided input for subsequent tethered experiments in much later flights.
- Zero-Gravity Experiment – The Lewis Research Center sponsored this experiment as an early investigation of the behavior of liquids in a weightless state. It was very simple and lightweight. A special glass sphere was made containing a capillary tube which extended from the interior surface to just past the center. The sphere was $3\frac{3}{8}$-inch in diameter, and the liquid represented hydrogen peroxide in a fuel tank. Carpenter took photographs that confirmed some predictions and were used in

the future construction of space fuel tanks. Even today, the International Space Station conducts experiments to study the effects of weightlessness on various fluids. Space travel will have to deal with this problem for all types of fluids in various tanks/containers. Remember that the Apollo 13 problem was an explosion of a tank holding oxygen in a "supercritical" state. Consider this issue for future very long duration missions, such as to Mars.

- Ground Flare Visibility Experiments – The new Manned Spacecraft Center sponsored this experiment in an effort to determine the capability of the astronaut to acquire and observe a ground based light of known intensity and to determine the attenuation of this light source through the atmosphere. At a prescribed time, Carpenter was to use a special photometer to measure ten-million-candle power flares located near Woomera, Australia. As it turned out, the weather was too cloudy and he didn't see them. The experiment was discontinued due to continued cloudiness.

- Photographic Studies – The Massachusetts Institute of Technology (MIT) had been hired to develop a navigational system for manned missions whereby an astronaut would use a sextant to measure the altitude of stars above the Earth's horizon. So it sponsored a study and supplied equipment to determine the definition of the horizon. Carpenter took several photographs and discussed these in his post-flight report. I have a personal connection to this study and problem. Five years later, during Apollo 7, MIT wanted the crew, headed by Wally Schirra, to attempt some additional experiments and in particular to employ the optical portion of the navigation system to "mark" on the visual horizon. As it turns out, there are several horizons when viewed from space! When Donn Eisele "marked" on the horizon, the onboard computer thought the mark was above the horizon and was unable to perform the necessary computation. The computer shut down, causing a lot of alarms and computer restarts. There were some harsh words from the crew to the ground. I was the Mission Staff Engineer on that mission, and in that capacity I had coordinated all of the flight test objectives. This was an extra test after we had finished all of the primary test objectives. I had to find the MIT experimenters in the support rooms of the Mission Control Center and get an explanation for the Flight Director. This incident apart, all in all it was a very successful mission with over 100% of the objectives accomplished.

13.4.3 MA-8 Science

The objectives for this mission addressed engineering issues rather than science. Schirra even named his spacecraft *Sigma 7* to reflect this point. The sigma symbol represents the summation of an engineering evaluation. Only four non-engineering scientific experiments were planned, two of which were completely passive, so his active involvement was required in only two of them.

The Ground Flare Visibility Experiment was carried over from MA-7. This time there were two locations. Woomera in Australia had high powered flares and Durban in South Africa had a xenon arc lamp. However, poor weather at both sites meant Schirra wasn't able to see them. He did report his ability to spot illuminated cities and lightning flashes.

Photographic experiments were to examine the spectral reflectance characteristics of cloud, land, and water areas using a 70-mm Hasselblad camera with a special filter for six spectral bands. These photographs were sponsored and analyzed by the National Weather Satellite Center to assist in designing future weather satellites. The results showed that the best wavelength for viewing the Earth might be the near-infrared portion of the electromagnetic spectrum, in which scattering by atmospheric particles is relatively weak. Although this filter photography was successful, the conventional color photography was not as a result of either overexposure or excessive cloud cover.

One of the passive science experiments involved carrying two photographic films that were sensitive to radiation. It was sponsored by the Goddard Space Flight Center and the U.S. Navy School of Aviation Medicine. The objective was to estimate the astronaut's overall exposure to radiation. Post-flight analysis showed that he received a minimal dosage. The other experiment involved six ablative materials that were attached to the capsule. They all survived the re-entry heating and were analyzed for future flight applications.

After this mission, it was determined that a significant amount of weight could be removed from the spacecraft to accommodate more consumables and scientific experiments for the final flight.

13.4.4   MA-9 Science

For this "full day" flight, the planners assigned Gordon Cooper much more to do. Some of the experiments on earlier flights that had been canceled for either weather or operational reasons were now rescheduled.

On the third orbit, Cooper began working on the experiments listed in the flight plan. They included:

- Ejection of a 6-inch sphere with xenon strobe lights
- Periodic urine samples
- Deployment of a tethered balloon with strain gauge
- Radiation monitoring
- Photography using different cameras and wavelengths
- Control systems checks
- Zodiacal light and night-time "air glow" observations
- Horizon definition observations
- Photography of the Earth's limb
- Radio frequency tests.

One experiment built by the Langley Research Center pertained to the future requirement for two spacecraft to rendezvous in space. A light which flashed once per second was ejected from the spacecraft and viewed by the astronaut at varying distances over several orbits. This sphere, mentioned below, added 10 lbs. to the experiment weight. It was a very successful experiment, and Cooper saw it several times at distances up to about 17 miles. As a personal note, I was in the MCC for the first active rendezvous, when Gemini 6 slid alongside Gemini 7 in December 1965.

Cooper used a 70-mm Hasselblad camera with special filters for horizon definition studies. This data, along with Carpenter's film, was analyzed by the MIT Instrumentation

Laboratory. Cooper made many observations of the Moon, Sun, night sky, "day glow" layer, and stars. This was of assistance to the atmospheric scientists. There was also interest in the appearance of the Earth at sunset and twilight, as well as the apparent flattening of the Sun; a phenomenon that is brief and difficult to see. Cooper confirmed the observations of Schirra regarding the scattering of light at the horizon during twilight; especially the blue band.

The Hasselblad camera was also used for weather photography, using filters to examine the spectral reflectance of clouds, land, and ocean areas for the National Weather Satellite Center. Because the weather for this flight was much better than the earlier flights, and it passed over more land areas, the pictures were more useful. The coverage of the African and Asian deserts and the Himalayan Mountains were analyzed by scientists at the Goddard Space Flight Center. Cooper took some thirty 70-mm photographs showing abundant topographical and geological detail. Those of the Tibetan plateau showed some unique geological detail. The synoptic nature of space photography is valuable in meteorology and oceanography applications. Cooper made interesting observations of lightning and correlated these with bursts of radio interference. This was useful to the Weather Bureau, as were his observations of cloud and snow cover.

The dim-light photography included the zodiacal light and night-time "air glow". They were sponsored by the School of Physics at the University of Minnesota. The zodiacal light pictures were not very useful because they were underexposed. The "air glow" images were valuable to the scientists' investigations.

The Langley Research Center went to a lot of effort to redesign, retest, and requalify the same balloon experiment that failed during Carpenter's MA-7 flight. Unfortunately the balloon didn't deploy, seemingly because of the failure of a squib firing circuit. The balloon was housed in the antennae canister that was jettisoned prior to landing.

13.5 MEDICAL

13.5.1 Space Flight Medicine

At the beginning of Project Mercury, the concept of monitoring the health and performance of an astronaut on-orbit was new. The flight surgeons (as these doctors were called) had different ideas about how to perform this monitoring. Previously, they had monitored a pilot before a flight in a high performance aircraft and again afterward, and a typical flight lasted only a few hours, not a day or more. In order to monitor a pilot for days during a space mission, they needed to come up with new methods. These methods could not be allowed to interfere with the pilot's execution of the mission. Furthermore, he was now in a full pressure suit connected to a complex system in a cramped capsule. New sensing devices had to be designed that would transmit biomedical data to the ground. The whole purpose was to allow the flight surgeons to assess the performance of the astronaut, to know whether he was capable of continuing the mission from a physiological point of view. That is, at any moment, should the mission be allowed to continue or should the flight surgeon recommend that it be terminated? This medical decision process now required mission rules just as other spacecraft systems did.

It was clear that people react quite differently to an event, and that hard and fast rules would not work. One pilot's reaction to a booster engine cutoff or a tumbling capsule may cause quite different physiological reactions from another pilot. Just like in a doctor's office, the diagnosis might just as well be made by talking to the patient and asking questions. Some flight surgeons wanted to talk frequently to the astronauts, but as missions became longer it was clear that the astronaut would be out of range much of the time due to the gaps between the tracking stations. When the spacecraft did come into contact with the station, the CAPCOM would be the one to talk to the astronaut and their conversation would necessarily be more oriented toward how the systems were performing and how closely the mission was following the flight plan.

As experience was gained, the bioinstrumentation selected gave the aeromedical monitors on the ground sufficient information, in addition to their monitoring of the voice communications, to assess the astronaut's status. They did have the prerogative to talk directly to the astronaut if they thought there was a medical emergency. Even so, the decision to terminate the mission had to take into account how far away the recovery forces and medical help were. Many recovery areas were very remote and many hours might elapse before help could reach a capsule which might be wallowing in rough seas. It was determined that the best course of action was to keep the astronaut in his air-conditioned capsule until he was able to get to a planned recovery area.

By the end of 1961 and early 1962 when John Glenn was preparing for his orbital flight, all the aeromedical and life science people had transferred to the new Manned Spacecraft Center. Dr. William K. Douglas was still the astronauts' flight surgeon, and he was supported by many others from the new MSC Life Systems Division, each of whom had his own specialty. There were others from Army, Navy, and Air Force aeromedical organizations with interests in this pioneering space program and they were involved in different phases such as pre-flight, flight, recovery, and post-flight analysis.

13.5.2   MR-3 and MR-4 Shepard & Grissom's Medical

By the time Alan Shepard flew on May 5, 1961 and Gus Grissom on July 21, 1961, two dogs, three primates and Yuri Gagarin had flown in space. It was known that the human pilot would survive and that many doctors were being overly cautious. Perhaps the lesson here is that for a totally new adventure that is without precedent, the managers, scientists, and engineers can be rather extreme in their thinking. Nowadays, the aeromedical studies are channeled toward the real problems of future long term exposure to weightlessness and radiation. The usefulness of having "a man in the loop" has been proven many times. Back in 1961, however, there were a great many unknown and unanticipated risks.

In the post-flight report on Alan Shepard's 15 minute suborbital flight, it was stated that "the remote monitoring on a noninterference basis of parameters such as temperature, respiration and the electrocardiogram, and blood pressure in active men fully engaged in prolonged and exacting tasks, is a new field. Hitherto, flight medicine has accepted the information concerning the well-being that could be derived from the pilot's introspection and conveyed by the invaluable voice link. For the rest, it has relied on performance to tell how close the man was to collapse. It is to be hoped that some of the developments in automation necessitated by Project Mercury will find application in clinical medicine."

13.5.3   MA-6 Glenn's Medical

No human was ever poked, probed, or prodded more than John Glenn when preparing for the first manned orbital Mercury mission. The life sciences objectives included the study of launch and re-entry accelerations and the intervening period of weightlessness. At almost 5 hours, this flight would be of sufficient length to understand prolonged weight-lessness better than the two manned suborbital flights which had provided only about 5 minutes each in this state.

Glenn was thoroughly monitored with leads feeding the telemetry system. He had two ECG leads, a respiration rate sensor, and a body temperature sensor. It was the first flight of a blood pressure measuring system that had gone through extensive design and testing for months prior to the flight. Glenn was monitored by the aeromedical teams at the remote tracking stations and at the MCC. The doctors also wanted him to report on eating, drink-ing, urinating, and carrying out his assigned tasks, and whether he suffered any spatial disorientation.

Occasionally, Glenn would report his condition during a pass over a station. Even his mood and voice were monitored and evaluated. He had no problem with weightlessness and actually found it helpful in performing some tasks. During the now famous Segment 51 problem, voice communications over the sites showed he was not alarmed by the situ-ation; merely questioning the ground because they had not fully involved him in the analy-sis except to just ask him a few unexpected questions. He dealt with the updates to the re-entry checklist and procedures in his usual professional manner, whilst still not fully informed of the situation.

The decisions and lessons learned on this flight weren't so much to do with physiology but with keeping the pilot in the loop; rather than in the dark. In his post-flight report, Glenn said:

> On the ground, some things would be done differently. For example, I feel it more advisable, in the event of suspected malfunctions, such as the heat shield and retro pack difficulties that require extensive discussion among the ground personnel, to keep the pilot updated on each bit of information rather than waiting for a clear cut recom-mendation from the ground. This keeps the pilot fully informed if there would hap-pen to be any communication difficulty and it became necessary for him to make all the decisions from on-board information.

Glenn proved that there were no deleterious psychological or physiological effects resulting from prolonged exposure to weightlessness, despite efforts undertaken to induce such effects.

13.5.4   MA-7 Carpenter's Medical

As Glenn's backup, Scott Carpenter underwent the same aeromedical preparation for his own flight. The flight surgeons now had an orbiting astronaut's data for comparison to Carpenter's while they performed the same tests and examinations. These tests included electrocardiogram, electroencephalogram, and audiograms. They also included X-rays, blood pressure, respiration and pulse rates. There were also extensive blood and urine chemistry tests.

During the flight, an attempt was made to control fluid and electrolyte balance but this was complicated by a high suit-inlet temperature and associated sweating and the increase in fluid intake by which Carpenter compensated for this situation.

In both flights, a xylose tolerance test was performed to measure intestinal absorption while the astronaut was weightless. Unfortunately, there were so many differences between Glenn and Carpenter's timing and amounts of hydration and urination, both during weightlessness and post-flight, that it was difficult to compare the results. Nevertheless, it was concluded that there were no abnormal gastrointestinal symptoms during their missions, and that both urine sensation and function were normal for both pilots.

Carpenter appeared to have tolerated the flight well, and all responses were considered to be within acceptable physiological ranges. No disturbing body sensations were reported as a result of the weightless flight.

13.5.5   MA-8 Schirra's Medical

Schirra's pre-flight examinations were similar to those for his predecessors. There were some changes to the blood pressure measuring system. These were principally to the position of the electrodes and the adhesives used, to provide better readings and reduce skin irritations. There were also changes to the gain settings in the controller for this system and this astronaut.

There were a few special studies to provide information about selected body functions and sensations. These produced biochemical and plasma enzyme determinations and three special measurements: a modified caloric test, radiation dosimetry, and retinal photography. Results of the retinal photography and the modified caloric tests showed no significant changes from prior to the flight. The dosimeters were located in the helmet and underwear and established that the radiation dose posed no hazard. The post-flight analysis of the plasma enzyme studies suggested that the elevations in some parameters were due to muscular activity rather than visceral pooling of the blood. A comparison of the MA-8 biochemical results showed that the astronaut's 9-hour exposure to weightlessness caused no biomedical changes that hadn't been noted after previous manned orbital flights. There were no medical reasons not to embark upon a longer mission.

13.5.6   MA-9 Cooper's Medical

The data obtained during this "day long" flight was developed on the foundation of knowledge gained from each of the preceding missions. Therefore, this flight was approached with a better understanding of the likelihood of a given physiological response occurring after exposure to the known stresses of a mission profile than had been previously possible. Taken together with four years' of pre-flight data on Cooper from centrifuge runs, altitude chamber runs, and his tests as the backup pilot for MA-8, the doctors had more data on Cooper than on any other astronaut.

It proved possible to load sufficient resources into the spacecraft to schedule a flight of over 34 hours, making 22 orbits of the globe. Cooper had two sets of electrocardiographic leads, an oral (rather than rectal) temperature thermistor, and a blood pressure measuring system. There was also a respiration sensor but it failed in-flight. The ECG intervals were well within normal physiological limits during the major portion of the flight. Later on it provided noisy readings. His blood pressure did not vary remarkably in-flight compared to before the flight. Post-flight readings of the film badges revealed the total radiation to be below the level of concern. And Cooper reported that the "g" forces were tolerable and weightlessness was an entirely pleasant experience to which he readily adapted.

Because of problems with the food containers and water nozzle, he was unable to properly reconstitute the freeze-dehydrated food and could only eat one-third of a package of beef pot roast. He only ate about one-fourth of the amount of calories made available. His water intake was also limited due to a problem with the condensate transfer system. It is estimated that he drank less than 1,500 cc during the entire flight.

Cooper had a good sleep on the night before the launch and was fully rested at the time he entered the spacecraft. Nevertheless, he briefly dozed off during a delay in the countdown. In space, he was able to sleep when his flight plan allowed; mostly this was in naps of 30 to 60 minutes, but sometimes he would fall asleep again some 30 to 45 minutes later. His total sleep time for the entire mission was only about 4½ hours.

On the advice of the MCC flight surgeon, Cooper took 5 mg of dextro-amphetamine sulfate about one orbit prior to retrofire. This made him feel more alert. There was a slight increase in the carbon dioxide partial pressure within his suit during the last two orbits, so he activated the emergency oxygen flow rate for 30 seconds. He also closed his face plate and kept it sealed for the final orbit and re-entry.

After the capsule had been recovered by the recovery ship and its hatch opened, the NASA flight surgeon attached an 8-foot extension cord to the biomedical cable. This cord was attached to Cooper's biosensor plug and blood pressure fitting and connected to the spacecraft onboard recorder to get his data before, during, and after egress. This system was extremely effective in deriving egress data. Upon standing on the deck, he swayed slightly and reported symptoms of impending loss of consciousness, dimming of vision, and tingling of his feet and legs. The post-flight examination continued for the next two days while the recovery ship steamed to port. He had lost 7¾ lbs. during the flight and was dehydrated.

Cooper's flight, and the others, had shown no evidence of significant degradation of pilot function attributable to making a flight into space. As a result of Project Mercury, the flight surgeons had a wealth of astronaut biomedical data with which to plan longer missions with Gemini and Apollo spacecraft.

Part III
Achievements

14

Facilities Created for Project Mercury

14.1 MERCURY CONTROL CENTER

The original Mercury Control Center (MCC), later called the Mission Control Center, was built in the 1956 to 1958 timeframe. The architects were the U.S. Army Corps of Engineers and the Burns and Roe, Inc., in the role of subcontractors to the Western Electric Corporation (WEC). The builder/contractor was Carlson-Ewell. As an Air Force facility, Pan American's technical contractor RCA was also involved. There were later additions for the Gemini Program but this section will just concentrate on Mercury. The control center was added to an existing Receiver Building #3 that had a large roof designed for telemetry equipment and a data processing area.

The final layout had operational input from Walter Williams, Associate Director for Project Mercury Operations, Christopher C. Kraft, Flight Director, Dr. Stanley White, Chief Flight Surgeon, G. Merritt Preston, Launch Operations Manager, and Scott H. Simpkinson, Capsule Operations Manager. Others also had operational input including Kraft's deputy Sig Sjoberg, Navy Lt. Paul Havenstein, and WEC's Paul Johnson. IBM engineers had input concerning the data interface to Goddard Space Flight Center.

In a speech given to the Society of Experimental Test Pilots in Los Angeles, California on October 9, 1959, Flight Director Kraft defined the major functions of the MCC as follows:

- Direction of all aspects of the capsules flight.
- Monitoring the aeromedical status of the astronaut and systems status of the capsule.
- Making all decisions to abort the mission.
- Determining the proper procedures following an abort decision.
- Commanding the re-entry of the capsule in both normal and emergency situations.

© Springer International Publishing Switzerland 2016
M. von Ehrenfried, *The Birth of NASA*, Springer Praxis Books,
DOI 10.1007/978-3-319-28428-6_14

- Keeping the astronauts and all tracking stations informed of the mission's progress.
- Coordinating and maintaining the flow of communications between all tracking stations.
- Informing the recovery forces subsequent to the decision to have the capsule re-enter the atmosphere.

These functions were facilitated by the following console positions:

- Operations Director, a NASA employee who supervised the overall operations for the flight and made the official decision to launch or scrub based on the recommendation of the Flight Director.
- Flight Director, a NASA employee who supervised and directed all the activity within the MCC. He also oversaw the pre-launch countdown, made recommendations to hold or scrub the launch, and made the decision to abort following launch or to have the capsule make an early re-entry.
- Flight Dynamics Officer, a NASA employee who advised the Flight Director as to pre-launch readiness and monitored the launch trajectory and orbital insertion.
- Capsule Communicator (CAPCOM), a NASA astronaut who monitored all pre-launch, launch, flight, and re-entry communications with the in-flight astronaut.
- Flight Surgeon, a military doctor employed by NASA who observed the condition of the astronaut in all phases of the mission and made recommendations to the Flight Director as to a launch abort or early re-entry.
- Capsule Environment Monitor, a NASA employee who monitored the environmental systems of the capsule during the pre-launch phase and during the flight, and advised the Flight Director as to a launch abort or re-entry.
- Capsule Systems Monitor, a NASA employee who observed the capsule systems (other than environmental) during all phases of a mission and made recommendation to the Flight Director as to a launch abort or early re-entry.
- Retrofire Controller, a NASA employee who advised the Flight Director as to the duration of the mission based on the orbital insertion conditions; established the time of retrofire for proper re-entry; advised the appropriate range station to command a reset of the retrofire time with the agreement of the Flight Director; and, in the case of a launch abort, determined the time of retrofire for capsule impact in a designated recovery area.
- Recovery Status Monitor, an officer from the Navy who reported to the Flight Director as to the readiness of the recovery force; monitored the force's status throughout the mission and advised the force of the mission's progress, the expected time of re-entry, and the predicted impact area.
- Missile Telemetry Monitor, an employee of Convair (Atlas manufacturer) who monitored the missile and advised the Flight Director of any situation that might require launch abort.
- Network Status Monitor, an employee of the DOD who reported to the Flight Director as to the readiness of the Cape; monitored the Range throughout the

mission; and advised the Range on the mission's progress. (This position was manned by Air Force Captain Henry "Pete" Clements.)

- Range Safety Observer, an employee of the Cape who observed the activity in the MCC throughout the flight; advised the Range Safety Control Center of any potential abort and advised the Flight Director of an imminent violation of safety criteria and a possible abort action by the Cape's Range Safety Control Center.
- Network Commander, an officer of the Air Force who commanded the Mercury Range and ordered the appropriate actions to rectify any Range Station malfunction.
- Recovery Task Force Commander, an officer of the Navy who commanded the Recovery Operations.

After some experience with simulations and the MR-1 mission, some changes were made to this initial list. The Range Safety Observer was no longer needed, as the Range Safety Officer had direct communications with the Flight Director and had responsibility to destroy the launch vehicle if it went out of limits.

The Recovery Status Monitor position was no longer needed in the flight control area, as the entire recovery operation was controlled in a separate room. Nevertheless, the Recovery Task Force Commander remained in the flight control area on the top row. This operation had direct communications to the Flight Director, the Retrofire Officer, and others involved with recovery forces. The console was then taken over by the Operations and Procedures Officer who assisted the Flight Director with the countdown, operational procedures, mission rules and coordination of all communications with the remote tracking stations. This position evolved into the role of the Assistant Flight Director.

The Missile Telemetry Monitor position was only used for MR-1 on November 21, 1960. The equivalent Redstone and Atlas monitoring positions were in the launch control areas with direct voice communications to the Flight Director. Later, a position called Booster Systems Monitor was added to monitor the launch vehicles.

The RCA Maintenance and Operations (M&O) support staff also had a position in the flight control area called the Support Control Coordinator (SUPPORT). He made sure all the systems supporting the flight controllers were functional including telemetry, displays, communications, power, lighting, data, voice, and teletype. This position was initially on the front row, then later moved to the side of the room, facing inward. He primarily interfaced with the Operations and Procedures Officer to assure MCC readiness to support the mission.

After Mercury, the Gemini Program made a number of changes to the flight control functions and names. The final mission to be controlled from the MCC was Gemini 3 in March 1965, but the center served as a backup to the new Houston MCC for the Gemini 4 mission. The original MCC functioned as a remote site for subsequent missions in this program by transmitting real-time voice, telemetry and data from the Atlantic Missile Range to the Houston MCC, and then retired with the completion of the program in November 1966.

Fig. 14.1 The Mercury Control Center during MA-6. (Photo courtesy of NASA)

Fig. 14.2 The restored MCC Flight Control Area at the KSC Early Space Exploration exhibit. The Operations and Procedures Officer and the Assistant Flight Director consoles are to the right and the Flight Surgeon is to the left. (Photo courtesy of NASA)

Fig. 14.3 The restored MCC front row of consoles. (Photo courtesy of NASA)

Fig. 14.4 The restored Environment and Systems consoles. (Photo courtesy of NASA)

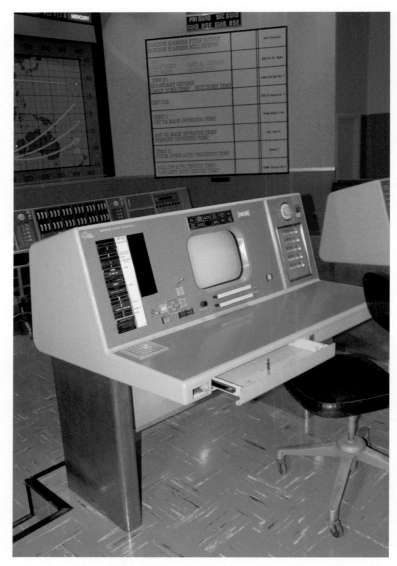

Fig. 14.5   The restored Flight Director console. (Photo courtesy of NASA)

Fig. 14.6 The restored upper row of consoles for the Navy Recovery Task Force Commander, Public Affairs Officer, Operations Director, and Air Force Network Commander. (Photo courtesy of NASA)

Fig. 14.7 The restored Retrofire Officer and Flight Dynamics Officer consoles and plot boards. (Photo courtesy of NASA)

The entire Mercury Control Center building was demolished in 2010. However, its contents, including the consoles and the wall map which had been restored in 1999, are now displayed in the Dr. Kurt H. Debus Conference Center at the Kennedy Space Center Visitors Complex.

Fig. 14.8   The MCC in 2010 prior to demolition. (Photo courtesy of NASA/Jack Pfaller)

Fig. 14.9   The end of an historic landmark. At least the Flight Control Area and Viewing Area survive at KSC. (Photo courtesy of NASA)

MISSION (MERCURY) CONTROL CENTER

Built on this site in 1957, the Mercury Control Center, later renamed Mission Control, was the United States' first mission control for unmanned and manned space programs under the leadership of the National Aeronautics and Space Administration. The center housed critical launch equipment for the Mercury and Gemini programs. These programs made crucial advancements in the development of spaceflight, including placing astronauts in suborbital and orbital space within and outside a spacecraft, and safely returning them to Earth. The Gemini program was the first American attempt in orbital rendezvous and docking, a critical maneuver used in future manned lunar landings. The control room was dominated by a world map with a miniature spacecraft that tracked the capsule's planned flight path. Teams at the center controlled all flights launched aboard Redstone, Atlas, and the first three Titan II vehicles. After mission control functions were transferred to Houston, Texas, the center provided backup for the initial launch and trajectory. Before the facility's demolition in 2010, its essential historic components were removed for preservation and are displayed at the Kennedy Space Center Visitor Complex.

A FLORIDA HERITAGE LANDMARK
SPONSORED BY THE NATIONAL AERONAUTICS & SPACE ADMINISTRATION
AND THE FLORIDA DEPARTMENT OF STATE

F-711   2011

Fig. 14.10   The MCC Historical Marker. (Photo courtesy of NASA)

14.2   BERMUDA CONTROL CENTER

The Bermuda Tracking Station was one of the sites in the Mercury Space Flight Network built on Cooper's Island and accessible by Kindley Air Force Base security. It was erected in 1960, and became fully operational in September 1961. It supported all of the Mercury flights from MA-4 through MA-9, as well as Gemini and Apollo flights. Owing to its location, it played a critical role in backing up the MCC at the Cape with at least a minute's worth of FPS-16 radar tracking prior to orbital insertion. It was also the only station that could track the last stages of the launch vehicle prior to insertion. The next station in the chain was on the Canary Islands. Bermuda would typically acquire the capsule about three minutes after liftoff and could receive data for about 12 minutes. It could compute the trajectory and provide the Goddard computers with radar tracking data and a trajectory solution during the final portion of powered flight. This data was used to confirm the orbital "Go/No-Go" decision. This data was crucial for a potential abort decision, in order to preclude a landing in Africa. An abort during the pass over Bermuda would land in a planned recovery area in the Atlantic. The Bermuda facility was configured as a miniature control center with consoles, displays, plot boards, and an IBM 709 computer.

The IBM 709 was an improved version of the IBM 704 at Langley. It used vacuum tubes and interfaced with the two IBM 7090 computers at Goddard, which were essentially transistorized IBM 704s with other improvements but whose reliability was, at first, not much better than their vacuum tube powered predecessors. The Goddard computers were also receiving data from the Cape radars during the early portion of the launch phase and transmitting data to the MCC plot boards. The Bermuda plots displayed the FPQ-6 C-Band radar plot, the Impact Prediction Plot, and the FPS-16 C-Band radar plot. Mounted above them were the digital mission clocks.

The Bermuda station was operated by NASA STG flight controllers and for the first mission, MA-4, the team included: Bermuda Flight Director John Hodge, Flight Dynamics Officer Glynn Lunney, and Capsule Communicator Deke Slayton. For subsequent manned missions other flight controllers were added, including a Capsule Systems Monitor, an Environment Systems Monitor, and a Flight Surgeon.

After the launch phase and during orbital passes, the station performed as a regular manned MSFN site with voice, TTY, telemetry, command and radar. It was supported by Bendix Field Engineering contractors and local Bermudian administrative and logistics personnel. Benjamin Gallup was the Bendix Maintenance and Operations supervisor. Bermuda was also used by the NASA/Navy recovery crews as a staging base and by NASA/Bendix MSFN station checkout crews.

The Bermuda Control Center and tracking station are long gone. All the facilities have been demolished, and the entire Cooper's Island is now a nature preserve. What was once a place to spot spacecraft is now a place to spot birds and whales from the old radar pedestal refurbished into an observation tower.

Fig. 14.11 The Bermuda Control Center (top left), with a control console (bottom left) and the plot boards (right).

Fig. 14.12 The NASA C-Band radar pedestal is now the Cooper's Island Wildlife Observation Tower. (Photo courtesy of the Bermuda Department of Conservation Services)

14.3 MERCURY/MANNED SPACE FLIGHT NETWORK

One of Goddard Space Flight Center's greatest contributions to Project Mercury was the creation of the world-wide network of remote tracking stations that provided communications links back to the Goddard computers and to the Mercury Control Center. See the discussion in Section 7.4 for their contribution. This section gives more technical details about this Mercury (later called Manned) Space Flight Network (MSFN).

Goddard was already involved in other unmanned satellite programs and a tracking network. The basis for an early system was the Minitrack Network that was built by the Naval Research Laboratory (NRL) for the International Geophysical Year (IGY) which ran for 18 months from mid-1957 through 1958. When this scientific endeavor was conceived in 1955 the highlight of the U.S. contribution was to be Project Vanguard, which would launch a satellite. Much of the initial research in tracking was done at the White Sands Missile Range in New Mexico. While this world-wide network wasn't suitable for tracking a manned spacecraft at orbital altitudes, it did provide the people and facilities that would be transferred to NASA in 1958. Nearly all of those people at NRL and White Sands involved in Minitrack were transferred to Goddard, even though that center was not yet built. Many would continue to work at existing sites, but others would move into temporary buildings in Greenbelt, Maryland. Along with the people from the Langley STG Tracking And Ground Instrumentation Unit (TAGIU), the Minitrack specialists began to define the very special real-time pilot safety requirements for a new network suitable for tracking a manned spacecraft. By the spring of 1959, four contracts were let. The Goddard team was given two years to complete the new Mercury network.

By 1960 new technology, requirements for supporting other satellite systems, modernization of existing systems, and the addition of more stations, led to what was then called the Satellite Tracking and Data Acquisition Network (STADAN). This system also supported satellites in polar orbits. Most of these stations were not applicable to Mercury. Another system called the NASA Communications Network (NASCOM) tied all the stations together with an eventual 2 million miles of voice and data circuits. Portions of this system did support Mercury. About this same time, the Jet Propulsion Laboratory (JPL) was building the Deep Space Network (DSN). Eventually this would be integrated into the overall system for Apollo.

By early 1960 the site surveys were finished and construction of the MSFN was under way. This network was specifically designed for manned flight with an orbit that was inclined to the equator at 32.5 degrees. It assessed possible abort landing and recovery sites, and attempted to minimize the periods during which the ground wouldn't have contact with the astronaut in the event of an emergency. International relations with those countries with ground stations had to be considered. This required the efforts of the State Department.

Wherever it was practical, proven and reliable military systems such as the FPS-16 and the Very Long Range Tracking (VERLORT) S-Band radars were employed, with improvements in acquisition aides. Because some systems weren't very reliable they either had to be modified or have a complete set of spares available; the latter option being complicated by the remoteness of some of the sites in Africa, Australia, and on oceanic islands. This network included the critical stations such as the MCC in Florida and the control center/ tracking station in Bermuda, as well as the Goddard computers that received all the tracking, telemetry and voice communications.

The original MSFN, configured for a three-orbit mission, included the following sites:

North America

- CNV – Cape Canaveral, Florida
- TEX – Corpus Christi, Texas

- EGL – Eglin AFB, Florida
- GYM – Guaymas, Mexico
- CAL – Point Arguello, California
- WHS – White Sands New Mexico

Australia

- MUC – Muchea, Western Australia
- WOM – Woomera, South Australia

Africa

- KNO – Kano, Nigeria
- ZZB – Zanzibar

Atlantic

- BDA – Bermuda, United Kingdom
- GBI – Grand Bahamas, British West Indies
- CYI – Grand Canary Island, Spain
- GTK – Grand Turk, British West Indies

Pacific

- CTN – Canton Island, Kiribati Republic
- HAW – Kauai, Hawaii

Ships

- ATS – Atlantic Ship, *Rose Knot Victor*
- CSQ – Indian Ocean Ship, *Coastal Sentry Quebec*.

Construction of the MSFN began in April 1960 and was completed in March 1961. On July 1, 1961, just 24 months after awarding contracts, NASA officially accepted the new MSFN. It was designed for a three-orbit mission. By relocating the two tracking ships, adding ships, and using some DOD sites it was able to support the later, much longer missions. DOD aircraft were used for voice and telemetry relays.

From concept to completion in only five years, the MSFN successfully supported all four of the orbital manned Mercury missions, and with modifications, it went on to support Gemini and Apollo.

For a fuller understanding of the NASA Spaceflight Tracking and Data Network, see *Read You Loud And Clear* by Sunny Tsiao, publication details of which are listed at the end of this volume.

14.4   MERCURY PROCEDURES TRAINERS

The Mercury Procedures Trainer created at Langley in 1960 was located in the Full Size Wind Tunnel Building 643. The training group partitioned off apparatus in an adjacent room to enable a remote site team to interact with the astronaut in the trainer in precisely

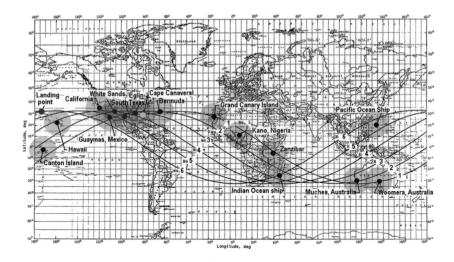

Fig. 14.13 The Manned Space Flight Network. (Photo courtesy of NASA)

the same way as they would during a pass over their station during a flight. Runners would carry simulated teletype messages between partitions as if these were instructions from the MCC. As the CAPCOM in each partition received the message, he would call the astronaut in the trainer and act upon the message just as if it were a real mission. This "paper drill" was crude, but it did show the team how the ground-to-capsule voice communications might work. As such, it was one of the first attempts at simulation as a training tool.

By 1961, the simulations were becoming ever more sophisticated, and a Flight Simulation Section was created under Harold Miller in the Spacecraft Operations Branch. The astronaut would sometimes suit up and enter the trainer. The simulation person could send faults to the trainer for the astronaut to act upon. McDonnell Aircraft set up an analog commutator in the trainer to enable the simulation person to send data about the capsule's systems to the control center so that the flight controllers there could act on that data. To ensure realistic data to the MCC, the data format and transmission system was identical to the real capsule.

Fig. 14.14 John Glenn in the Langley Mercury Procedures Trainer NASA STG engineer Charles Olasky is running the Simulation console. (Photo courtesy of NASA)

Fig. 14.15 The Mercury Simulation Control Room at the Cape.

15

Mission Designs and Concepts

15.1 MISSION RULES

Developing "mission rules" or "flight rules" as a concept was not really new in 1958. Aircraft flight test programs employed checklists for certain situations, with procedures for what to do in the event of something happening. In a way, these emergency procedures for specific events had the connotation of "rules." In Project Mercury, this methodology became a discipline. The space environment naturally leads you to think, "What if this happens?" In spaceflight operations you think about these situations in the context of your job, your position, and your responsibility, as well as from the perspective of the launch vehicle, the spacecraft, the control center, the tracking station, etc. The rules that you operate by depend upon your point of view and where you are in the overall system.

The rules are related to individual systems, but it falls to the operators of the overall system to implement them. Since the nature of man is to have different points of view, and for some to be more vocal about their opinions than others, it becomes necessary to have an ultimate manager to decide what the rules should be and when they must be implemented. This manager also carries the dreaded responsibility of "overruling the rule" when a situation arises. Experience has shown time and time again that you can never foresee all the possibilities, and therefore can never have a complete set of rules. In the case of Project Mercury, the environment for executing the rules was the world of operations. This meant that the MCC Flight Director Christopher C. Kraft and his boss Mercury Operations Director Walter C. Williams bore the overall responsibility for the conduct of a mission. They would confer if time allowed, but the real-time decisions had to fall to the Flight Director, who could "overrule the rules" if necessary for the safety of the astronaut or the mission.

The development of the rules began at the system level; perhaps the component level if the failure of that component was a single point failure that could have catastrophic consequences. The design of the Mercury spacecraft, as well as the design of the Redstone and

© Springer International Publishing Switzerland 2016
M. von Ehrenfried, *The Birth of NASA*, Springer Praxis Books,
DOI 10.1007/978-3-319-28428-6_15

Atlas missiles, had failure modes that were documented and fed into the mission rules process. Likewise, other supporting systems had to be considered; e.g. tracking, communications, and computers. Even non-hardware or software needed to be factored into the decision making process; e.g. weather conditions for launch photography or the weather at the recovery location, including sea states.

Military missile test conductors were ahead of NASA in terms of launch rules. They knew what they needed to do to launch, but weren't concerned about a manned capsule or even the weather. Their only concerns for life and limb were if the rocket blew up or strayed towards a populated area; which did occur occasionally. In that situation, the rule was to have the Range Safety Officer transmit the destruct command to the missile.

When the STG became involved in launches with Project Mercury, they needed to concern themselves with the total system. That meant all systems, all vehicles, and all of the supporting elements including the world-wide tracking network. The task of developing the mission rules for Mercury was daunting. In 1959 the McDonnell capsule systems engineers began designing the systems with some input from only a few engineers at Langley. It was some time before the STG was able to assign more engineers to work with the company to understand the systems in more detail, and in particular how they might fail.

By October 9, 1959 Kraft had finished the specification of how the MCC would be manned, and who would be responsible for which systems. The concepts of mission operations began to take shape. In October, Kraft sent Gene Kranz to the new MCC to learn about how the military counted down the Redstone, and how the MCC would interface with those operations. And oh, while you're at it, write some mission rules! Kranz had been onboard for just two weeks. This was his trial by fire. He was introduced to the control center by Western Electric Engineer Paul Johnson and Navy Cdr. Paul Havenstein, who was on Gilruth's staff for the MCC. This began the slow process of developing the original mission rules. Kranz took the lead for the STG and MCC on mission rules and became the first Operations and Procedures Officer. His call sign in the MCC was PROCEDURES. His call sign when communicating with the MSFN was CAPE PROCEDURES. When communicating with the Atlantic Missile Range his call sign was DEVIL FOX BRASS ONE.

MR-1 was the first mission supported by the new MCC. Each mission contributed new rules. Between flights, the teams returned to Langley. Kranz worked with engineers from other STG Branches and Sections to get their input. The process for gathering up everyone's input became more formal with periodic reviews. On returning to the MCC for a mission, the formal mission simulations and post-mission debriefings would facilitate a lot of interaction between the flight controllers and astronauts. This lengthy process included the results of simulations and could be contentious, but as everyone came to know why things were decided the way they were, and by whom, it ultimately delivered the teamwork that enabled real-time "Go/No-Go" decisions to be made with confidence and conviction.

The mission rules considered the overall mission objectives, and as they became ever more formal they evolved into a document that considered all possible events and contingencies. Just as significant was the timing of the events, and the degree to which some systems were critical for mission success. The rules would dictate who (e.g. astronaut, Surgeon, Systems, etc.) would take an action based upon each condition (e.g. abort, failure to separate, loss of radar, etc.) and when (e.g. ASAP, or end of orbit). The rules would be noted as mandatory, highly desirable, or desirable. The mandatory rules would be in red. The document could run to 30–50 pages for the early Mercury missions and hundreds of pages for later missions. They became the operational "Bible" and copies would be on hand at each console in the MCC and at all the remote sites. As the missions grew in complexity, the mission rules process took up a lot of people's time before and after the flight simulations to ensure that both the astronauts and the flight controllers were ready for a mission.

The prime objective of the mission rules is to identify equipment configuration for mission support and formulate a series of basic ground rules based upon systems analysis and mission planning considerations that will provide for the safety of the flight crew, optimize chances for mission success, and provide guidelines to expedite the decision process. Having analyzed the conditions and malfunctions prior to their occurrence will offer those personnel involved with pre-thought, pre-arranged and pre-planned actions which are known from experience to be the best solution to the malfunction. Each person would analyze their responsibilities, and provide their proposed rules. For example, the Flight Dynamics team would propose rules for RETRO, FIDO and, later, GUIDO. Similarly, the SYSTEMS and SURGEON people would do the same. All the inputs were provided to and prepared by the Flight Control Branch and signed off by the Operations Division and the Flight Director. After each simulation the rules would be reviewed and modified as required.

One of the greatest operational achievements of Project Mercury was its establishment of the mission rules process. The STG created a spaceflight operational management tool. The lessons learned about real-time, operational decision-making permeate operations to this day aboard the International Space Station as well as the space programs of other countries. This process is so well thought out and vetted that it is sure to be used, in some form, for a future Mars mission. It has also been used in other application areas, including the nuclear industry.

15.2   OPERATIONAL PROCEDURES

As each MCC and MSFN flight operations position became defined, there was a need to define what, where, and when each person would perform an action. This varied considerably for each position and it was up to them to write up their procedures, specifying their fundamental duties and responsibilities as well as their "normal" and "emergency" actions. During flight operations, there are many interactions with people you don't even know, or may never see, but who either need information from you in a timely manner or, conversely, you need to inform of something in a timely manner. For example, in Mercury, there were people in the blockhouse conducting launch operations who needed to know if the MCC was ready to pick up the count or why they were having a problem and required to hold the countdown. Likewise, there were people in the Range Safety Officer's area at an Atlantic Missile Range, Bermuda control center, or a remote site who needed to communicate a problem.

Often the "countdowns" were the mechanism for communicating the "Go/No-Go" calls from the MCC to the launch conductor, who was the driving force during the pre-launch phase. The Bermuda control center and the remote sites were in a supporting role but they could hold up a count if their site was temporarily unable to provide that support. Similarly the Atlantic Missile Range had their procedures, as did the Computing Complex at Goddard and the GE/Burroughs guidance computer complex.

Eventually, an "integrated" countdown was prepared for the MCC to ensure that every flight controller knew who was doing what to whom, and when they themselves would be required to take an action. It turned out that the Operations and Procedures Officer (PRODEDURES) was the one to understand the entire operation and document the count-down and procedures for the Flight Director and his team of flight controllers in the MCC. The countdown would drive each console position's procedures. When the launch vehicle was changed from the Redstone to the Atlas, the countdown was revised, as were some of the procedures for the MCC. Things would change again with the introduction of the Titan launch vehicle for Gemini and the Saturn types for Apollo.

Each remote site had their own procedures integrated with the MCC's countdown and these would later become more formalized and documented into console handbooks. The astronauts' procedures were handled by their training support people, taking into account their work in the Mercury Procedures Trainers at the Langley Research Center and at the Cape. Their procedural actions were documented in the mission rules. One of the achievements of Project Mercury was the formalization of operational procedures. See Appendix 7 for my experience in this area.

15.3   SIMULATIONS

Closely related to both mission rules and operational procedures are the flight simulations that drive the flight control team to undertake given actions at certain times. Many of the rules and procedures were tested in simulations and found to be undesirable or in need of modifications. The concept of simulating an action as a training tool wasn't new in 1958–1961; it was used in aircraft flight testing. However, the STG (and subsequently the MSC) took the concept to new levels of sophistication and efficacy for manned space missions. Typically, training a pilot is a one-on-one relationship; perhaps only an instructor pilot and the trainee. Certainly, the Mercury astronauts had plenty of training, but not that which they would need in order to interface with the operations teams in a launch complex, the Mercury Control Center, and the Mercury Space Flight Network. Conversely, the flight controllers needed training in their respective areas and systems in order to support an astronaut.

In Robert Gilruth's original staffing memo of August 3, 1959, there wasn't an organization with "Simulation" in the title. There was an Operational Analysis Section under John Mayer's Mission Analysis Branch that included many people who began to work on simulations. This small group was initially focused on training the new flight controllers in the Mercury Control Center. They included Jack Cohen, Stan Faber, Harold Miller, Richard Hoover, Arthur A. Hand, Glynn Lunney, Dick Koos, Charles Olasky, and Bob Eddy. Sometimes people were assigned to one organization and then temporarily assigned to another. In their individual bios, they would say that they worked for a given organization, but when you checked that organization chart or the phone book you would find there wasn't an organization by that name! Nevertheless, there were sometimes informal names of groups that consisted of people temporarily borrowed from different organizations.

The Operations Analysis Section headed by Jack Cohen provided input to Western Electric, which was building the control center and the new tracking stations. This group also did some requirements work for the company regarding the interface to the planned Mercury Procedures Trainer, two of which were to be built. See Section 14.4. This effort began to clearly show that operational requirements for the engineering design were critical. On March 31, 1960 the group (informally called the Simulation Group) published the first document called "Plan for Control Center Training Simulations."

As the simulation capability increased, radar and guidance data was fed in. The increase in capabilities required more people to create the tapes to drive the trajectory displays in the MCC. It was also considered important for the MCC RCA technicians to support the simulations. They were required to verify various settings to help in diagnosing data problems. At times they were "in" on the simulations, and helped to trigger flight controller faults and subsequent actions. In addition, some of the simulation tapes were sent to Goddard so that engineers there could send simulated data to the RETRO and FIDO displays in the MCC. As the astronauts were presented with ever more complex failures, their frustration at what they regarded as "unrealistic" failures was strongly voiced. Later, however, when some of these failures actually occurred they would thank the simulation people for this training. Losing the MCC or an entire tracking station was thought unrealistic, but both actually occurred. A bulldozer actually cut through a power cable and rendered the MCC dead for several hours; fortunately not during a real mission. They even simulated a critical flight controller becoming sick and having to be replaced during a mission.

Eventually, many simulations were conducted world-wide, and the MSFN and MCC flight controllers began to function as a professional flight control team. They were challenged many times in Mercury, and ever more sophistication was introduced into the simulation process for Gemini and Apollo. Exploiting the rapid pace of development in the digital realm the Houston MCC had systems, consoles, and displays for a whole new Branch that designed and produced the simulations necessary to deal with new space hardware.

The main lesson learned from this experience was the importance of integrating the training element early in the design of the overall spaceflight system; both in space and on Earth. While the industry making the hardware may think they can handle the operational requirements, it is imperative that the trainers know the hardware in order to simulate its failures and the hardware manufacturers need to bring the operators in on the design process. Both hardware and software changes were made as a direct consequence of simulations. Likewise from the point of view of the astronauts, the often tricky simulations were a major factor in giving all the Mercury Seven the utmost confidence in the ground's ability to "have their back."

The culture of simulation developed over time with Mercury, and would become even more sophisticated in later programs like Gemini and Apollo. However, there were times when things happened so rapidly that the training simulations were of no immediate help. The shutdown of Wally Schirra's Gemini 6-A Titan and his decision not to eject from the capsule (contrary to the mission rules) was a result of years of "seat of the pants" experimental flight test experience. No one on the ground had any input to that decision. Likewise, soon after Gemini 8 docked with its Agena target vehicle, the combination started to tumble. Thinking that it was the Agena that was malfunctioning, astronauts Neil Armstrong and David Scott promptly undocked only to find that the fault was one of *their* thrusters that was firing when it shouldn't and that the rate of tumbling was accelerating. Because they were between tracking stations, they had to resolve the situation on their own, which they did by shutting down their primary maneuvering system and switching on the system which was intended only for controlling the attitude of the vehicle during re-entry. This development came as a considerable surprise to the MCC when the spacecraft came within range of the next tracking station. The mission rules required that the mission be terminated and the capsule returned to Earth as soon as possible. Given the urgency, the Flight Director chose a contingency area in the Pacific where there was a single destroyer on station. These actions were further justification of the decision to select experimental test pilots as the early astronauts. The Apollo 1 fire was a catastrophe that occurred so rapidly that the launch team could do nothing to save Gus Grissom, Ed White, and Roger Chafee who were unable to save themselves. Ironically for Grissom, this was another hatch problem.

But there are times when the years of teamwork honed with literally hundreds of simulations saves lives and missions. The best example is Apollo 13 in April 1970. The problem was never simulated, but the knowledge of the entire ground team was able to save the lives of Jim Lovell, Fred Haise and Jack Swigart. Books and movies have been made about this mission. There are several videos about it on YouTube and on the Discovery Channel. Almost every flight control discipline was involved in the work-arounds that first got the astronauts into a safe posture and then brought them home. Other astronauts in Houston worked in the crew procedures trainer to simulate and evaluate situations. Engineers worked on how to use the Lunar Module as a "life boat," worked on how to fully power down the Command Module, and analyzed the source of the explosion in the Service Module. Others worked on a work-around for the carbon dioxide filter problem using only items that were available to the crew; and yes, their use of what was referred to as "Duck Tape" was the best use of duct tape ever! Trajectory analysts assessed the situation and devised timely maneuvers to get the crew into the proper position for the burn to return to Earth.

All of this work, much of which was improvisation, was orchestrated by the Flight Directors over a period of six days. The flight control teams had many missions and hundreds of different simulations under their belts, and this supplied the knowledge base and competence to work on any problems. Some of this emergency work had been simulated during "routine" training, but what saved the astronauts was the utter professionalism of the entire team and the leadership of the Flight Directors. Over the course of Mercury, Gemini, and Apollo, simulation training had instilled an absolute trust between astronauts and flight controllers. They did not achieve their planned mission, but the Apollo 13 crew has the distinction of having been farther from Earth than anyone; 248,655 miles!

15.4 SPACECRAFT DESIGNS

The Mercury capsule design was driven by three fundamental constraints. Foremost, of course, was the safety and survivability of the astronaut. A close second was the fact that, at that time, only two missiles were capable of lifting a capsule weighing approximately one-and-a-half tons to the desired height and velocity. Obviously weight was a major part of that problem. This was more of a manufacturing concern, but it drove the internal systems designs. The Redstone could only propel the capsule on a suborbital flight, but the Atlas was capable of inserting it into orbit. The third constraint was the expected aerodynamic heating of the capsule during re-entry. This could generate sufficient heat to kill the astronaut if insufficiently isolated and protected by the heat shield and heat resistant afterbody shingles. There were many other concerns related to the safety and survivability of the astronaut, such as "g" loads and weightlessness but these were in the process of being solved either by ourselves or by the Soviet Union. The "g" load issue was addressed by tests of the astronauts in their customized couches in the centrifuge at Johnsville. This concern had a lesser influence on the design of the capsule than the other constraints. The couch and the crushable honeycomb landing impact material did not represent a major weight penalty. The weightlessness concern was a big one for some of the aeromedical people but the flights by our chimpanzees and Soviet cosmonauts minimized that problem, which didn't have much of a design impact anyway.

As discussed in Section 13.2, some of the basic concerns were in work even before NASA was created. NACA wind tunnels had addressed some of the fundamental issues that drove the design and shape of the capsule. Shortly after the STG was formed, the Little Joe tests gave the STG actual flight data for some of the critical issues, including the launch escape problem. The Big Joe mission was critical to solving the re-entry heating problem and the appropriateness of the protective shielding.

So less than a year after NASA's establishment on October 1, 1958, not only was the STG organization created but the design of the capsule was well understood and underway. Its size, shape, systems, and approximate weight were known, and on January 26, 1959 the contract was let to build it. That was certainly a major programmatic achievement not only involving NASA Headquarters and the three research centers, but the aerospace industry as well. Furthermore, it took managers with vision, creative engineers including many wind tunnel and shop technicians, and administration and contracts people. Bear in mind that there were few computers (and most of those used vacuum tubes) and only secretaries equipped with typewriters to generate all the necessary documentation. There was no word processing software. Even engineering drawings were handmade. Can you imagine a major program of any kind today being able to make such progress?

15.5   LAUNCH VEHICLE DESIGNS

Although NASA didn't have to design the Redstone or Atlas, they certainly had a hand in man-rating both those vehicles. This was a contentious process at times. After all, it was the Army's Redstone and the Air Force's Atlas. They were designed for warheads, not capsules fitted with escape towers. They were also not designed to put a payload into orbit. The new space agency was even blamed for the structural failure of MA-1. Fortunately, NASA had very experienced structural engineers who could go toe-to-toe with the military's engineers. NASA's impact on the missile designs was discussed in Section 13.2.5.

15.6   MERCURY FULL PRESSURE SUIT

When the astronauts came onboard in 1959, there had been about 25 years' of experience and development of pressure suits. Robert Gilruth, running the STG, was well aware of the need to have experts available in the STG to help with the selection of the Mercury pressure suit and to work with the astronauts during testing and evaluation. He recruited Dr. James Paget Henry, a pioneered in the development of the Army S-1 partial pressure suit in 1945. The suit was even called the "Henry Suit." It was placed into limited production by the David Clark Company in 1948; a company which would subsequently bid for the Mercury suit. When the Air Force was made an independent military service in 1947 they adopted full pressure suits, as did the Navy.

When the NASA X-15 started flying in 1959 its pilots wore the MC-2 full pressure suit that David Clark supplied. This later evolved into the A/P 22 suit that had -2, -4, and -6 versions. I wore these in the NASA RB-57F in the late 1960s and early 1970s. I also spent a lot of time in Apollo suits. The U-2 pilots preferred partial pressure suits until the larger super U-2Rs came into production in the late 1960s. U-2 and WB-57F pilots now employ David Clark S1034 full pressure suits.

Meanwhile, the Navy worked with B. F. Goodrich and the Arrowhead Rubber Company to produce their full pressure suits, known as the Mark III and Mark IV. Malcom Ross and Victor Prather wore the Mark IV on the Strato-Lab V unpressurized gondola that rose to an altitude of 113,740 feet on May 4, 1961, which was the day preceding Alan Shepard's suborbital mission.

All of this technology and experience was available to the pressure suit manufacturers when, in July 1959, the STG let a request for proposals for the Mercury suit to three companies: David Clark, B. F. Goodrich, and the International Latex Corporation. Several weeks later, on July 27, Goodrich was awarded the contract to create the Mercury suit using a modified Mark IV design. The decision was in keeping with the Mercury Design Objectives and supporting Guidelines to use off-the-shelf hardware and technology whenever applicable.

Several design changes were required in order to interface with the Mercury capsule and to carry out the mission:

- A "closed loop" system was integrated into the capsule's life support system.
- The outer nylon layer was aluminum-coated for launch and re-entry thermal reasons.
- Safety boots were also aluminum coated for the same reason.

- Various refinements were made to enhance mobility.
- Special gloves would improve dexterity in handling switches.
- An interface for biomedical connections to the telemetry system.

Further modifications were made after almost every flight; for example:

- A urine collection device was added for MR-4 after Shepard's need to urinate during an extended launch delay.
- Improved wrist bearings and ring locks were added.
- A convex mirror was added to the astronaut's chest to enable a single camera to observe both him and the instrument panel.
- After Gus Grissom almost drowned, a small inflatable life vest was added.
- Cooper's boots were incorporated into the suit and there were improvements in his helmet, microphones, and thermometer (he had an oral thermometer whereas his predecessors had rectal thermometers).

The B. F. Goodrich contract called for three suits for each astronaut; a training suit, a flight suit, and a backup. All the suits were tailor-made for each man and weighed only 22 lbs. Since the Mercury capsule never lost pressure, the suits never were inflated during a flight, but they were inflated for testing and training purposes. No suit ever failed.

For a full history of pressure suits, see *Dressing for Altitude, U.S. Aviation Pressure Suits-From Wiley Post to Space Shuttle* by Dennis R. Jenkins, published by NASA as SP-2011-595 and available online.

15.7   MISSION ANALYSIS AND TRAJECTORY PLANNING

It is now hard to believe that nearly six decades ago few people had any experience planning a manned orbital flight with any detail. In 1957, the Soviets proved they could launch something into orbit and set about planning a manned orbital mission. While the military people knew the equations of ballistic and orbital flight, much more is required for mission planning, trajectory analysis, and contingency planning for a human flight. When NASA was created, the Mercury Control Center and the Mercury Space Flight Network didn't exist. The words "astronaut" and "Mercury" were not in use in the context of human flight. We still used the words "pilot" and "capsule."

Certainly, mission concepts existed, and there were visionaries who contemplated flights in space, but they were just that; concepts. There was much more to be done. There were rockets aplenty in 1958, but we had only one to lift a capsule to orbit; and even then just barely. What was required was a group of experienced flight people and mathematicians equipped with the most modern computer (even if it did have vacuum tubes) to run launch, abort, orbital, and re-entry simulations with variable constraints and situations.

This challenge was given to a former NACA Flight Research Engineer, John P. Mayer. He was one of Gilruth's STG Core Team and the first Chief of the Mission Analysis Branch in the Operations Division. Mayer broke down the work into several phases:

- Mission analysis supporting the capsule design
- The Mercury mission design based on requirements and objectives

- Operational analysis of each planned flight
- Formulation of the computer mission logic for real-time flight control.

This work had to consider structural loads, heating, and propulsion performance. Once the capsule design became better known, the analysis shifted to flight operations. Then the launch vehicle constraints (initially the Redstone and later the Atlas), the capsule constraints, and the operational constraints were factored into the analysis. Only then could a specific mission be planned, including mission rules such as when to abort, when to re-enter, etc.

Once into an actual flight, the work of real-time mission planning began. This was the time when computer calculations were performed using the logic and equations developed during the pre-mission phase. Although every effort was made to predict all possible situations this initially proved almost impossible; nevertheless, experience evolved over time. Both on-line and off-line computers were used to analyze situations. Once the flight was completed, a post-flight mission analysis was carried out and the results fed into preparations for the next flight. Sometimes, the results actually changed the launch vehicle, the spacecraft, or the ground systems. Examples of mission constraints are as follows:

Spacecraft

- Performance, e.g. propulsion
- Guidance and Control, e.g. attitude accuracies
- Systems limitations, e.g. consumables, "g" loads

Control Center, Tracking Stations, Recovery Areas

- Performance, e.g. acquisition
- System limitations, e.g. position accuracies, range safety, weather

Launch Vehicle

- Performance, e.g. thrust, weight, abort conditions, "g" loads
- Guidance and Control, e.g. accuracies, radar elevation
- System Limitations, e.g. heating, loads.

Over 90% of the computer simulations studied abort conditions at various times during the launch phase. Even the angle of offset for the escape rocket of the Mercury capsule involved a compromise between high lateral loads and low miss distances between the spacecraft and the launch vehicle in the high dynamic pressure of the abort phase. This also involved probability analysis for different conditions.

The determination of the insertion orbit depended on the performance of the launch vehicle, and in particular on the cutoff conditions; e.g. velocity, altitude, and staging time. A family of orbits was needed because the number of orbits possible for a mission depended on the cutoff conditions. The effects of atmospheric drag had to be considered in the orbit determination. A "Go/No-Go" criteria was developed, and eventually real-time data gained over several flights provided ever more data to increase confidence in these numbers. Some computer simulations indicated that there were higher orbits that would impose excessive re-entry heating, but the Mercury flights never exceeded those limits.

Operational analysis also involved the orbital inclination of the Mercury missions, and the positioning of the tracking stations. Owing to the inclination angle, selected re-entry locations were optimized for recovery forces, taking into account the constraints imposed by the safety, health, and comfort of the astronaut while he awaited recovery. Also, the atmospheric densities at orbital altitudes were not well known at the time, nor was the manner in which the gravity of the Earth varies with geographic location. These factors would affect both orbital lifetimes and acquisition times at the tracking stations.

The Mission Analysis disciplines (including mathematical, trajectory, and operational) that were developed during Project Mercury were new in 1959. The entirely new methodology is a credit to the people in the STG Mission Analysis Branch working under the leadership of John Mayer. See Section 9.4.1. This development continued when the group moved to the Manned Spacecraft Center in Houston, where it continued for Gemini, Apollo, Space Shuttle, and now the International Space Station.

16

The Impact of NASA and the STG on History

16.1 ORGANIZATIONAL EXCELLENCE

In retrospect, it is difficult to believe the level of excellence that was evident in many federal government organizations in the years 1957–1961. Of course, an organization is made up of people, and there couldn't have been a better group to start the space program. These people were exceptional. Here is an example by organizational area.

16.1.1 President and Congress

President Eisenhower's response to the launch of Sputnik on October 4, 1957 was to control the narrative and to control the civilian space program's future. His direction was specific and, by today's standards, both rapid and effectual. His vision and leadership after the Soviet challenge provided clarity of purpose to those wondering how to respond to Sputnik, which was perceived more as a threat to national security than a scientific satellite. But Eisenhower knew the Soviets well. He had already responded on the military front with ICBM and IRBM developments, and he was now prepared to respond on the civilian front. On November 21, 1957 he organized the President's Scientific Advisory Committee (PSAC), named James R. Killian, Jr. of MIT as the chairman of an 18 member committee, and relocated the committee to the White House. It was Harvard physicist James B. Fisk who led one subcommittee which included NACA Chairman General James H. Doolittle and which, with input from NACA's Director Hugh L. Dryden and Associate Director of NACA's Lewis Flight Propulsion Laboratory Abraham Silverstein, along with others, proposed a comprehensive national program in astronautics, emphasizing peaceful, civilian-run research and development. These recommendations were acted upon by others, but these men were the driving force and carried the load through Congress. See Chapter 3.

© Springer International Publishing Switzerland 2016
M. von Ehrenfried, *The Birth of NASA*, Springer Praxis Books,
DOI 10.1007/978-3-319-28428-6_16

It is hard to believe the speed with which others acted to implement the plan for a civilian space program. The Bureau of the Budget responded to Eisenhower on March 5, 1958, and he approved the plan on March 25, just five months after Sputnik. On April 14 Eisenhower sent a Bill to create the new agency to the 85th Congress. House and Senate special committees held hearings on it in May. The House passed it on June 2 and then the Senate did likewise on June 26. Eisenhower signed the National Aeronautics and Space Act on July 29 and NASA became effective on October 1. Now that's how things were done in Washington in 1958!

Only a few years later, in 1961, John F. Kennedy, a new President with vision and leadership qualities, initiated the Apollo program when this country had yet to put a man in orbit. His Vice President, Lyndon B. Johnson became Chairman of the National Aeronautics Space Council and was instrumental in getting NASA's budget through Congress.

16.1.2   NACA and NASA

With the creation of the National Aeronautics and Space Administration (NASA), the National Advisory Committee on Aeronautics (NACA) ceased to exist. The first NASA Administrator, T. Keith Glennan, made sweeping changes. The existing three NACA laboratories and their test facilities were renamed as NASA research centers and given a new focus. Other space related organizations were incorporated. Parts of the Naval Research Laboratory were brought into the newly created Goddard Space Flight Center. Some of the DOD and ARPA satellite programs and lunar probes were transferred to the Jet Propulsion Laboratory, which was run by the California Institute of Technology. Glennan also transferred the Army Ballistic Missile Agency to the new Marshall Space Flight Center.

The new NASA leadership in 1958 for the space program included:

- NASA Headquarters
  T. Keith Glennan, Administrator
  Hugh L. Dryden, Deputy Administrator
  Abe Silverstein, Director of Space Flight Programs
- Langley Research Center
  Henry J. E. Reid, Director
  Floyd L. Thompson, Deputy Director
  Robert R. Gilruth, Director of the STG and Project Mercury
- Lewis Research Center
  Dr. Edward R. Sharp, Director
- Ames Research Center
  Smith J. DeFrance, Director
- Marshall Space Flight Center
  Dr. Wernher von Braun, Director
- Goddard Space Flight Center (1959)
  Harry J. Goett, Director.

16.2   MERCURY MISSION ACCOMPLISHED

The concept of the manned space mission was being studied by NACA long before NASA was created on October 1, 1958. The STG was formed soon thereafter to undertake Project Mercury. The word "Mercury" was first proposed by Abe Silverstein during the fall of 1958. There were many other suggestions but the Olympian messenger was familiar to Americans and finally, on November 26, 1958, it was approved by Administrator Glennan.

From the inception of Project Mercury to its declared conclusion with the launch on May 15, 1963 of Gordon Cooper's MA-9 was barely 4 years and 8 months!

There were only three project objectives:

- To place a manned spacecraft in orbital flight around the Earth.
- To investigate man's performance capabilities and his ability to function in the environment of space.
- To recovery the man and spacecraft safely.

There were only four guidelines to achieve those three straightforward objectives:

- Existing technology and off-the-shelf equipment should be used wherever practical.
- The simplest and most reliable approach to system design would be followed.
- An existing launch vehicle would place the spacecraft into orbit.
- A progressive and logical test program would be employed.

Of course ever more detail followed, but fundamentally the project had three objectives and four guidelines!

In early 1959 a complete flight schedule for the capsule and launch vehicles was drawn up, including development, qualification, ballistic, and orbital flights. There were 25 major flight tests, including eleven additional flights made in response to lessons learned on earlier flights. There were only six manned flights; two suborbital and four orbital. One could consider John Glenn's flight to have achieved the three objectives, but the remaining flights drove home the lesson that man was able to cope with all the previous concerns and was an integral part of the spacecraft system. The additional flights also honed the management, engineering, operations, science, and medical skills of the entire team. NASA was clearly ready to carry out President Kennedy's visionary challenge.

16.3   FUTURE PROGRAMS

The writing of this book was essentially completed in October 2015, exactly 57 years after the start of the American space program in 1958. In that time, there have been several generations of launch vehicles and spacecraft. We've been to the Moon, but it has been over 40 years since humans have been that far into space. And the Moon isn't really all that far away. Humans are driven to explore; it is our nature. The spinoff from that urge to explore has changed the world. Always remember, it was the NASA STG that paved the way for the American space program. So, where are we going now?

NASA's plans are laid out for the future and include the development of the necessary tools and technologies to enable us to return to the vicinity of the Moon and even to Mars. But there are things that must be done first, because we aren't currently capable of achieving these goals with humans. This year was the 50th anniversary of the flyby of the Mariner 4 probe that began our exploration of Mars. Follow-on flyby probes, orbiters, and landers have taught us a lot, but we will need to learn a great deal more in the coming decades before we will be able to attempt to send a crew to Mars.

Today, the International Space Station (ISS) is yielding answers to some of the fundamental science, technology, and life support questions, and NASA has several programs and projects to build up our capabilities for deep space travel, including:

- Commercial Crew Program to develop the space transportation capability for safe, reliable and cost effective access to and from the ISS. This includes the Space Launch System and the Orion spacecraft.
- The Asteroid Initiative that includes both the Asteroid Redirect Mission and the Asteroid Grand Challenge. It includes the development and use of a Solar Electric Propulsion (SEP) system and a vehicle to capture an asteroid and redirect it to a location in the vicinity of the Moon, where it will be accessible to a crew flying an SLS/Orion mission. NASA estimates these missions will occur in the 2020s.
- Although NASA doesn't yet have a fully defined Mars program, it is planning to develop the technologies necessary for such a mission. These will include radiation shielding and mitigation techniques, advanced life support systems, advanced pressure suits, advanced propulsion systems, a Mars landing system, plus crew habitats for the 6–9 month cruise to Mars and a month on the surface of the planet before the long haul back to Earth. NASA is committing only to the 2030s. As yet, no one is brave enough to make a prediction such as "before this decade is out…"

NASA has created a series of videos that describe many of the above mentioned programs, projects, and technologies for the journey to Mars. These can be found at: http://www.nasa.gov/topics/journeytomars/videos/index.html.

The Orion Multi-Purpose Crew Vehicle (MPCV) has been designed and built by Lockheed Martin, but it isn't capable of going any great distances without a service/habitat module. The European Space Agency will provide the Orion service module, developed by Airbus Defense and Space. The first unmanned orbital test flight of an Orion capsule occurred on December 5, 2014, using a Delta IV launch vehicle; you can watch that launch on YouTube. There are plans to launch an unmanned Orion on a Block 1 Space Launch System into a circumlunar trajectory on or about September-November 2018 as Exploration Mission-1. The first manned mission is scheduled for 2021 and it will fly in lunar orbit. The flight to a captured asteroid is planned for 2026. These longer duration missions will require a deep space habitat module and a logistics module to support a crew of four. The modules will be built in three variants, depending on the requirements of the missions. Further information is available at www.nasaspaceflight.com.

These new launch vehicles will be state-of-the art in terms of technology, but will be built on the experience of past missions. The Block 1 will have about 10% more thrust at liftoff than the Saturn V that sent Apollo to the Moon. It will be able to place 154,000 lbs.

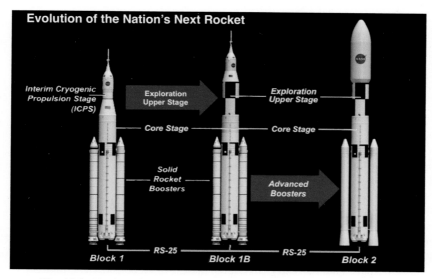

Fig. 16.1 The Space Launch System vehicles. (Photo courtesy of NASA)

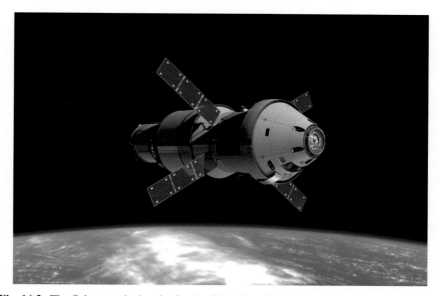

Fig. 16.2 The Orion attached to the Service Module and booster upper stage. (Artist concept courtesy of NASA)

into low Earth orbit. The Block 2 will have 20% more thrust and carry a payload of 286,000 lbs. They will both use advanced solid rocket boosters, and advanced RS-25 engines similar to the main engines of the Space Shuttle.

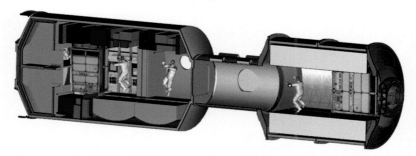

Fig. 16.3 The ISS derived concept for a deep space habitat module. (Artist concept courtesy of NASA)

I predict that if technologies related to long duration habitat modules, extended life support systems, radiation protection and mitigation, and advanced pressure suits all progress at a faster pace than currently envisaged, there will be considerable thought given to a mission to Phobos, the larger of the two Martian moons. It would most likely involve a rendezvous and perhaps a remote soil sampling operation, although a landing on such a small body would be feasible. It would be far more exciting than an asteroid mission, engender more public and Congressional support, provide valuable scientific results, and demonstrate a lot of Mars mission technology. Depending on when that mission decision is undertaken, it may be possible for the mission to occur by 2028. I also predict that the first manned landing on either Phobos or Mars will occur on or about March 30, 2036! That of course is just my dream to have the landing on my 100th birthday. When you consider this Nation's great debt and the fact that 66% of our GDP goes to pay off this debt and the out-of-control entitlement programs, with Defense consuming another 17%, there is precious little left for all the other programs, let alone space exploration. Perhaps 2050 or later is more realistic date for a Mars mission.

In the meantime, the Russian Soyuz will continue to ferry astronauts to the ISS. It has gone through many generations since its introduction in the 1960s and the current version is state-of-the-art. This is reflected by the fact that the Chinese and the Indian designs borrow much from the latest Soyuz design. The Soviets lost four cosmonauts on two missions in the early days but the current Soyuz is considered to be safe and cost-effective.

The first Chinese manned mission occurred on October 15, 2003 and they have launched five crews to date. They have evolved their Shenzhou (Divine Vessel) version of the Soyuz design as well as their Tiangon (Heavenly Palace) laboratory design and the Tianzhou (Heavenly Vessel) cargo craft. They hope to man their 60 ton space station in the 2020–2022 timeframe. They also have ambitions for lunar flights. They soft-landed a robotic craft on the Moon on December 14, 2013 that deployed a rover. In the longer term, they have plans for a human landing on Mars in the 2040–2050 timeframe.

The Indian space program has been successful with unmanned satellites, including launching 10 simultaneously, and on September 24, 2014 it inserted a probe into orbit around Mars on its first try. It hopes to launch a human into orbit in 2017.

The Japanese manned space program is primarily linked to the ISS. It doesn't currently have plans to develop its own spacecraft, but they have tentative plans to send a robot to the Moon.

16.4   TECHNOLOGY TRANSFER

The National Aeronautics and Space Act of 1958 that created NASA called for the new agency to disseminate its technology for public benefit. While the average person today doesn't give it much thought, the technology "spinoffs" from the space program are now part of everyday life, not only in the U.S. but nearly everywhere, including the Third World. Initially, the public had the wrong impression about what NASA provided society. Myths about NASA's contributions abound, even today. NASA did NOT invent Tang, Velcro, barcodes, smoke detectors, and the MRI. They did much, much more than that.

The Space Act obliged NASA to make the widest practicable and appropriate dissemination of its results to the public. Whilst the agency was too busy during the formative days of Project Mercury to devote much effort to this task, it began to do so immediately afterwards. In 1962 it created the Industrial Applications Program. This evolved into the current Technology Transfer Program, designed to carry out the responsibilities of sharing NASA's research with the public. Over the past 50 years there have been at least 13 major laws and executive orders designed to enhance the legal authority of NASA to facilitate the transfer of its technologies to industry and the public.

As technology spinoffs emerged in the Apollo era NASA started to send reports to Congress to demonstrate the results of its Industrial Applications Program. It followed up with its annual *Spinoff* publication.

Today, Technology Transfer is a major NASA effort at Headquarters and at each of its field centers. It covers the following areas:

- Health and Medicine
- Transportation
- Public Safety
- Consumer Goods
- Energy and Environment
- Information Technology
- Industrial Productivity.

In the Project Mercury era it would have been impossible to predict the future benefits of the space program to society. There were people then, as there are now, who expect a prediction of the benefits from spending national treasure on the "harebrained scheme" of space travel. Even the spinoffs from the current Mars rovers have resulted in software imaging technology that can detect heart disease earlier than was previously feasible. One trip to a hospital emergency room will open your eyes to space age spinoffs. There are books elaborating thousands of spinoffs.

NASA is tapping into the imagination of American youth to see what they envisage for the world by sponsoring essay contests. The Technology Transfer Program is the responsibility of the NASA Office of the Chief Technologist. For more information see http://spinoff.nasa.gov and http://technology.nasa.gov.

16.5   NATIONAL PRIDE

Why do a million people crowd the roads to Cape Canaveral to see a rocket launch? Why do a million people watch astronauts parade down New York's Canyon of Heroes? Although there may be some who do so on the off-chance of seeing a rocket blow up, most are there to join in the celebration of a human being going into space. They celebrate because they are proud of the astronaut, the team, and their country. Why do people remember where they were and who they were with when Armstrong and Aldrin walked on the Moon? For sure American pride is part of the celebration. Americans, Russians, Chinese, and Europeans all have an emotional attachment to their astronauts, cosmonauts, taikonauts and spationauts; it's a matter of national pride.

The success of a "ticker-tape" parade in New York is officially measured by the Department of Sanitation that has to clean up the mess. John Glenn's parade generated 3,474 tons of paper, making it the largest parade clean up since V-J Day. No parade since has broken this record; for many reasons. Nobody uses "ticker-tape" anymore and with more televisions in homes, people watch the parade there rather than stay at work in New York; or worse, commute into the city. Also, some the windows of some modern office buildings don't open because they have closed air conditioning systems.

Fig. 16.4   John Glenn's parade down the Canyon of Heroes. (Photo courtesy of Wikipedia)

Project Mercury certainly instilled pride in America. After half a century, flying in space is commonplace. Indeed, the ISS has been permanently occupied since the turn of the century and most members of the public don't know the names of the astronauts. Although people are more blasé now, they're still very proud of the astronauts and of NASA, which is probably the most respected of all the government agencies.

16.6   GENERATIONAL IMPACT

There are many definitions of a "generation" but for this discussion let's assume it is 25 years; the reasonable time for one generation to have offspring. If we start in 1958, the beginning of NASA and Project Mercury, here are just a few of the things which space-flight has brought to the future in exploration and technology:

- $1958 + 25 = 1983$ (1 generation):
  Lunar landing, Viking on Mars, Space Shuttle, PCs
- $1983 + 25 = 2008$ (2 generations):
  Internet, Hubble, Windows XP, iPhone, GPS, ISS
- $2008 + 25 = 2033$ (3 generations):
  Mars robots, Pluto, robotic surgery, new materials
- $2033 + 25 = 2058$ (4 generations):
  World Peace? World Destruction? More of the same? How about a crew on Mars?

The space program instilled wonder and excitement into people all around the world. "If we can do that; we can do anything" became the positive "can do" spirit. "It's not rocket science" became the comment for when we thought we couldn't do something, meaning it couldn't be as difficult as going to the Moon. But alas, there are things that do seem impossible, like balancing the budget, or getting Congress to agree on societal, energy, and economic solutions. However, when it comes to science, we have accomplished much because science is truth; uncorrupted by human frailties, except for those who have warped their data to assure the continued funding for their research. The advancement of science drives the technologies that improve our lives and advanced technologies drive exploration. Science has not only improved our lives, but has also increased our life spans. The fact that we are living longer than previous generations becomes a problem for governments whose planning was based on the promise to pay social security to an increasingly aging population. How do you modify a promise?

NASA routinely informs the public, businesses, and industries of the many "spinoffs" from the space program. These have occurred in almost every field. In many cases you can actually see it if you look. Next time you go to the hospital, look at all the bioinstrumentation, sensors, telemetry, emergency room equipment and monitors. And then add in the space age spinoff of microelectronics in phones and computers. Add to that the advances in medicines, and it is no wonder we are living longer. The impact of these advances also makes us accept them without question. We have come to expect, indeed almost demand technological advances. Just look at the youngsters standing in line to buy the latest phone! In contrast, my generation is inclined to be satisfied with their old phones, in some cases still of the type which plugs into the wall by a cable.

NASA set up an organization called the Space Technology Mission Directorate (STMD) to develop the crosscutting, pioneering, new technologies and capabilities required to achieve its current and future missions. This work takes place in all the NASA centers, academia, industry, and both U.S. and international partnerships. The program seeks to identify and rapidly mature innovative and high impact capabilities and technologies. By stimulating breakthroughs, these programs and activities could transform future missions. Because many of the research results have non-space applications as well, there are many cross applications to technology.

Although many of us would like to see humans exploring the Red Planet, the stay-at-homes on the Blue Planet may be having our own problems by 2050, if not actually earlier. The world population in 1958 was about 2.8 billion people. One estimate by the United Nations for 2050 is 9.3 billion. That's a lot of people to feed, clothe and shelter. But space age technology spinoff is helping farmers to be more productive even today.

In summary, there is no question that Project Mercury had a positive and long lasting impact on American society that has been felt for generations. While the Atlas and the Mercury capsule are primitive by today's standards, one could think of them as the Model T's. And although the STG people who started it all are now old or gone, they have inspired at least two generations of space workers. Hopefully, the spark of imagination for youngsters will keep the space program going; be it manned or unmanned spaceflight. I fear that I will never live to see humans walking on Mars, but you might!

Appendix 1

STG Organization Lists, Charts and Manning

The summer of 1958 was very busy. The "birth of NASA" was close at hand, and so was the birth of the Space Task Group. As soon as the National Aeronautics and Space Act was signed on July 29, 1958, memorandums began to fly around Washington. On September 3, before the establishment of the STG, Paul E. Purser wrote to Robert R. Gilruth describing the number of positions that Dr. T. Keith Glennan would authorize for the nascent "Space Center." Glennan wasn't formally the Administrator yet, but he was taking charge. He authorized 200 positions and Purser detailed the understanding that Langley would provide for, or hire, 14 aeronautical positions, 6 propulsion, and so on. No names were mentioned in the memo. However, back at Langley, names were being mentioned in organizations such as the Pilotless Aircraft Research Division (PARD), the Instrument Research Division (IRD), etc. Hence the question was: Who would volunteer and who would be selected to join this new space projects effort?

The very next day, September 4, Edmond C. Buckley of the IRD sent a memorandum to the Langley Associate Director Floyd Thompson describing how his Division would support the on-going projects and what kind of people would be needed for the "Space Center." Still no names were mentioned; only technical disciplines, e.g. communications, tracking, computing, etc.

Once Gilruth was officially designated as Director of Project Mercury by Glennan on January 26, 1959, the basic outline of the Space Task Group began to take on official form. On the same day, Gilruth issued a "Memorandum for all Concerned." The circulation list specified 71 names, counting himself. Over the next six months the STG grew. The astronauts and their training and aeromedical people came onboard. Administration people were identified. So too were those in technical services and scheduling, and those in capsule requirements and contracting. While the organization went through several iterations over the summer, the organization dated August 10, 1959 proved to be the basic STG organization for Project Mercury with slight changes along the way, and by the end of 1961 included about 750 people.

© Springer International Publishing Switzerland 2016
M. von Ehrenfried, *The Birth of NASA*, Springer Praxis Books,
DOI 10.1007/978-3-319-28428-6

NASA - Space Task Group
Langley Field, Virginia
August 10, 1959

MEMORANDUM for Staff

Subject: Organization of Space Task Group

1. Effective this date, the following organization of the Space
Task Group is established:

OFFICE OF DIRECTOR

Director of Project Mercury, Robert R. Gilruth
Assistant Director of Project Mercury, Charles J. Donlan
    Jacquelyn B. Stearn, Secretary to Director

Staff Assistants

Paul E. Purser, Special Assistant
James A. Chamberlin, Technical Assistant
Raymond L. Zavasky, Executive Assistant
Col. Keith G. Lindell, USAF
Lt. Col. Martin L. Raines, USA
Cdr. Paul L. Havenstein, USN
W. Kemble Johnson, Langley Research Center

Public Affairs Officer

Lt. Col. John A. Powers, USAF

Staff Services

Burney H. Goodwin, Personnel Assistant
    Betty S. Knox, Secretary
Guy W. Boswick, Jr., Head, Administrative Services
    Jo Ann S. Fountain, Receptionist
    Ann W. Hill, Alternate Receptionist

Files and Library

Norma L. Livesay, Head
    Carole A. Howard
    Mary C. Morrison
    Doris W. Reid
    Shirley P. Watkins
    Samuel S. Westbrook, Jr.

Stenographic Pool

Margaret B. Burcher, Head
    Nancy C. Alexander
    Norma C. Eacho
    Irene J. Helterbran

- 2 -

Ann W. Hill
Patricia H. Murphy
Rita F. Pereira
Genevieve B. Taylor

Technical Services

Jack A. Kinzler, Technical Services Assistant
David L. McCraw
Orrin A. Wobig

Astronauts and Training Group

Col. Keith G. Lindell, USAF, Head
Lt. Col. William K. Douglas, USAF, Flight Surgeon
Lt. Robert B. Voas, USN, Training Officer
(Warren J. North will continue to participate in all astronaut
  activities as representative of the Office of Space Flight
  Development)

Astronauts

Lt. Malcolm S. Carpenter, USN
Capt. Leroy G. Cooper, Jr., USAF
Lt. Col. John H. Glenn, Jr., USMC
Capt. Virgil I. Grissom, USAF
Lt. Cdr. Walter M. Schirra, Jr., USN
Lt. Cdr. Alan B. Shepard, USN
Capt. Donald K. Slayton, USAF

Training Office

George C. Guthrie
Raymond G. Zedekar

FLIGHT SYSTEMS DIVISION

Maxine A. Faget, Chief
Robert O. Piland, Assistant Chief
J. Thomas Markley, Executive Engineer
   Betsy F. Magin, Secretary

Computing Group

Katherine S. Stokes, Head
   Patricia D. Link
   Mary W. McCloud
   Anne F. Wilson

- 3 -

Systems Test Branch

William M. Bland, Jr., Head
H. Kurt Strass, Assistant Head
  Lawrence W. Enderson, Jr.
  Edison M. Fields
  Louis R. Fisher
  Jerome B. Hammack
  Jack C. Heberlig
  Robert A. Hermann
  Walter J. Kapryan
  Ronald Kolenkiewicz
  Owen E. Maynard
  James T. Rose
  Rodney G. Rose

Performance Branch

Aleck C. Bond, Head
  Betty M. Breault, Secretary

Aerodynamics Section

Alan B. Kehlet, Head
  Steve W. Brown
  David D. Ewart
  William H. Hemby
  Dennis F. Hasson
  Bruce G. Jackson
  William C. Moseley, Jr.
  Herbert G. Patterson
  William W. Petynia
  Edward F. Young

Loads Section

George A. Watts, Head
  Joseph E. Farbridge
  May T. Meadows
  Robert P. Smith
  William Rogers

Heat Transfer Section

Leonard Rabb, Head
  Eugene L. Duret
  Richard B. Erb
  Joanna M. Evans
  Archie L. Pitzkee

- 4 -

    Stephen Jacobs
    John S. Llewellyn, Jr.
    Robert O'Neal
    Emily W. Stephens
    Kenneth C. Weston

Life Systems Branch

    Lt. Col. Stanley C. White, USAF, Head
    Richard S. Johnston, Assistant Head
    Lt. Col. James P. Henry, USAF, Specialist for Biological Flight
    Gerard J. Pesman, Specialist for Stress Tolerance
        Capt. William S. Augerson, USA
        William H. Bush, Jr.
        H. Jack Grames
        James R. Hiers
        Dorothy B. Histand
        George T. Jenkins, Jr.
        Lee N. McMillion
        Jack A. Prizzi
        Frank H. Samonski, Jr.
        Charles D. Wheelwright
        Frank C. Robert
        William E. Thompson

On-Board Systems Branch

    Harry H. Ricker, Jr., Head
    William T. Lauten, Jr., Assistant Head
        Stella R. McDonnell, Secretary

Electrical Systems Section

    Harry H. Ricker, Jr., Acting Head
        Robert E. Bobola
        Norman B. Farmer
        John H. Johnson
        Milan J. Krasnican
        Harold R. Largent
        Robert E. Munford
        Thomas E. Ohnesorge
        Ralph Sawyer
        James E. Towey

Mechanical Systems Section

    John B. Lee, Head
        Philip M. Deans
        James K. Hinson
        Witalij Karakulko
        David L. Winterhalter, Sr.

- 5 -

<u>Dynamics Branch</u>

Robert G. Chilton, Head
Jean S. Saucer, Secretary

<u>Flight Controls Section</u>

Richard R. Carley, Head
Thomas V. Chambers
Stanley Galezowski
Paul F. Horsman
Thomas E. Moore
Fred T. Pearce
Donald J. Jezewski
Thomas N. Williams

<u>Space Mechanics Section</u>

Robert G. Chilton, Acting Head
Robert C. Blanchard
Harold R. Compton
Thomas F. Gibson, Jr.
Jack Funk
Morris V. Jenkins
Frank A. Volpe

OPERATIONS DIVISION

Charles W. Mathews, Chief
G. Merritt Preston, Assistant Chief (Implementation)
Christopher C. Kraft, Jr., Assistant Chief (Plans and Arrangements)
Chris C. Critzos, Executive Engineer
John D. Hodge, Assistant to Division Chief
Cdr. Paul L. Havenstein, USN, Assistant for Operational Planning
Robert D. Harrington, BMD Coordinator
Edith K. Spritzer, Secretary

<u>AMR Project Office</u>

Elmer H. Buller, Head
Richard G. Arbic
B. Porter Brown
Sarah H. Greenfield
Emory F. Harris
Phillip R. Maloney
Charles S. Murray
Loyd C. Shelton
T. Bradley Curry, Jr.

- 6 -

Mission Analysis Branch

John P. Mayer, Head
Jack Cohen, Assistant Head
    Shirley J. Hatley, Secretary

Trajectory Analysis Section

John P. Mayer, Acting Head
    Charlie C. Allen
    John A. Behuncik
    James A. Ferrando
    Jack B. Hartung
    Claiborne R. Hicks, Jr.
    Carl R. Huss
    John W. Maynard, Jr.
    John C. O'Loughlin
    Ted H. Skopinski

Operational Analysis Section

Jack Cohen, Acting Head
    Paul G. Brumberg
    Robert E. Davidson
    Arthur A. Hand
    John H. Lewis, Jr.
    Glenn S. Lunney
    Harold G. Miller

Mathematicial Analysis Section

Stanley H. Cohn, Head
    Jerome N. Engel
    John N. Shoosmith
    Mary S. Burton
    Nancy K. Carter
    Shirley A. Hunt
    Elizabeth P. Johnson
    Pattie S. Leatherman
    Catherine T. Osgood

Flight Control Branch

Gerald W. Brewer, Head
    Nancy C. Lowe, Secretary

Control Central and Flight Safety Section

Gerald W. Brewer, Acting Head
Howard C. Kyle, Assistant Head
    Arnold D. Aldrich
    Alfred Amar

- 7 -

John H. Dabbs
Frank J. Chalmers
Robert E. Ernull
James F. Dalby
Dennis E. Fielder
Elmer A. Horton
John K. Hughes
C. Frederick Matthews
Joan B. Maynard
David T. Myles, Jr.
Leonard E. Packham
Tecwyn Roberts
Richard F. Schultheiss
Rafael Cosme
John T. Koslosky

Training Aids Section

Harold I. Johnson, Head
Bruce A. Aikenhead
Stanley Faber
Arthur E. Franklin
Rodney F. Higgins
Alfred J. Meintel, Jr.
Charles C. Olasky, Jr.

Launch Operations Branch

G. Merritt Preston, Acting Head
Scott H. Simpkinson, Assistant Head
Maxwell G. Christopher, Ground Support
Arthur M. Busch
Alfred L. Conner
Kenneth E. Glover
Harold G. Johnson
Robert G. Mungall
Carlos L. Springfield
John J. Williams
(The following personnel on the Lewis Space Task Group
    complement are also assigned to the Launch Operations
    Branch)
Francis Bechtel
Joseph Bender
Dugald O. Black
Joseph M. Bobik
Jack A. Campbell
Robert L. Carlson

- 8 -

Donald M. Corcoran
Frank M. Crichton
Edward J. Cudlin
William R. Dennis
A. Martin Eiband
Emily M. Ertl
Vernon E. Fisher
Clifford J. Haight
Charles J. Heckelmoser
John Janokaitis, Jr.
Elmer H. Karberg
Frank A. Maruna, Jr.
William R. Meyer
Jacob C. Moser
Donald H. E. Priebe
Howard R. Roe
Armand Sanvido
Robert L. Sorg
Michael A. Wedding
Donald F. Wilfert
Donald J. Woods

Recovery Operations Branch

Robert F. Thompson, Head
Peter J. Armitage
John B. Graham, Jr.
Exiaslav N. Harrin
Enoch M. Jones
Carl J. Kovitz
Charles I. Tynan, Jr.
Julia R. Watkins
Milton L. Windler

ENGINEERING AND CONTRACT ADMINISTRATION DIVISION

James A. Chamberlin, Acting Chief
Andre J. Meyer, Jr., Assistant Chief
Norman F. Smith, Executive Engineer
John C. French, Technical Assistant
Acquilla D. Saunders, Secretary

- 9 -

Field Representative at McDonnell

Wilbur H. Gray, Head
    William J. Nesbitt
    Louise E. Kase

Capsule Coordination Office

James A. Chamberlin, Head
Andre J. Meyer, Jr., Assistant Head
    Norman F. Smith
    Barbara K. Lawrence
    Margaret M. Nichols

Contracts and Scheduling Branch

George F. MacDougall, Jr., Head
Joseph V. Piland, Assistant Head

    John R. Bailey, Transportation
    Richard F. Baillie
    Jack Barnard
    Lanell B. Beavers
    Wanda Carter
    Joyce A. Lloyd
    William B. Long
    Margaret Marshall
    William C. Muhly
    Frank L. Parmenter
    Albert J. Saecker
    Floyd M. Saxton
    Edward L. Snoddy
    Douglas E. Steele
    Lester A. Stewart
    Paul M. Sturtevant

Engineering Branch

Caldwell C. Johnson, Jr., Head
    Norman R. Bailey
    Douglas R. Broome, Jr.
    John E. Canady
    Richard A. Colonna
    Joe W. Dodson
    Douglas J. Geier
    John E. Gilkey
    R. Bruce Grow
    Robert T. Gunderson
    John B. Hall, Jr.
    Nicholas Jevas
    Joe S. Hunter

- 10 -

William H. Keathley
Clarence O. Keffer
William C. Kincaide
William E. Lane
John H. Langford
Robert H. McDonnell
Norma C. Medina
Earnest F. Perkins
Harry C. Shoaf
Edward E. Shumilak
John B. Slight
Richard F. Smith
Joseph E. Walz
Lawrence G. Williams
Charles H. Wilson
Thomas F. Woods

    2. There are attached organizational charts dated August 10, 1959
reflecting the organization established by this memorandum.

Robert R. Gilruth
Director of Project Mercury

Enc:
5 charts

WKJ.bsk

Copies to: Each Space Task Group employee
          NASA Headquarters (3)
          Goddard Space Flight Center (20)
          Langley Research Center (40)
          Space Files (3)

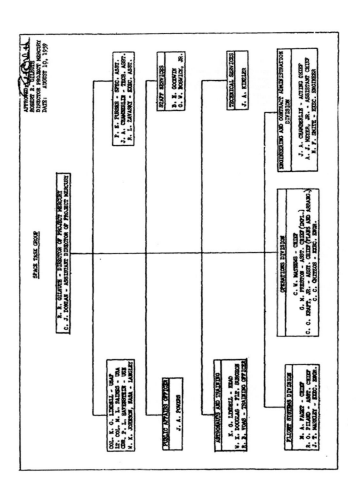

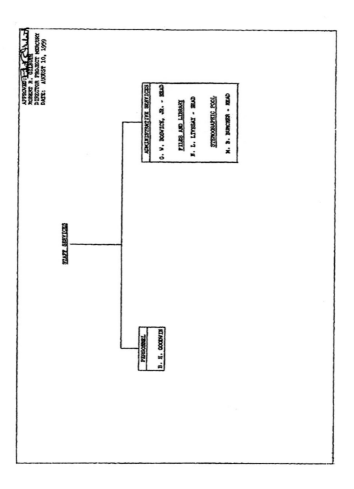

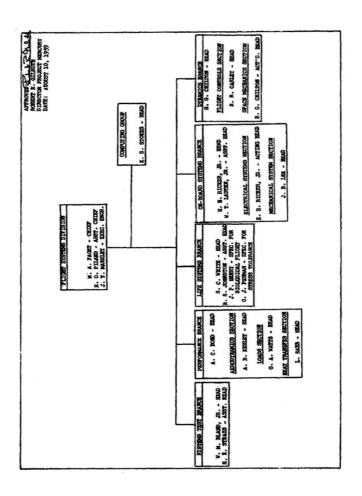

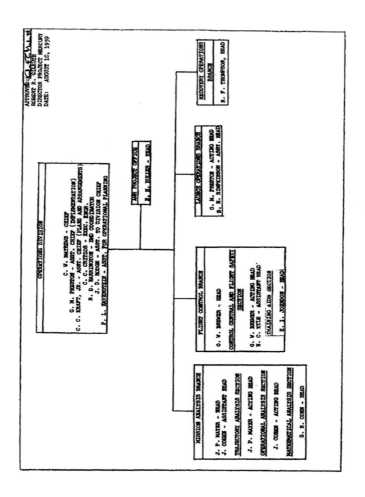

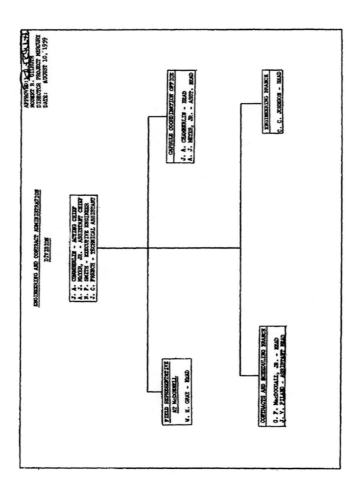

The following are the manning lists for the Mercury Control Center and remote sites, kindly provided by Gene Kranz and Arnie Aldrich.

MERCURY

MERCURY CONTROL CENTER AND REMOTE SITE MANNING

SITES/POSITION	MA-1	MR-1	MR-1A	MR-2	MA-2 "HAM"	MA-3	MR-3 SHEPARD	MR-4 GRISSOM	MA-4
	######	######	12/19/1960	1/31/1961	2/21/1961	/ 4/24/1961	5/5/1961	7/21/1961	9/13/1961
TEST NUMBER						#835	#108		
FLIGHT DIRECTOR	KRAFT	KRAFT	KRAFT	KRAFT	KRAFT	KRAFT	KRAFT / KYLE	KRAFT	KRAFT
PROCEDURES		KRANZ / JOHNSON	KRANZ	KRANZ / JOHNSON	KRANZ	KRANZ	KRANZ	KRANZ	KRANZ
SYSTEMS							HAVENSTEIN / ARABIAN		ARABIAN
CAPCOM							SLAYTON / COOPER-B	SHEPARD / SLAYTON-B	SHEPARD / GLENN / GRISSOM
FLIGHT DYNAMICS		ROBERTS	ROBERTS	ROBERTS	ROBERTS	ROBERTS	ROBERTS	ROBERTS	ROBERTS
RETRO		HUSS	HUSS	HUSS	HUSS	HUSS	HUSS	HUSS	HUSS
ENVIRONMENT							SCHLER		SCHLER
SURGEON							WHITE		
NETWORK							CLEMENTS		
SCC							SHARKEY		
SIM SUP		FABER	FABER	FABER	FABER	FABER	FABER	FABER	FABER

BDA-FLT

SITES/POSITION	SIM TASK GROUP 1959	MA-2		MA-4
BDA-FLT		BDA-FLT		HODGE
CC	FABER	CC		SLAYTON
FDO	KOOS	FDO		LUNNEY
SYS	HOOVER	SYS		
ENV	LUNNEY / MILLER	ENV		
SURGN		SURGN		

Remote Sites

Site / Position	MA-2	MA-3	MR-3	MR-4	MA-4
CYI-CC	CYI-CC	ERNULL		CYI -CC	BECKMAN
SYS		BARKER			STENFORS
A/M		WARD/BERRY			WARD
KNO-CC	KNO-CC	WAFFORD		KNO- CC	VOLPE
SYS		N/A			HUBER
A/M		N/A			KRATCHOVIL
ZZB-CC	ZZB-CC	AIKENHEAD		ZZB - CC	LLEWELLYN
SYS		N/A			WAFFORD
A/M		N/A			FLOOD
IOS-CC	IOS-CC	N/A	LLEWELLYN		BRUMBERG
SYS		N/A	RUMBAUGH		RUMBAUGH
A/M		N/A	HALL		AUSTIN
MUC-CC	MUC-CC			MUC-CC	CARPENTER
SYS					ROSENBLUTH
A/M					BISHOP/TURNER
WOM-CC	WOM-CC	GUTHRIE		WOM-CC	GUTHRIE
SYS		HOPP			BARKER
A/M		LANE			LANE
CTN-CC	CTN-CC	VOLPE		CTN-CC	MOORE
SYS		ROSENBLUTH			T. WHITE
A/M					SMITH
ATS-CC	ATS-CC	MOORE	MOORE		LANGFORD
STS		STENFORS	STENFORS		REMBERT
A/M		AUSTIN/GORDON	AUSTIN/GORDON (Port CNV)		HAWKINS
HAW-CC	HAW-CC			HAW-CC	HIGGINS
SYS					DELUCA
A/M					MOSER
					COOPER?
CAL-CC	CAL-CC			CAL-CC	DURET
SYS					LOONGAN
A/M					PRUETT/MARCHBKS
		CARPENTER			SCHIRRA?
GYM-CC	GYM-CC	ALDRICH		GYM-CC	ALDRICH
SYS		HUNTER			HUNTER
A/M					DAVIS
TEX-CC	TEX-CC	BECKMAN		TEX-CC	FABER
SYS		LONGAN			CROSS
A/M		SMITH/HALL			HALL/HOLMSTROM

Page 1

MERCURY

SITE/POS'N	"ENOS" MA-5 11/29/1961	GLENN MA-6 2/20/1962	CARPENTER MA-7 5/24/1962	SCHIRRA MA-8 11/3/1962	COOPER MA-9 5/15/1963	
FLIGHT DIR	KRAFT	KRAFT	KRAFT	KRAFT	KRAFT	HODGE
PROCEDRS	KRANZ	KRANZ	KRANZ	KRANZ	VONEHREN	KRANZ
SYSTEMS	ARABIAN	ARABIAN	ARABIAN(?) ALDRICH	ALDRICH	ALDRICH LOCKARD GULL	BROOKS
CAPCOM	GRISSOM GLENN CARPENTER-B	SHEPARD SLAYTON CARPENTR-B	GRISSOM	SLAYTON	SCHIRRA	SHEPARD
FLIGHT DYN	ROBERTS	ROBERTS-LUNNEY	LUNNEY		LUNNEY	CHARLESWTH
RETRO	HUSS		LLEWELLYN-HUSS		LLEWELLYN	HUSS
ENVIRON	SCHLER		HUGHES-SAMONSKI		SAMONSKI	HUGHES
SURGEON			WHITE BERRY		BERRY	CATTERSON
NETWORK					BURWELL	
SCC					HATCHER	SHARKEY
	FABER	HOOVER	MILLER/SULLIVAN	KOOS/BROOKS	BROOKS/FERGUSON	
BDA-FLT	HODGE	HODGE				
CC	SHEPARD	GRISSOM		TOMBERLIN		
FDO	LUNNEY	LUNNEY			BROWN	
SYS		TOMBERLIN STRICKLAND			MOSER	
ENV		SAMONSKI HUGHES		BROWN	SHEA	
SURG'N		BERRY COONS		KELLY		
CYI-CC	LANGFORD	LLEWELLYN	OLASKY	LEWIS	PLATT	
SYS	STENFORS	ROSENBLUTH	ROSENBLUTH	WAFFORD	REMBERT	
A/M	WARD/BURWELL	WARD/BURWELL	HAWKINS/UNGER	UNGER	KRATCHOVIL	
KNO-CC	LLEWELLYN	DURET	MOORE	HANNIGAN	PENDLEY	
SYS	ROSENBLUTH	RUMBAUGH	DELUCA	CASSELLI	BARKER	
A/M	KRATOCHVIL/MARCHBK	KRATOCHVIL/MARCHBKS	BECKMAN	BURWELL/BIDWELL	BLACKBURN	
ZZB-CC	BECKMAN	LANGFORD	LEWIS-PARKS	PARKS	TOMBERLIN	
SYS	LONGAN	REMBERT	WAFFORD	AMOS	ROSENBLUTH	H. SMITH
A/M	FLOOD/FOX	FLOOD/FOX	N/A		GRAVELINE	
CSQ-CC	HIGGINS	BECKMAN	DURET	LOCKARD	GLENN	
SYS	L. WHITE	HOPP	RUMBAUGH	T. WHITE	RUMBAUGH	
A/M	HALL/HANSEN	AUSTIN/BENSON	BENSON	KRATOCHVIL	BECKMAN	
		COOPER	SLAYTON			
MUC-CC	SCHIRRA-HOOVER	FABER	ERNULL	DURET	LEWIS	
SYS	BARKER	L. WHITE	HUNTER	HUNTER	STENFORS	
A/M	BECKMAN/BISHOP	BECKMAN/BISHOP	AUSTIN/BISHOP		BISHOP	
WOM-CC	VOLPE	VOLPE	BRUMBERG	FERGUSON		
SYS	WAFFORD	WAFFORD	HUBER	DUNBAR		
A/M	LANE/OVERHOLT	OVERHOLT/LANE	LANE	LANE		
CTN-CC	OLASKY	HIGGINS	GUTHRIE	PLATT	PARKS	
SYS	DELUCA	HUBER	CROSS	RUMBAUGH	AMOS	
A/M	GRAVELINE/HOLMSTROM	GRAVELINE/HOLMSTROM	FLOOD	SHEA	LUCHINA	
RKV-CC	ZEDEKAR	PRIM		PCS-SHEPARD	HUNTER	
SYS	HUBER	CROSS		STRICKLAND	WALSH	
A/M	KELLY/HAWKINS	R. KELLY/ F/ KELLY		BECKMAN	BRATT	
HAW-CC	ERNULL	ERNULL-OLASKY	PRIM	GRISSOM	CARPENTER	ROACH
SYS	T. WHITE	LONGAN	REMBERT	BARKER	STRICKLAND	DAVIS
A/M	MOSER/AUSTIN	MOSER/HALL	MOSER	MOSER	MOSER	
	COOPER	SCHIRRA		GLENN		
CAL-CC	ALDRICH/PRIM	ALDRICH	SHEPARD	PRIM	T. WHITE	
SYS	REMBERT	T. WHITE	T. WHITE	REMBERT	DUNBAR	
A/M	BRATT/PRUETT/BENSON	BRATT/PRUETT	GRAVELINE/REED	HALL/TUCHINA	KELLY	
	SLAYTON			CARPENTER		
GYM-CC	MOORE	MOORE-BRUMBERG	COOPER	KAWALKIEWICZ	GRISSOM	
SYS	HUNTER	HUNTER	BARKER	ROSENBLUTH	WAFFORD	
A/M	DAVIS/TURNER	DAVIS/TURNER	REED/HOLMSTROM	GRASVELINE/MUSGRAVE	G. SMITH	
TEX-CC	KUEHNEL		LANGFORD/WHITE	MUSE-BATES-ROACH	BRAY	
SYS	RUMBAUGH		MUSE/BROWN	L. WHITE	L. WHITE	
A/M	SMITH/KELLY/LAWSON		PLATT/GULL KAWALWIEWICZ	BLACKBURN	BENSON	
			Page 2	WATERTOWN-GULL		

Appendix 2

Biographies

Introduction

It was not possible to include a biography for everyone who was at NASA and the Space Task Group when the space program started on October 1, 1958. The following are many of the key players as well as those who were fortunate enough to be a pioneer in America's civilian space program; specifically Project Mercury. My apologies if I've missed someone. The eight people from the Joint NASA-ARPA Manned Satellite Panel are included; six were former NACA and two from ARPA. I also include the ARPA director because he had much to do with the creation of NASA. These men were among the first to make specific recommendations and start to plan the manned satellite project.

Other agencies and organizations such as the President's Scientific Advisory Committee (PSAC) are not included, but certainly played significant roles in the beginning of the space program. Specifically, Dr. James R. Killian, Chairman of the PSAC, with General James H. Doolittle, Chairman of the NACA and a member of the PSAC, provided the conceptual and formative direction for the new space agency. They could easily be called "Architects of the Space Program." Doolittle helped to navigate the Space Act through Congress. Neither man became part of the agency that this produced. Their biographies aren't included, but Killian's influence is related in Section 3.4 and Doolittle's powerful speech to Congress is included in Appendix 5.

This book focuses on the NASA employees, and especially those of the Space Task Group. Therefore, the list of "Founding Fathers" is my own and shouldn't be considered as an official NASA list. The agency never officially used that term. The history books do not make it very clear as to who was behind the beginning of NASA, and one would have to read a lot of books and reports in order to determine that point. For example, one would think that James E. Webb should be included in this list, but he was an oil man when NASA was formed. He was not the NASA Administrator until February 14, 1961; two years and four months after the agency was formed. So I think my list may be helpful in understanding who were "the attending physicians behind the birth of NASA."

© Springer International Publishing Switzerland 2016
M. von Ehrenfried, *The Birth of NASA*, Springer Praxis Books,
DOI 10.1007/978-3-319-28428-6

The following biographies are rather brief, but of sufficient detail to make you aware of who they were, where they came from, and what they did for the program. In a few cases, I couldn't locate any details or pictures of the people in 1958. The Headquarters list mostly includes those in the temporary offices of the Dolley Madison House, near the White House. There are some for whom I could find very little detail. The field centers' list primarily includes those from Langley, Ames, and the Flight Research Station. The STG Core Team list includes those people chosen by Robert Gilruth in the beginning of the program in October and November 1958. The third list has other members of the STG who came later in order to work on Project Mercury; myself included.

The biographies are organized as follows:

- The Founding Fathers from NACA, NASA Headquarters, the field centers and ARPA
- The Core Team of the Space Task Group (including Langley and Lewis)
- Other members of the STG, primarily during the 1959–1961 timeframe.

A.2.1   The Founding Fathers

From Headquarters and the Joint NASA-ARPA Manned Satellite Panel:

- T. Keith Glennan – Administrator
- Hugh L. Dryden – Deputy Administrator
- John L. Johnson – General Counsel
- Abe Silverstein – Director of Space Flight Development
- Newell D. Sanders – Assistant Director for Advanced Technology
- George M. Low – Chief, Office of Manned Space Flight
- Warren J. North – Chief, Office of Manned Satellites
- Ira H. Abbott – Director of Advanced Research
- John W. Crowley – Director of Aeronautics and Space Research
- Roy W. Johnson* – Director of ARPA and member of the Joint NASA-ARPA Panel
- Samuel Batdorf* – Head of the ARPA Man in Space Panel
- Robertson C. Younquist* of ARPA and member of the Joint NASA-ARPA Panel.

From the NACA/NASA field centers (1958, prior to NASA and the STG):

- Henry J. E. Reid – NACA Langley Director
- Floyd L. Thompson – Acting NACA Langley Director, STG Center Support
- Robert Gilruth – Director of the STG and Project Mercury
- Maxime Faget – STG Chief, Flight Systems Division, early capsule designer
- Charles J. Donlan – STG and Associate Director of Project Mercury
- Charles H. Zimmerman – STG Executive Counsel
- Joseph G. Thibodaux, Jr. – Langley Head, High Temperature Materials Branch
- Charles W. Mathews – Pre-NASA space concept work and STG Chief, Operations Division

---

*The ARPA Biographies are not included in the following section.

- Paul E. Purser – Pre-NASA PARD concept work and STG Special Assistant to the Director
- Alfred J. Eggers, Jr. – Ames hypervelocity expert
- Harvey Allen – Ames "blunt body" aerodynamicist
- Walter C. William – High Speed Flight Research Station and STG Associate Director.

The first nine biographies were members of NASA Headquarters, and are in approximate order of rank.

T. Keith Glennan
Founding Father and the first NASA Administrator

Born: September 8, 1905
Died: April 11, 1995 (age 89)
Age at NASA: 53
Spouse: Ruth Haslup Adams
Children: one son and three daughters

BACKGROUND

Glennan was born in Enderlin, ND. He attended Eau Claire State Teacher College in Wisconsin, then went to Yale and graduated in 1927 with a B.S. in electrical engineering. He spent the next 14 years working in the film industry in the U.S. and Britain. He was the operations manager for Paramount Pictures and studio manager of the Samuel Goldwyn Studios.

In 1942, he joined the Navy's Underwater Sound Laboratory in New London, CT, where he became the director. After the war, he worked briefly for General Analine and Film Corporation. In 1950, he became President of the Case Institute of Technology in Cleveland, OH. And from October 1950 to November 1952 he was a member of the U.S. Atomic Energy Commission.

NASA

With the encouragement of Abe Silverstein, Glennan met with President Eisenhower and was invited to become the first Administrator of the newly formed NASA, which he did on October 1, 1958. NASA's initial, temporary offices were in the Dolley Madison House near the White House. Many of the original members of NASA worked there. Glennan consolidated the new agency from many different agencies including NACA, Wallops Station, the High Speed Flight Station, the Jet Propulsion Laboratory, and elements of the

Army Ballistic Missile Agency. He also picked up several satellite programs, two lunar probes, and the research efforts of a single-chamber rocket engine. In 1960 he created the Marshall Space Flight Center.

POST-NASA

Glennan left NASA in January 1961 and returned to the Case Institute, serving as its president until 1966, during which period he negotiated the merger of the Institute with Western Reserve University, creating the Case Western Reserve University. He was a trustee from 1970 to 1978 and an honorary trustee until his death in 1995. Glennan was also a U.S. representative on the International Atomic Energy Agency from 1970 to 1973. Thereafter, he spent two years as the President of Associated Universities, Inc. He occasionally consulted with the State Department until 1980. After 20 years of retirement, he died at age 89.

Hugh L. Dryden
Founding Father and the first NASA Deputy Administrator

Born: July 2, 1898
Died: December 2, 1965 (age 67)
Age at NASA: 60
Spouse: Mary Libbie Travers
Children: one son and two daughters

BACKGROUND

Dryden was born in Pocomoke City, MD, the son of Samuel Isaac and Nova Hill Culver Dryden. He was raised in Baltimore and attended public schools, graduating with honors. He earned his way through the Johns Hopkins University and graduated with honors. There he was influenced by Dr. Joseph S. Ames, a pioneer in aerodynamics; the Ames Research Center was named in his honor. Dryden then continued his graduate studies and also worked at the Bureau of Standards in the area of fluid dynamics. This work was credited toward his Ph.D. in physics and mathematics.

In 1920, he was promoted to head of the Bureau's Aerodynamics Section. He made some of the earliest studies of airfoil characteristics close to the speed of sound. In 1934 he became the Chief of the Mechanics and Sound Division. In 1946 he was made the Assistant Director of the Bureau, and six months later became its Associate Director. In 1938 he was selected to deliver the Wright Brothers Lecture before the Institute of the Aeronautical Sciences on the subject of "Turbulence and the Boundary Layer."

In WW-II he served on several technical groups advising the Armed Forces on aeronautical matters and guided missiles. In 1945 the Army presented him with its Medal of Freedom for "an outstanding contribution to the fund of knowledge of the Army Air Force with his research and analysis of the development and use of guided missiles by the enemy."

NACA

In 1947 Dryden resigned from the Bureau of Standards to become the Director of Aeronautical Research at NACA and two years later was appointed the overall Director. In 1948 he won the Presidential Certificate of Merit for work undertaken during the war and was also appointed an honorary officer of the Order of the British Empire.

Dryden was a member of several scientific advisory committees. Most of his research dealt with the properties of airfoils at high speeds, wind tunnel investigation, boundary layer effects and turbulence, noise suppression, and other aeronautical matters. He was involved in much of the post-war research of high speed aircraft and also involved with many Air Force studies for sending a man into space. He was involved with the research at the Wallops Island Station and the Pilotless Aircraft Research Division of the NACA Langley laboratory. After the launch of Sputnik by the Soviet Union he chaired a meeting of all of the NACA laboratory directors and associate directors in December 1957 to discuss the future role of NACA; this resulted in each laboratory proposing ideas for a space initiative.

Dryden's work at NACA led him to Washington to guide the planning that would extend its traditional role in aeronautics to the new field of astronautics. He participated in all the high-level meetings in Washington that were planning the space program, including the President's Scientific Advisory Committee.

NASA

Shortly after President Eisenhower signed the Space Act into law on July 29, 1958, Dryden met with ARPA and the DOD to discuss the management of space efforts and negotiate the transfer of programs and funds from the military to the new agency. On August 15, 1958, Dryden was confirmed by the Senate as the Deputy Administrator of what would become NASA on October 1, providing continuity as that absorbed and expanded NACA. He and Administrator Glennan began working on the proposed organization. Dryden would function as NASA's scientific and technical overseer. He also testified before Congress in order to obtain the first funding for the new agency. In the early months, Dryden and his staff drafted a Ten Year Plan which included Apollo. He negotiated the transfer of the Air Force's F-1 rocket engine to the Apollo effort at the new Marshall Space Flight Center. He also worked on the creation of the Goddard Space Flight Center, which had to set up the tracking stations around the world. When James Webb became the new NASA Administrator in January 1961, Dryden was retained as his Deputy. Dryden also served on the United Nations Ad Hoc Committee on the Peaceful Uses of Outer Space which led to joint space research projects of mutual international interests.

Dr. Dryden died in office on December 2, 1965 at the young age of 67, after maintaining a rigorous schedule despite a lengthy illness. His greatest honor was the renaming of the NASA Flight Research Center on March 26, 1976 as the Hugh L. Dryden Flight Research Center.

For a more complete history of this great man, read Michael H. Gorn's 1996 book *Hugh L. Dryden's Career in Aviation and Space*, which is available on line from the NASA History Office.

John A. Johnson
Founding Father and the first NASA General Counsel

Born: 1915
Died: 2005 (age 90)
Age at NASA: 43
Spouse: Harriet Nelson

BACKGROUND
Johnson was born in Milwaukee to a family of Norwegian descent. He was raised in Hammond, Indiana, graduated from DePauw University in 1937, then attended the University of Chicago's Law School and gained a law degree in 1940. He practiced law in Chicago until he was drafted into the Navy during the war and served aboard the USS *Robert H. Smith* (DM-23).

After the war he went to Harvard for a master's degree. In 1946 he joined the Department of State as an assistant for international security affairs. In 1949 he joined the Air Force, becoming the General Counsel of the Air Force. While living in Falls Church, Virginia, he was active on the school board, argued in favor of desegregation and opposed the call by Senator Harry Byrd for resistance to school desegregation.

NASA
In October 1958 Johnson became the first NASA General Counsel. He was involved in the controversial decision to allow the astronauts to sell their personal stories to Time-Life, Inc.

POST-NASA
Johnson left NASA in 1963 and was appointed as director of international arrangements at the Communications Satellite Corporation (COMSAT), where he eventually became the CEO. He retired in 1980.

Abraham "Abe" Silverstein
Founding Father and Lewis Center Director

 Born: September 15, 1908
Died: June 1, 2001 (age 92)
Age at NASA: 50
Spouse: Marion Croster (deceased)
Children: Two sons and one daughter

BACKGROUND
Silverstein was born in Terre Haute, Indiana in 1908. He attended college in his home town; the Rose Polytechnic Institute where he gained his B.S. in mechanical engineering in 1929 and then his M.S. in engineering in 1934. He was awarded an honorary doctorate from Rose in 1959 and later honorary doctorates from Yeshiva University and from John Carroll University.

NACA
After college, Silverstein joined the NACA Langley Memorial Aeronautical Laboratory in 1929, where he helped to design and later was put in charge of the Full Scale Wind Tunnel. He carried out aerodynamic research that led to increased high speed performance of combat aircraft which saw service in WW-II.

During the war, he was transferred to the Lewis Aeronautical Laboratory in Cleveland, Ohio as Chief of the Wind Tunnel and Flight Division. He directed work in propulsion aerodynamics in the Altitude Wind Tunnel that led to significant improvements in both reciprocation and early turbojet aircraft engines. He also pioneered research on large-scale ramjet engines.

After the war, Silverstein was responsible for the conception, design, and construction of our Nation's first supersonic propulsion wind tunnels, which greatly contributed to the development of present day supersonic aircraft.

In 1949, Silverstein was placed in charge of all research at Lewis, and in 1953 was appointed Associate Director of the Laboratory.

NASA
In the summer of 1958, Silverstein helped to organize NASA and plan its initial programs. On October 1, 1958, when NASA was established, he became the Director of the Office of Space Flight Programs. He was involved with mission planning, spacecraft design and development, and in-flight research and operations. He was also involved with the

development of the Space Act and briefed Congress on the nascent space program. In addition to Project Mercury (which he named, just as he did in the case of Apollo), the programs included unmanned satellites and early deep space missions.

From 1961 until his retirement in October 1969, he was the Director of the Lewis Research Center that is nowadays named the Glenn Research Center. There he directed the research and development of the hydrogen-powered Centaur upper stage which has sent probes out to Mars, Jupiter, Saturn, Uranus, and Neptune.

OTHER
On retirement, Silverstein joined the Republic Steel Corporation as Director of Environmental Planning. He was also a Technical Advisor for the Lake Erie International Jetport Task Force.

During his many decades with NACA/NASA, Silverstein gave many formal papers including the 49th Wilbur Wright Memorial Lecture in London in 1961; the International Council of the Aeronautical Sciences Fight Congress in London in 1966; and also the Biennial Theodore Von Kármán Memorial Lecture at the Tenth Israel Annual Conference on Aviation and Astronautics in Tel Aviv, Israel, in 1968. Amongst his many awards and honors were the following: the Air Force Exceptional Civilian Service Award (1960); the NASA Medal for Outstanding Leadership (1961); the National Civil Service League's Career Service Award (1962); the Sylvanus Albert Reed Award of the AIAA (1964); the Louis W. Hill Space Transportation Award of the AIAA (1967); the Boy Scout Silver Beaver Award; the NASA Distinguished Service Medal; and the Rockefeller Public Service Award (1968).

Ira H. Abbott
Founding Father

Born: July 18, 1906
Died: November 3, 1988 (age 82)
Age at NASA: 52
Spouse: Martha Leola Streeter
Children: Three

BACKGROUND
Abbott was born in Wolfeboro, New Hampshire and graduated from Newburyport High School in Massachusetts.

NACA

On graduating from MIT in 1929 with a B.S. in aeronautical engineering, Abbott immediately joined the NACA Langley Memorial Aeronautical Laboratory at Langley Field,
Virginia, as a junior aeronautical engineer. His early work included writing technical
reports on aerodynamics. He was instrumental in setting up programs for high speed
research. By the early 1930s Abbott was involved in designing the Two-Dimensional Low
Turbulence Tunnel, and then in 1933 he worked on the design of the 24-inch High Speed
Tunnel which was used to investigate airfoils and fuselages. In 1941 he was made Assistant
Chief of the Airflow Research Division. In 1944 he became the Assistant Chief of the Full
Scale Research Division and the next year he became the Assistant Chief of the Research
Department.

   In 1948 Abbott was transferred to the Headquarters of NACA in Washington to serve
as an aeronautical consultant. By 1950 he was the Assistant Director of Research for
Aerodynamics. He supervised the X-15 rocket plane, supersonic transport aircraft, nuclear
rocket engine, and advanced re-entry programs. From 1957–1958 he was Chairman of the
Wind Tunnel Panel of NATO's Advisory Group on Aeronautical Research and
Development.

NASA

When NASA was established on October 1, 1958, Abbott was appointed Assistant Director
of Research for Aerodynamics and Flight Mechanics. He was involved with the appointment of members to the NACA steering committee that worked with the Air Force to
cooperate in the joint manned satellite venture. Abbott was very much involved with
addressing how the 8,000 NACA employees would organizationally transform into
NASA. It was Dryden, Abbott, and John W. Crowley who gave Robert Gilruth the authority to proceed with Project Mercury and form the STG. Abbott retired from NASA
Headquarters in 1962 after 32 years of service with the NACA and NASA. Abbott was a
Fellow of the Institute of the Aeronautical Sciences and a member of the New York
Academy of Sciences.

John W. Crowley
Founding Father

Born: 1899
Died: 1994 (Age 95)

BACKGROUND
"Gus" Crowley graduated from MIT with a B.S. in mechanical engineering in 1920 and joined the Langley Memorial Aeronautical Laboratory in 1921.

NACA
Crowley was involved in flight research and became Head of the Research Department in 1943. In 1948 he was made Associate Director of Research under Hugh Dryden. During that period he published many papers on a variety of aerodynamic problems. He is referenced in many NACA publications and technical reports. In 1945, he transferred to NACA Headquarters as Associate Director of Research.

NASA
When NASA was formed in 1958, Crowley was appointed Director of Aeronautics and Space Research. He was on the team headed by Abe Silvestein which drafted the initial plans for the new space agency. On his retirement in 1959, Crowley became the first recipient of the NASA Distinguished Service Medal.

Newell D. Sanders
Founding Father

Born: October 29, 1914
Died: October 10, 2011 (Age: 96)
Spouse: Thelma Burrows

BACKGROUND
Newell and his twin brother John were born in the southwestern town of Saltville, Virginia. As they grew up they were known as the "Mad Scientists" of Saltville because of their experiments and antics. Sanders later attended the Virginia Polytechnic Institute and earned a B.S. and M.S. as well as the rank of second lieutenant in the ROTC, then he joined NACA.

NACA Langley/Lewis

Sanders joined the Langley Memorial Aeronautical Laboratory in 1938, then he followed Abe Silverstein to the Lewis Flight Propulsion Laboratory when that was established in 1941. By 1945, he was an assistant to Silverstein, who was the Division Chief of the Wind Tunnels and Flight Division. In 1949 Sanders became the Head of the Physics Division and worked on fuels for military aircraft.

NACA Headquarters

In 1957, Sanders followed Abe Silverstein to Washington to help transform NACA into NASA. He worked on the development of early communications and weather satellites, and assisted in the planning of future concepts. He also served on the President's Special Committee to Study Aircraft Noise. During the transition from NACA to NASA, he was the Assistant Director for Advanced Technology. In 1962, he returned to Lewis and worked on a "quiet engine" program.

OTHER

After retiring from NASA in 1972, Sanders and his wife traveled to six of the seven continents and joined several societies at the Natural History Museum. Among his many honors and awards are the NASA Exceptional Service Medal and the Greater Cleveland Growth Association Career Service Award.

George M. Low
Founding Father

Born: June 10, 1926
Died: July 17, 1984 (age 58)
Age at NASA: 32
Spouse: Mary Ruth McNamara
Children: Three boys and two girls

BACKGROUND

George Michael Low was born in Vienna, Austria, to Artur and Gertrude Burger Low. In 1940 the family moved to the U.S. After graduating from Forest Hills High School in New York he entered the Rensselaer Polytechnic Institute in Troy, New York. He became a naturalized U.S. Citizen while in the Army between 1944 and 1946. After the war he returned to Rensselaer and received a B.S. in aeronautical engineering in 1948. He briefly worked for General Dynamics-Convair in Fort Worth, TX, then gained an M.S. in aeronautical engineering from Rensselaer in 1950.

NACA

Armed with his M.S. degree, Low joined NACA as an engineer at the Lewis Flight Propulsion Laboratory in Cleveland, OH. He was made Head of the Fluid Mechanics Section in 1954 and Chief of the Special Projects Branch in 1956. He specialized in experimental and theoretical research in the fields of heat transfer, boundary layer flows, and internal aerodynamics. He also worked on space technology problems such as orbit calculations, space rendezvous techniques, and re-entry paths. In the summer of 1958 Low worked on a planning team on the organization of NASA.

NASA

In October 1958 Low transferred to NASA's initial Headquarters in the Dolley Madison House, where he served as Chief of Manned Space Flight. He was also a member of the NASA-ARPA Manned Satellite Panel. He served as the primary interface between Headquarters and the field centers for planning the space program and worked closely with Robert Gilruth at the STG. He also worked with the Administrator Keith Glennan, the Director of Space Flight Development Abe Silverstein, and the Assistant Director for Advanced Technology Newell Sanders.

During the astronaut selection criteria process, Low advocated the test pilots' point of view that the success of the mission might well depend on the actions of the pilot, and hence jet test pilots were best suited for the role. Low was involved with the decision to direct McDonnell to design the Mercury capsule to use both a beryllium heat shield and an ablative heat shield. He was involved with reporting to Administrator Glennan on the progress of technical issues and flight schedules for Mercury. He attended design reviews, inspections at McDonnell, and the capsule coordination panel.

Low was also involved with the status and ongoing issues concerning the various rocketry aspects of the program, including the Little Joe, Big Joe, Scout (which he recommended to be canceled), Redstone, and Atlas. He participated in the debates on program reliability, systems analysis for the capsule, and man-rating the launch vehicles. He was involved in how failures were analyzed and worked. He often attended launches at Cape Canaveral.

POST-MERCURY

Low remained at NASA supporting Mercury, Gemini, and Apollo; transferring to the Manned Spacecraft Center in Houston in February 1964 as Deputy Center Director. After the Apollo 1 fire in 1967 he became the Manager of the Apollo Spacecraft Program Office, responsible for directing the changes to the Apollo spacecraft in order to make it flight worthy.

Low became the NASA Deputy Administrator in December 1969, serving Administrators Thomas O. Paine and James C. Fletcher. He was involved with the development of the Space Shuttle, Skylab, and the Apollo-Soyuz Test Project.

OTHER
On retirement in 1976 after 27 years with NACA/NASA, Low became President of his alma mater, the Rensselaer Polytechnic Institute, a position that he held to his early death in 1984.

Among Low's many awards is the Presidential Medal of Freedom. There is also an award named after him; the George M. Low Award which is given by NASA to its contractors and subcontractors for quality and excellence.

Warren J. North
Founding Father and Member of the STG

Born: April 28, 1922
Died: April 10, 2012 (89)
Age at NASA: 36
Spouse: Mary
Children: one boy and one girl

BACKGROUND
North attended the University of Illinois from 1940 to 1943 and enlisted in the ROTC. During the war he attended engineer officer's training school and was then sent to Officer Candidate School. In November 1944 he received a commission as a fighter pilot. He was an instructor pilot and flew B-17s, twin P-51s, and B-29s. After the war he went to Purdue University and received a B.S. in aeronautical engineering in 1947.

NACA
North joined the NACA Lewis Flight Propulsion Laboratory and worked with Abe Silverstein for several years. He then became an engineering test pilot and flew most of the WW-II fighters and bombers. North also dropped experimental ramjets over Wallops Island to be monitored by Langley's Pilotless Aircraft Research Division.

While at Lewis, North attended night school at the Case Institute of Technology and received instruction from Keith Glennan, who later became the first NASA Administrator. He obtained an M.S. in theoretical aeronautical engineering in 1954. With an AIAA fellowship, North attended Princeton and graduated with an M.S. in applied aeronautical engineering in 1956. He was then made Assistant Chief of the Aerodynamics Noise Branch at Lewis and worked on a hydrogen-powered second stage for the Sergeant solid rocket. He led the Stability Group of the Mission Design Panel.

It was in the summer of 1958 that Abe Silverstein was transferred to NACA Headquarters to initiate the planning for the civilian space program. North was a member of a group from Lewis that commuted to Washington to assist Silverstein in that task.

NASA/STG

When NASA was established in October 1958, North moved to NASA Headquarters as Chief, Manned Satellites. As a former test pilot, he was a member of the group led by Charles Donlan that selected the Mercury astronauts. He not only closely monitored that process, he went to the Lovelace Clinic and subjected himself to the same physical evaluations. North was initially the Headquarters representative on Robert Gilruth's STG staff for astronaut training and activities. He permanently moved over to the STG in 1962 as Chief of the Flight Crew Support Division, and was involved in the selection of the second and third groups of astronauts, was involved in the development of advanced astronaut simulators, and was involved in all aspects of flight crew activities including the lunar landing simulators. In 1971 he was made the Assistant Director for Space, a post that he held until 1985.

POST-NASA

North took a special interest in his aunt Romalda Spalding and her work in an effective method of teaching reading, and supported the Spalding Educational Foundation for advanced literacy. He was a member of many professional associations and received many awards for his work on the space program, including the Institute for Advance Study Flight Test Engineering Award and the AIAA De Florez Training Award.

– O –

The following biographies of Founding Fathers are those from the NASA field centers, listed in alphabetical order.

Dr. H. Julian Allen
Founding Father

Born: April 1, 1910
Died: January 29, 1977 (age 66)
Age at NASA: 48

BACKGROUND
Allen was born in Maywood, Illinois. The family moved to Palo Alto, California. As a young man in high school, he was interested in electrical engineering. Attending Stanford University, he discovered his life-long love of aerodynamics. After earning his B.A.in engineering in 1932 he graduated with an aeronautical engineering degree in 1935. He worked as a junior engineer for Shell Chemicals for six months while looking for a job in aerodynamics.

NACA
Allen joined the NACA Langley Memorial Aeronautical Laboratory in 1936 as a junior engineer in the Variable Density Tunnel Section working on laminar flow low-drag airfoils. His personal interest was making wind tunnels more efficient and versatile. In 1940, this work led him to the newly established Ames Aeronautical Laboratory in California.

Throughout WW-II, Allen was Chief of the Theoretical Aerodynamics Section at Ames. This work and his experience at Langley included the application of laminar-flow airfoils to the P-51 Mustang. He suggested a new type of wind tunnel in which to test airfoil shapes for aircraft that was approaching the speed of sound. By 1945, Allen (whose nickname was "Harvey" from the invisible rabbit in the Broadway Play) was Chief of the High Speed Research Division and was not only thinking of how to break the sound barrier but also about hypersonic flight. This led to the aerodynamics and thermodynamics associated with ballistic missiles. As an advisor on the ICBM program, Allen worked on the problem of a warhead re-entering the atmosphere. As the warhead slowed, much of its kinetic energy would be converted into heat that was sufficient to melt a pointed nose cone. This led to the "blunt body" theory that is associated with his name, and which applied directly to the re-entry of a capsule as well as a warhead. Allen teamed with fellow Ames aerodynamicist Dr. Alfred J. Eggers to refine the theory and then they published a formal paper called "A Study of the Motion and Aerodynamic Heating of Missiles Entering the Earth's Atmosphere at High Supersonic Speeds" in 1953.

In 1955 the Air Force chose the "blunt body" shape for the warhead of its Atlas ICBM. But this was classified work and didn't become public until 1957. Allen's work in hypersonics was applied to the X-15, and picked up by NACA Langley engineer Max Faget when investigating the design of a manned capsule.

NASA

In 1959 Allen became the Assistant Director for Astronautics responsible for finding ways for Ames experts and facilities to help get America into space. In October 1965 Allen became the Ames Center Director, a post he held until retirement in 1969. His "blunt body" showed up in Mercury, Gemini, Apollo, Soyuz, and now the new Orion spacecraft.

OTHER

Among Allen's many awards and honors are the NACA Distinguished Service Medal and the NASA Medal for Exceptional Scientific Achievement.

Charles J. Donlan
Founding Father and Member of the STG Core Team

Born: July 15, 1916
Died: September 28, 2011 (age 95)
Age at STG: 42
Spouse: Marguerite Phelan
Children: Two sons

BACKGROUND

Donlan was born in Lawrence, Massachusetts. He graduated from MIT in 1938 with a B.S. in aeronautical engineering. He received an honorary doctorate of aeronautical engineering from Ohio Northern University in 1972.

NACA

Donlan worked at the NACA Langley Memorial Aeronautical Laboratory for 30 years, rising to become the Deputy Director of the NASA Langley Research Center. In the late 1930s and 1940s, the period leading up to and during WW-II, he worked in the Stability Analysis Group using the Langley Spin Tunnel. His work in the 7–10 foot 500 MPH tunnel included jet aircraft. This work coupled with that of other aerodynamicists led him to work on the F-102, F-106 and D-558-I. By 1955 he was the Technical Assistant to the Associate Director of Langley.

STG

Even before the official creation of the STG, many people at Langley were getting prepared to work with industry to build the Mercury capsule. It wasn't the type of work typically done by Langley researchers and engineers. Upon the 1958 decision to go ahead with a manned space program, Robert Gilruth chose Donlan as his Assistant Director of Project Mercury in the new STG organization. The STG would have to have an Engineering and Contract Administration Division. One of Donlan's first major efforts was to oversee the procurement process for the Mercury capsule. This required reviewing proposals and setting up a Source Selection Board. Donlan chaired the Technical Assessment Committee and appointed eleven teams to perform assessments. He coordinated the Board's progress with NASA Headquarters. The selection of McDonnell Aircraft was made and on February 5, 1959 the contract was signed.

Another of Donlan's major contributions was setting up the criteria for astronaut selection, a task that he did with former NACA test pilot Warren North of NASA Headquarters and Allen O. Gamble, a physiologist from the National Science Foundation. This process involved examining the records of hundreds of military test pilots for consideration as astronaut candidates and then conducting interviews. With the help of Dr. Stanley White, Warren North, and Robert Voas, the selection process narrowed the field down to the final seven. Robert Gilruth approved Donlan's recommendation.

After completion of the STG work and the creation of the Manned Spacecraft Center in1961, Donlan returned to Langley as Deputy Director with major responsibility of its aeronautical and space activities.

NASA HEADQUARTERS

In 1968 Donlan moved to NASA Headquarters and became the Deputy Associate Administrator in the Office of Manned Space Flight. Then from 1970 through to 1973 he was the first Director of the Space Shuttle Program.

OTHER

After retiring from NASA in 1976, Donlan provided consultation to the Institute for Defense Analysis and then served on the NASA Aerospace Safety Advisory Panel from 1983 to 1994. His many awards include the NASA Outstanding Leadership Medal; the Exceptional Service Medal; and the NASA Distinguished Service Medal (twice). His other awards may be seen in various NASA publications.

Dr. Alfred J. Eggers, Jr.
Founding Father

Born: June 24, 1922
Died: September 22, 2006 (age 84)
Age at NASA: 36

BACKGROUND
Eggers was born in Omaha, Nebraska and graduated from the University of Omaha with a B.A. in 1944. He completed the V-12 Navy College Training Program, but didn't serve in the Navy. After graduation he joined the NACA Ames Aeronautical Laboratory.

NACA
Eggers' early work included hypersonics and the development of new wind tunnels. In 1954 he was made Chief of the Vehicle Environment Division, which consisted of the Physics Branch, Structural Dynamics Branch, 3.5 foot Hypersonic Wind Tunnel Branch, Hypervelocity Ballistic Range Branch, and Re-entry Simulation Branch. He became known for supersonic flight and re-entry physics. He worked with Harvey Allen on the "blunt body" theory. In 1956 he earned his Ph.D. from Stanford University and also headed the Manned Satellite Team that was to review design problems and propose a practical system for a satellite while recommending a suitable research program. This ultimately led to Ames developing and managing the highly successful Pioneer planetary missions.

NASA
Eggers designed and built the Atmospheric Entry Simulator which was used to test the re-entry heating of the Mercury capsule. He also worked on supersonic interference lift, work which led directly to the XB-70 Valkyrie supersonic bomber. In 1959 he became the Assistant Director at Ames.

In 1964 Eggers was appointed Deputy Associate Administrator for Advanced Research and Technology at NASA Headquarters. In 1968 he became the Assistant Administrator for Policy and served until retirement in March 1971. During this period he was also visiting professor at MIT.

OTHER
After NASA, Eggers was made the Assistant Director for Research Applications at the National Science Foundation. His many awards included the Arthur S. Flemming Award in 1957 and the Outstanding Alumni Award from the University of Omaha in 1959.

Maxime A. Faget
Founding Father and Member of the STG Core Team

Born: August 26, 1921
Died: October 9, 2004 (age 83)
Age at STG: 37
Spouse: Nancy Carastro
Children: 3 girls and 1 boy

BACKGROUND
Faget was born in Stann Creek, British Honduras. His father, Dr. Guy Henry Faget, who was a pioneer in the treatment of leprosy, served as the director of the National Leprosium in Carville, Louisiana, where the family lived. "Max" Faget attended the San Francisco Junior College from 1939–1940 and then obtained a B.S. in mechanical engineering from Louisiana State University in 1943. He joined the Navy and served on submarines for three years during WW-II, and saw combat. In 1946 he joined the NACA Langley Memorial Aeronautical Laboratory as a research scientist. He holds an honorary doctorate of engineering from the University of Pittsburgh and Louisiana State University.

NACA
Faget worked in hypersonic aerodynamics under Robert Gilruth at Langley's Pilotless Aircraft Research Division. This was the era of research and testing at the Wallops Island Station which was created in 1945. He became Head of the Performance Aerodynamics Branch in 1951, and remained there until the establishment of the STG in 1958.

Faget's research at Langley dealt with high Mach numbers, high temperature structures and high dynamic loads. This was in the era of the X-Series of aircraft, as well as various rockets, most of which were tested at Langley either in the wind tunnels or with models launched from Wallops Island Station. Faget worked on a variety of hypersonic design concepts including the X-15. He studied the pioneering work of H. Julian "Harvey" Allen, including his "blunt body" theory for re-entry vehicles. Allen had preceded Faget at Langley and transferred to Ames in 1940, where he was now Chief of the Theoretical

Aerodynamics Branch. Faget applied Allen's theory to the Mercury capsule, and after wind tunnel testing he devised the final design for its overall shape. Faget and several other Langley/STG people took out the patent on the "Space Capsule." Faget also holds a patent for a Mach Number Indicator.

Concerned about the dynamic loads that the pilot would endure during launch and re-entry, Faget conceived the idea of an astronaut's couch that would support the body to minimize the effects of the anticipated "g" loads. A prototype couch was built by the West Model Shop with the assistance of NACA test pilot Robert Champine. Tests were then undertaken at the Navy's Johnsville centrifuge with several pilots. After the evaluations, the West Model Shop designed couches for all seven Mercury astronauts. Faget and other Langley/STG people hold the patent on the "Survival Couch."

Shortly prior to the creation of the STG, Faget joined Robert Gilruth and others to write the plan for the space agency that would later be presented to the President's Scientific Advisory Committee.

STG

After the creation of the STG in 1958, Faget was selected by Gilruth to serve on the initial Core Team of the STG. He became Chief of the Flight Systems Division, a group that included many from this initial list. It was this group that developed the escape tower system and launched the test articles on Little Joe rockets from Wallops Island Station. Faget holds a joint patent on the "Aerial Capsule Emergency Separation Device." He supported the rigorous flight test program throughout the three year period of the STG's existence.

POST-STG

After the establishment of the Manned Spacecraft Center in November, 1961, Faget became the Director of Engineering and Development, with responsibility for Gemini and Apollo. By 1972 he had filed a patent for the design of a Space Shuttle. He retired from NASA in 1981 but went on to become a co-founder of Space Industries, Inc., co-inventing five patents for that company between 1988 and 1992. Faget's honors and awards are too numerous to mention but the can be read in various NASA history books and on the Wikipedia website.

Robert R. Gilruth
Founding Father and Director of the STG and Project Mercury

Born: October 8, 1913
Died: August 17, 2000 (age 86)
Age at STG: 45
Spouse: Jean Barnhill
Children: one daughter

BACKGROUND

Gilruth was born in Nashwauk, Minnesota. Both his parents were educators. He graduated from Duluth Central High School in 1931. During the Depression, Gilruth first went to a local junior college and then to the University of Minnesota. After gaining a B.S. in aeronautical engineering in 1935 and an M.S. in 1936 he joined the NACA Langley Memorial Aeronautical Laboratory in early 1937. Most of his work prior to and during WW-II was on the subject of supersonic flight. He received an honorary doctorate of laws from the University the New Mexico State University in 1970.

NACA

Early in his career with NACA Langley, and throughout the war, Gilruth worked with research pilots and aircraft designers to document flying characteristics. This required pioneering work in the data recording of instrumented flight tests. In 1945, Gilruth organized a group to research transonic and supersonic flight using rocket-powered models, and this led to the creation of the Pilotless Aircraft Research Division and the launching facility at Wallops Island, Virginia. This effort had a significant impact on what would become the Mercury capsule. Many future STG employees got their early experience in this organization.

In 1952 Gilruth was appointed Assistant Director of NACA Langley and directed research in hypersonic aerodynamics at the Wallops "Station" (as it was called then), as well as researching high temperature structures and dynamic loads. The military exploited this research for missile and payload applications, and it was directly applicable to the design of re-entry vehicles.

STG

Even prior to Sputnik, Gilruth was deeply involved in the NACA Headquarters studies about a manned space capsule. In the wake of Sputnik, he was involved in the highest level discussions about the creation of a new space agency. Shortly after NASA was created in October 1958 he was appointed Director of the Space Task Group.

Immediately after NACA was transferred to NASA, Gilruth changed from a basic research engineer into a major program manager. He and his technical staff invented the organizational structure of the STG. He hired the best engineers and scientists from Langley, Lewis, Ames and the High Speed Flight Station (as it was known back then). All three military services had their own space-related projects, so Gilruth added military representatives to his staff in order to take advantage of their experience in implementing Project Mercury. He also hired people from the AVRO aircraft company in Canada, as well as other British and Canadian flight test engineers. In less than a year, the STG was comprised of three major organizations: the Flight Systems Division, the Operations Division, and the Engineering and Contract Administration Division. Astronauts and other staff and administration functions reported to the Office of the Director. When new hires came into the STG over the next three years, they were integrated into these three Divisions. See Appendix 1 for the STG's organizational structure.

POST-STG
In November 1961 the STG was folded into the new Manned Spacecraft Center, with Gilruth remaining in charge. During a tenure of over 10 years Gilruth directed 25 manned spaceflights starting with the first manned suborbital Mercury flight in 1961 through to Apollo 15 in 1971, with the highlight being the Apollo 11 lunar landing in 1969.

A little known fact about Dr. Gilruth is his conversation with President Kennedy, when the latter was quite frustrated by all the Soviet "firsts" in space. Gilruth pointed out that if we were to beat them, we would need to do something which was so "big" that they wouldn't be able to achieve it first by virtue of having more powerful rockets; like going to the Moon.

Charles W. Mathews
Founding Father and Member of the STG

Born: 1921?
Died: September 10, 2001 Age 80
Age at STG: 37
Spouse: Marietta Gaynelle Short
Children: 1 son and 1 daughter

BACKGROUND
Charles "Chuck" W. Mathews was born in Gainesville, Florida, and his family later moved to Duluth, Minnesota. He graduated from the Rensselaer Polytechnic Institute in 1942 with a B.S. in aeronautical engineering, then joined the NACA Langley Memorial Aeronautical Laboratory in 1943.

NACA
Mathews was an aeronautical engineer in the Flight Research Division at Langley. His area of research was primarily stability and control. In 1947 he was appointed Assistant Head of that Section.

During WW-II, Mathews worked on drag measurements of freely falling bodies, and drag measurements of straight and sweptback airfoils.

In early 1957, still in the Flight Research Division, he became interested how an astronaut could safely return from suborbital and orbital flights. He studied the "blunt body" research of Harvey Julian Allen at Ames and calculated the "g" loads that would be placed on a crewman flying various re-entry trajectories.

After Sputnik, and in early 1958, Mathews' group worked on the delta-wing, high angle of attack approach to re-entry. This work showed that the vehicle would weigh too much for the launch vehicles available at that time. This favored the "blunt body" design being studied by Max Faget. In the spring of 1958 (long before the establishment of the STG) Mathews' group pulled together all the research material on the capsule, and developed a specification for the procurement process. This was subsequently presented to the President's Scientific Advisory Committee, and later served as the basis of the formal Statement of Work for the Request for Proposals to industry to develop the Mercury capsule. In the summer of 1958 Mathews was a member of the team that met at the Dolley Madison House to plan the new space agency.

STG

The STG was established on November 3, 1958; a month after NASA itself. Mathews was part of the Core Team and was made Chief of the Operations Division, where he remained until the organization of the Manned Spacecraft Center. His STG work during Project Mercury included managing four major Branches that were involved in all mission planning, pre-launch checkout, flight operations, and recovery of the capsule. He also obtained many of the AVRO engineers and Lewis engineers. In addition, he was responsible for coordinating with the Army Ballistic Missile Agency, the Atlantic Missile Range, McDonnell, and Convair. During this period, the Division was also responsible for training flight controllers for the Mercury Control Center and the remote tracking stations around the world.

OTHER

Mathews was Chief of the Operations Division throughout Project Mercury. After the move to the Manned Spacecraft Center in 1962 he became Chief of the Spacecraft Technology Division for about one year before becoming Deputy Assistant Director, Engineering and Development. When he was made Manager of the Gemini Program Office he worked on the unique problems of that program, including spacecraft fuel cells, rendezvous thrusters, ejection seats, the "pogo" vibrations of the Titan launch vehicle, and the engine instability of the Agena target vehicle.

In 1966 he moved to NASA Headquarters as Director of the Apollo Applications Program (which eventually developed the Skylab space station), and then became the Deputy Associate Administrator of the Office of Manned Space Flight in 1968 and Associate Administrator of the Office of Applications in 1971.

Mathews retired in 1976. Amongst his many awards and honors were the NASA Exception Service Award (1965) and the Distinguished Service Medal which he received twice (1966 and 1969).

Paul E. Purser
Founding Father and Member of the STG Core Team

Born: 1919
Died: September 8, 2003 (age 84)
Age at STG: 39
Spouse: Dottie

BACKGROUND

After high school, Purser graduated from Louisiana State University with a B. S. in aeronautical engineering in 1939; Walter C. Williams was one of his classmates. Purser briefly
worked for the Glenn L. Martin Company, but joined the NACA Langley Memorial
Aeronautical Laboratory in that same year. While at Langley, he attended the University
of Virginia Extension as a graduate student from 1942–1944.

NACA

Purser's early work at Langley was in the area of high temperatures on various metals,
and he tested the effects of temperature on aircraft using both low speed and supersonic
wind tunnels. During WW-II he tested models of most of the aircraft of that era. When
the Pilotless Aircraft Research Division was created in 1945, he worked with Robert
Gilruth, Max Faget, Caldwell Johnson, Guy Thibodaux, and others in testing model
aircraft on rockets at the Wallops Island Station. He gained experience in general aerodynamics and structures and analyzed telemetry from the flights. He analyzed the impact
of aerodynamic heating at high Mach numbers; work that was directly applicable to
spaceflight. He also worked on the Little Joe solid rocket which was used for Mercury
capsule tests.

   In 1957, Purser attended a major – now historic – NACA conference at the Ames
Research Laboratory that discussed the X-15 rocket plane, the Dyna-Soar space plane, and
putting a man into space in a capsule. The launch of Sputnik during this conference had a
major influence on everyone present. It was there that the three NACA laboratories and the
Air Force exchanged ideas about spaceflight.

   Purser served on many committees while at NACA, including the Army Picatinny
Arsenal Fin-Stabilized Ammunition Committee 1955–1956, the NACA Special Advisory
Group on the Navy Polaris Missile 1957, and the NASA Research Advisory Committee on
Structures 1960.

STG

During several months in the summer of 1958, before the creation of NASA, Purser joined the team at NACA Headquarters that defined the basic elements of Project Mercury. This was the work that went before the President's Scientific Advisory Committee and later to Congress for the Space Act. When the STG was formed in November 1958, he became the Special Assistant to Robert Gilruth. In this role, he was a member of the team which went to AVRO to interview the recently laid off engineers. Purser became the "Go To" guy for Gilruth for any issue which needed to be resolved.

POST-STG

In 1962 Purser joined the Manned Spacecraft Center, where he remained until 1971. He took a year off in 1969 to be the Special Assistant to the President of the University of Houston. After retiring from NASA in 1971 he did some consulting work for industry and also for the National Academy of Engineering in Washington, DC.

With Max A. Faget and Norman F. Smith, Purser co-edited *Manned Spacecraft: Engineering Design and Operation*, published by Fairchild Publications in 1964. He published a great many papers while with NACA/NASA. His awards include the NASA MSC commendation for work on the space program and also the NASA Group Achievement Award (twice, 1964 and 1967).

Henry J. E. Reid
Founding Father and NACA Langley Director

Born: August 20, 1895
Died: July 1, 1968 (age 72)
Spouse: Mildred Woods
Children: one boy and one girl

BACKGROUND

Born in Springfield, Massachusetts, Reid attended a technical high school there. This was a world without flying machines and before there was an aeronautical discipline or a research center. Reid graduated from Worcester Polytechnic Institute in 1919 with a B.S. in electrical engineering. In 1946 he was awarded an honorary doctorate in engineering.

NACA

After a brief stint in private industry Reid joined the NACA Langley Memorial Aeronautical Laboratory in 1921. His principal field of research was the design and improvement of basic instruments for flight research. He was promoted to Engineer-in-Charge in 1926, from which position he managed the rapid growth that came with the development of the laboratory as a leading research facility. In 1929 Langley's Fred Weick won the Collier Trophy for low-drag engine cowlings, and Langley was involved in three more during Reid's tenure.

During WW-II, Langley focused upon specialized research to ensure Allied air supremacy, directed a major expansion in facilities and personnel, and selected and trained staff who were destined to establish and lead new NACA laboratories. At the request of the Army, Reid toured France and Germany as Chief of the Alsos Mission; a scientific intelligence group charged with seeking information on Axis progress in scientific research and development. It was part of the Manhattan Project. Reid's focus was on the German aeronautical programs.

In 1948 Reid's title was changed by Public Law 167 to "Director," in response to Congress' authorizing ten professional and scientific positions to supervise and direct the scientific study of the problems of flight with a view to practical solutions. Reid resided over the extensive growth that accompanied Langley's development as a leading aeronautical and space research facility.

NASA

Reid served as Director of the Langley Research Center until 1960. Even as NASA was being created in 1958, he turned over many of his duties to Floyd L. Thompson in preparation for his retirement, which he took in 1961. He monitored the transition of Langley assets and personnel to the STG for Project Mercury. He saw the fruits of Langley's work benefit Mercury, Gemini, and a few Apollo flights. He died in 1968.

HONORS

Reid's honors include the War Department's Medal of Freedom for exceptional service during WW-II; the Medal of Merit awarded by President Truman for assisting the U.S. in attaining and maintaining air supremacy; and the National Safety Council Award of Honor in 1953.

Joseph Guy Thibodaux, Jr.
Founding Father and Consultant to the STG

Born: 1921
Age at STG: 37
Spouse: Mary Jo Goliwas
Children: Four

BACKGROUND
"Guy" Thibodaux grew up in the New Orleans, Louisiana area. After high school he attended Louisiana State University, graduating with a B.S. in chemical engineering in 1942. He was a classmate of Max Faget and Walter C. Williams. During WW-II he was an officer in the Army Corps of Engineers and participated in building a road from India to China, gaining first-hand knowledge of the use of explosives.

NACA
Thibodaux joined NACA Langley in 1946 as a propulsion engineer in the PARD. This was just after the time that Wallops Island was built. He worked with the engineers to provide them with the types of rockets that they needed in order to conduct their research. He would find the best rockets available and, if necessary modify them for the test launch. He was Head of the Model Propulsion Section from 1949–1955 and Head of the High Temperature Materials Section from 1955–1958.

STG
In the summer of 1958 Thibodaux was a member of the team at NACA Headquarters that was tasked with planning an agency to run a civilian space program. When the STG was formed in November, he stayed at Langley as Head of the High Temperature Materials Section but was a consultant to the STG Director Robert Gilruth for matters related to propulsion; these included the Mercury capsule retrorockets, posigrade rockets, escape tower, and pyrotechnics. When the STG became the Manned Spacecraft Center, Thibodaux was made Chief of the Propulsion and Power Division; a post he held until his retirement in 1980.

OTHER

In addition to holding four patents on various rocket motors, Thibodaux's honors and awards include the NASA Exceptional Service Medal (twice); membership of the agency's Research Advisory Committee on Chemical Propulsion; various positions at the American Institute of Aeronautics and Astronautics (Associate Fellow, membership of the Technical Committee on Solid Rockets, and Chairmanship of the Houston Section); and the James H. Wyld Propulsion Award.

Floyd L. Thompson
Founding Father and Acting Director of the Langley Research Center

Born: November 25, 1898
Died: July 10, 1976 (age 78)
Age at STG: 60

BACKGROUND

On enlisting in the Navy in 1918, Thompson spent a year in aviation mechanic's school and then joined a torpedo squadron based in Pensacola, FL. Because WW-I was over, he was able to leave with 3½ years of military service in 1922. He graduated from the University of Michigan in 1926 with a degree in aeronautical engineering.

NACA

On graduation, Thompson joined the NACA Langley Memorial Aeronautical Laboratory in the Flight Research Section under John W. "Gus" Crowley. (At that time, there was only one NACA laboratory and its total population was about 250 people.) Much of Thompson's early work was in testing various propeller designs. Later he was involved in instrumentation improvements for aerodynamic measurements in wind tunnels. In the early 1930s Thompson worked on pressure instrumentation for airships and seaplanes. Some of the many people that he worked with over the years were Fred Weick, Max Munk, Elton Miller, and R. T. Jones.

The NACA laboratories at Lewis and Ames were established during WW-II and much of the engine research and propulsion work at Langley was move to the Lewis. People involved with aerodynamic research went to Ames. Thompson worked with local colleges and universities to supply engineers to replace those that left and organized in-house training.

He worked with the supersonic wind tunnel and tested models to study the overpressure that would cause sonic booms. He worked with Dr. Busemann, an expert in supersonic theory, and Tony Ferri, an expert on supersonic tunnels. And he worked with the military services to study their aviation problems. He worked on problems with high wing loadings, and on the stability and control of aircraft diving at high speeds. He was also involved with the selection of Walter C. Williams to head up the High Speed Flight Center which was to test aircraft.

Thompson became Chief of the Flight Research Division in 1940 and Chief of Research in 1945. He was appointed Associate Director of the laboratory in 1952.

NASA/STG

Thompson was not a member of the STG but provided Robert Gilruth, the Director of Project Mercury, with significant support from the newly named Langley Research Center. The STG was a separate organization that reported to NASA Headquarters but needed Langley resources, offices, and people. Thompson worked with Gilruth to select Langley people to transfer to the STG. He received an honorary doctorate from William and Mary University and the University of Michigan in 1963.

OTHER

Thompson was made Director of Langley in 1968. After retiring in 1979 he temporarily served as a special advisor to the NASA Administrator. And over the years he also sat on many boards and committees, including the Apollo 204 Review Board that investigated the Apollo 1 fire. He published scores of reports while at Langley, including one to the Royal Aeronautical Society in 1949 on "Flight Research at Transonic and Supersonic Speeds with Free-Falling and Rocket-Propelled Models." Among his many honors was his appointment as President of the American Institute of Aeronautics and Astronautics.

Walter C. Williams
Founding Father and Member of the STG

Born: 1919
Died: 1995 (76)
Age at STG:39

BACKGROUND
On graduating with a B.S. in aeronautical engineering from Louisiana State University in 1939, Williams joined the Glenn L. Martin Company in Baltimore, Maryland but was there for only a few months.

NACA
He joined the NACA Langley Memorial Aeronautical Laboratory later in 1939, and primarily worked on stability and control designed to improve the handling, maneuverability, and flight characteristics of WW-II aircraft. In 1946 he was assigned as the Project Engineer on the X-1 experimental aircraft program that resulted in Chuck Yeager's breaking of the sound barrier in October 1947. When Langley's Muroc Field Test Unit was expanded into a permanent facility, Williams became the first Chief of the NACA High Speed Flight Station. In this assignment he directed the flight research programs for the Century series of military aircraft and the X-Series of research aircraft through to the X-15.

Williams attended the NACA conference on High Speed Aerodynamics at Ames in March 1958. This was his initial in-depth look into aeronautical research applicable to spaceflight. In August, he was a member of the NASA-ARPA Manned Satellite Panel which drew up specific recommendations for the manned satellite project.

NASA
When NASA was created in October 1958, Williams stayed on at Muroc, now named the High Speed Flight Station, for another year. On September 1, 1959, at the request of NASA Deputy Administrator Hugh L. Dryden, Williams joined the STG along with two others from the High Speed Flight Station; Kenneth S. Kleinknecht and Martin A. Byrnes. His initial position was Associate Director for Project Mercury.

STG
In his first assignment at the STG, Williams assisted Robert Gilruth with the staffing of critical management positions and in establishing working relationships with the military services at the general staff level. By January 1960, Williams and Charles Mathews had organized coordination committees with the Army, Air Force, and NASA for launch operations. The Launch Operations Branch of the STG was assigned capsule checkout responsibilities. By March, all the operational functions and positions were given to the Mercury Control Center and the Navy was assigned the recovery role for the MA-1 flight. Williams also oversaw the development of the Mercury Space Flight Network from the standpoint of operations.

As Operations Director for Project Mercury, Williams was involved with almost every aspect of flight preparations, as well as flight operations. During the flights, he manned the position for the Operations Director in the Mercury Control Center.

OTHER

In 1963 Williams became Deputy Associate Administrator in the Office of Manned Space Flight at NASA Headquarters. He left in 1964 to become Vice President of the Aerospace Corporation. He returned to NASA Headquarters in 1975 as Chief Engineer; a post which he retained until his retirement in 1982. Amongst Williams' many awards was the 1963 John T. Montgomery Award by the National Society of Aerospace Professionals.

Charles H. Zimmerman
Founding Father and Member of the STG Core Team

Born: 1908
Died: May 5, 1996 (Age 88)
Age at STG: 50
Spouse: Beatrice
Children: 2 sons

BACKGROUND

Shortly after graduating from the University of Kansas with a B.S. in electrical engineering in 1929, Zimmerman joined the NACA Langley Memorial Aeronautical Laboratory.

NACA

Zimmerman worked on various research topics, including loads, airfoils, and aircraft stability and design. He developed the theory of "kinesthetic control," in which the natural balancing reflexes of a pilot are deemed adequate to control small flight vehicles. During the 1930s, he worked on aerodynamic forces and moments which act on spinning aircraft. He examined the characteristics of different airfoils and of different aspect ratios, and also how the location of stabilizers influenced spins. In 1952 Langley Director Henry Reid appointed Zimmerman and two colleagues to draft a proposal for research in the upper atmosphere and spaceflight. This study led to the X-Series of aircraft. In 1954 he earned a master's in aeronautical engineering from the University of Virginia.

In the 1950s Zimmerman explored the novel aspects of flight, leading to the testing of the "flying platform" built by Hiller Aircraft with the Office of Naval Research. Another unusual aircraft of his design was the use of circular bodies as the lifting surface; as can be seen in the design of the Vought XF5U "flying pancake." He also worked on Vertical/Short Takeoff and Landing aircraft.

STG
Zimmerman assisted Robert Gilruth and Charles Donlan in developing the concept of Project Mercury, from its origins in 1958 through to the implementation of the STG organization. He was also involved with the process for selecting the contractor to build the capsule, which was ultimately awarded to McDonnell Aircraft. He led the inspection team that examined the first full-scale mockup of the capsule in March 1959. With the STG up and running, he returned to Langley, where he spent a total of 33 years.

OTHER
Among Zimmerman's awards were the Wright Brothers Medal in 1956 and the Dr. Alexander Klemin Award from the American Helicopter Society.

A.2.2   The STG Core Team

The following biographies are those discussed in Sections 7.2 and 7.3; first for the Langley team and then the Lewis team. The Core Team is considered to comprise the 36 people from Langley and the 10 from Lewis; 46 in all.

Appendix A.2.1 included biographies for the Founding Fathers, many of whom were also part of the Core Team; they will not be repeated here. In November and December 1958, an informal STG executive committee was formed consisting of:

- Robert R. Gilruth (PARD)
- Charles J. Donlan (OAD)
- Maxime A. Faget (PARD)
- Paul E. Purser (PARD)
- Charles W. Mathews (FRD)
- Charles H. Zimmerman (Stability).

The following people (listed in alphabetical order) round out the initial Langley members of the Core Team:

- Melvin S. Anderson (Structures)
- William M. Bland (PARD)
- Aleck C. Bond (PARD)
- William J. Boyer (IRD)
- Robert G. Chilton (FRD)
- Edison M. Fields (PARD)
- Jerome B. Hammack (FRD)
- Shirley Hatley (Steno)
- Jack C. Heberlig (PARD)
- Alan B. Kehlet (PARD)
- Claiborne R. Hicks (PARD)
- Ronald Kolenkiewicz (PARD)

- Christopher C. Kraft, Jr. (FRD)
- William T. Lauton (Dynamic Loads)
- John B. Lee (PARD)
- Norma L. Livesay (File Clerk)
- Nancy Lowe (Steno)
- George F. MacDougall, Jr. (SRD)
- Betsy F. Magin (PARD)
- John P. Mayer (FRD)
- William C. Muhly (Planning)
- Herbert G. Patterson (PARD)
- Harry H. Ricker, Jr. (IRD)
- Frank C. Robert (PARD)
- Joseph Rollins (Files)
- Ronelda F. Sartor (Fiscal)
- Jacquelyn B. Stearn (Steno)
- Paul D. Taylor (Full Scale Tunnel)
- Julia R. Watkins (PARD)
- Shirley Watkins (File Clerk).

The designations of the Langley organization given in parenthesis in the above list were:

DLD	Dynamic Loads Division
FSRD	Full Scale Research Division
FRD	Flight Research Division
IRD	Instrument Research Division
OAD	Office Associate Director
PARD	Pilotless Aircraft Research Division
Fiscal	Fiscal Division
Files	Office Services Division
Planning	Technical Services
Stability	Stability Research Division
Steno	Office Services Division.

The initial group of 10 people from Lewis who joined the STG at Langley were:

- John. E. Gilkey (assigned to the Engineering Branch)
- Milan J. Krasnican (assigned to the Flight Systems Division, Electrical Systems)
- Glynn Lunney (assigned to the STG Mission Analysis Branch)
- Andre J. Meyer, Jr. (assigned to the STG Engineering and Contracts Division)
- Warren J. North (Headquarters and the STG Astronauts and Training Group)
- Gerald J. Pesman (assigned to the Flight Systems Division in Life Systems)
- G. Merritt Preston (assigned to the Cape Canaveral for the launch of Big Joe)
- Leonard Rabb (assigned to the STG Flight Systems Division)
- Scott Simpkinson (assigned to the Cape Canaveral for the launch of Big Joe)
- Kenneth Weston (assigned to the Flight Systems Division in Heat Transfer).

The following biographies include those for whom details could be found. My apologies to those for whom I couldn't find sufficient detail; at least their names are included for posterity.

Aleck C. Bond
Member of the STG Core Team

Born: January 11, 1922
Died: September 9, 2015
Age at STG: 36
Spouse: Anastasia Marinos
Children: Two girls

BACKGROUND
After graduating from the Georgia Institute of Technology in 1943 with a B.S. in aeronautical engineering, Bond joined the Bell Aircraft Company in Marietta, Georgia as a liaison engineer who worked on the wing design of the B-29 bomber and as a problem solver on the production line, where he made necessary changes to replace damaged materials and overcome issues of quality control. In late 1945 he was inducted into the Army for a short period, but managed a transfer to Wright-Patterson Air Force Base, where he worked in the wind tunnel. Returning to Georgia Tech after the war, he graduated in 1948 with an M.S. in aeronautical engineering.

NACA
On being hired by Langley in 1948 as an aeronautical research scientist, Bond was assigned to the Pilotless Aircraft Research Division (PARD). His supervisor was Paul Hill and he worked with Max Faget, Guy Thibodeaux, and Robert Gilruth. He worked on aerodynamic heating and various ablative and heat sink materials, and launched a variety of rockets from Wallops Island including the Navajo cruise missile. During his 10 years with PARD, he held several positions including Head of the Performance Section and Head of the Structural Dynamics Section.

NASA STG
When the STG was formed in 1958, Bond was on the original team selected by Robert Gilruth. He became Head of the Performance Branch under Max Faget's Flight Systems Division. This Branch included many people from the PARD. Bond worked on the heat shield of the Mercury capsule and coordinated work with General Electric; the contractor

that built the ablative heat shield. He also coordinated the work on the beryllium heat shield with its supplier. The ablative heat shield design work was immediately applied to the Big Joe capsule. Bond worked with the Test and Checkout Team in Hanger S at the Cape to check out the instrumentation for the heat shield, as well as other capsule components. He conducted post-flight analysis to confirm the design of the heat shield. He supported the checkout and post-flight analysis of all the thermal protection systems of the early Mercury flights.

POST-STG

When the STG was dissolved and moved to the Manned Spacecraft Center in Houston, Bond stayed at Langley to complete Project Mercury. In 1962 he moved to Houston as Chief of the Systems Evaluation and Development Division; part of Faget's Engineering and Development Directorate. Bond was instrumental in getting state-of-the-art facilities at the center to support Apollo and Space Shuttle. These facilities included:

- Space Environment Simulation Laboratory
- Structures and Mechanical Laboratory
- Arc Jet Test Facility
- Vibration and Acoustics Test Facility.

During his 20 years in Houston, Bond held many management positions including Assistant Director of Chemical and Mechanical Systems and Assistant Director of Program Support. He retired in 1982.

OTHER

Amongst Bond's honors and awards are the following by NASA: Group Achievement Awards for Mercury, Apollo, Skylab; the Distinguished Service Medal (1973); the Exceptional Service Medal (1969); the Outstanding Leadership Medal (1981); plus various MSC and JSC Special Achievement Awards and Commendations.

Jerome B. Hammack
Member of the STG Core Team

Born: May 2, 1922
Died: August 20, 2007 (age 85)
Spouse: Adelin Worrill
Children: Two sons

BACKGROUND

"Jerry" Hammack was born in Randolph County, Georgia, the son of William D. and Gussie Beauchamp Hammack. After growing up on a farm he graduated from the Georgia Institute of Technology in 1943 with a B.S. in aeronautical engineering. He joined the Douglas Aircraft Company but was soon drafted by the enlisted reserve of the Army Air Corps and sent to the NACA Langley Memorial Aeronautical Laboratory.

NACA

After WW-II, Hammack was discharged by the Army Air Corps with three years' service. He elected to stay at Langley as a project engineer in the Flight Research Division. He worked on propellers for high speed aircraft, testing them at transonic and supersonic speeds, and he also worked on the XF-88B, which was the first aircraft driven by a propeller to exceed Mach 1. At one point he also worked on turbofans.

NASA STG

In 1958 Robert Gilruth selected Hammack to be a member of the STG Core Team. He initially worked in Max Faget's Flight Systems Division in the Systems Test Branch headed by William M. Bland. Hammack's topic was the capsule recovery system. He conducted various parachute drop tests and worked on landing systems. He shares the patent for the Mercury space capsule. Hammack later moved to the Engineering Division as Head of the Project Engineering Branch. This required coordination with the Army Ballistic Missile Agency in Huntsville, Alabama for the Redstone rocket; the NASA STG liaison officer there was Lt. Col. Martin Raines.

POST-STG

As Project Mercury was ending in 1962, Hammack transferred to the Gemini Program as the Deputy Manager of Vehicles and Missions. In 1966 he became Chief of the Future Programs Division until 1971, then Chief of the Landing and Recovery Division, and then from 1973 to 1987 he was Chief of the Safety Division. He obtained an M.S. on the topic of Studies in the Future from the University of Houston at Clear Lake, Texas in 1980. He retired from NASA in 1987 after over 40 years in aviation and spaceflight.

OTHER

In retirement, Hammack was a consultant with Hernandez Engineering for several years. His honors and awards include NASA Exceptional Service Medals (Apollo 8 and Apollo 11) and NASA Superior Achievement Awards (Gemini, Skylab, Space Shuttle).

Alan B. Kehlet
Member of the STG Core Team

Born: 1929
Age at STG: 29
Spouse: Lois
Children: Two

BACKGROUND
Kehlet gained a B.S. in aeronautical engineering from the University of Illinois in 1951 and an M.S. in aeronautical engineering from the University of Virginia in 1961.

NACA
Armed with his B.S., Kehlet was hired by NACA as an aerodynamics research intern with the Langley Memorial Aeronautical Laboratory's PARD, where he worked for John H. Parks and Clarence L. Gills in Paul E. Purser's Branch. Joseph Shortal was the PARD Division Chief at the time. Kehlet conducted secret studies of an ICBM nose cone and early studies on the F-104 Starfighter. Later he became an aerodynamic research scientist with the Aircraft Configuration Branch until 1958.

STG
Kehlet was selected by Robert Gilruth as a member of the STG Core Team, and headed up the Aerodynamics Section in the Performance Branch that was led by Aleck Bond, who was also from PARD and on the Core Team. Many of the former PARD people settled in Max Faget's Flight Systems Division. Kehlet worked on the launch aerodynamics of the capsule and escape tower for aborts at various altitudes. He also worked on the capsule stability for re-entry. They worked with "boiler plate" models and confirmed their results by wind tunnel tests. He worked with the source selection panels to review industry proposals. He worked with McDonnell at its factory in St. Louis, Missouri on many occasions, checking on their work and progress. He also followed the capsule to Cape Canaveral for pre-flight testing and post-flight analysis.

Kehlet worked on the escape tower aerodynamics and its separation system. He also worked on the roll control system that would offset any misalignment of the capsule's center of gravity during re-entry. He worked with John Glenn on the instrument panel configuration and layout.

POST-STG
After the creation of the Manned Spacecraft Center, Kehlet moved there to work in the Apollo Project Office of the Flight Systems Division. He, Robert Piland and Kurt Strauss worked on early Apollo spacecraft configurations. He worked with three companies that had Apollo study contracts.

POST-NASA
Kehlet left NASA in 1962 to work for North American Aviation as Technical Assistant to the Chief Engineer, and by 1967 he was Assistant Program Manager on the spacecraft that would ultimately fly the Apollo 11 mission. He gained an M.B.A. from California State University in 1976. Over the years he worked on many of North American's projects and became President of the Sabreliner Division and the Fairchild Aviation Division in 1983. He left in 1984 to work for McDonnell Aircraft as Vice President on many projects including the Tomahawk cruise missile, the Delta launch vehicle and Advanced Programs and Technology. He retired in 1996.

OTHER
Among Kehlet's many honors and awards were: the NASA Citation for Contribution to Project Mercury; the NASA Certificate of Appreciation for Apollo 11; the Who's Who in Aviation; the Distinguished Alumnus Award by the University of Illinois; and the Distinguished Engineering Achievement Award by the Los Angeles Council of Engineers and Scientists. In addition he was elected to Fellowship of the AIAA and the IAE.

Claiborne R. Hicks, Jr.
Member of the STG Core Team

Born: November 10, 1935
Age at STG: 23
Spouse: Former Frances Marie Vaughan
Children: 3

BACKGROUND
Clay Hicks was born in the old MacArthur Hotel, which was managed by his grandparents in Durham, North Carolina. His parents were Claiborne Sr. and Leila Ruth Keller, who also lived with the Hicks family at the hotel. Clay was raised by his grandparents who

moved to Newport News, Virginia in 1938 and resided in the North End Section. He graduated from the Newport News High School in 1953 and then attended the Virginia Polytechnic Institute, from which he emerged in June 1958 as an aeronautical engineer *cum laude*.

NACA

Seeking a way to fund his college education, Hicks found that he would be able to "co-op" with the NACA Langley Memorial Aeronautical Laboratory between semesters as one of the original members of a new cooperative engineering program which was being developed by the Virginia Polytechnic Institute. This gave him an insight into many aspects of Langley's research programs and activities. He actually became a Langley employee in 1953 as a Training Student (GS-2) and got some real "hands-on" experience with flight hardware and engineering design. While still a student, Hicks worked in the Fabrication Shops, the Wood Model Shop, the Wind Tunnels, the Electronic Instrument Research Division, and the Pilotless Aircraft Research Division (PARD), giving him insight into the differences between researchers, engineers, and operations people. Hicks therefore had a unique "leg up" in experience even before applying for his first job out of college, and he was known to many of the people who would shortly manage the newly created STG; he couldn't have been better prepared.

Shortly after the Sputnik launch on October 4, 1957 and during Christmas break, he met with Max Faget, Bill Bland, Jack Heberlig, and Tom Markley to discuss his new job assignment upon graduation; this was six months before he was to graduate. He was to be assigned to work on the configuration of the pilot's couch for the Mercury capsule. (At that time the pilot was not called an "astronaut," nor was the capsule called a "spacecraft.")

After graduation and before the STG was formed, Clay worked on many aspects related to launch trajectories, the various "g" loads at launch and during aborts, atmospheric density, and better determination of the Earth's shape as it relates to trajectories. He had supported launches at Wallops Island and at Cape Canaveral and gathered data on re-entry heating which led to the choice of an ablative material for the capsule's heat shield. From June 1958 until the STG was created in November he worked with the PARD, which was later reorganized by having some people and functions transferred over to the STG while others remained in place, as indeed did the PARD organization and name.

STG

Hicks became part of the STG's Trajectory Analysis Section headed by Carl Huss, which was in the Mission Analysis Branch headed by John Mayer. The Branch was in the Operations Division headed by Charles Mathews. In addition to John Mayer and Carl Huss, Clay's colleagues and co-workers at the Mission Analysis Branch who transferred to the Manned Spacecraft Center in Houston in late 1961 and early 1962 included Ted Skopinski, Charlie Allen, John O'Loughlin, Mary Shep Burton, Shirley Hunt Hinsen, Richard Koos, Harold Miller, Frances Vaughan Hicks, John Shoosmith, James Ferrando, Ed Behuncik, Lynn Dunsieth, Morris Jenkins, and Jerry Engel.

Hick's trajectory work was the beginning of the discipline of "Flight Analysis" as well as the work of the future Mission Planning and Analysis Division in Houston. This work also began to define the positions in the Mercury Control Center needed to monitor the spacecraft trajectories and abort conditions.

Once the astronauts came onboard in April 1959, Project Mercury got focused on the crew interface with the Redstone and Atlas launch vehicles. With his trajectory experience, Clay was assigned as the Atlas Ballistic Trajectory Project Engineer. He held this position when the STG transformed into the MSC.

POST-STG

Hicks went on to become the MSC Flight Analysis Branch Chief for both Gemini and Apollo. He also worked on the Apollo Applications Program which developed the Skylab space station. As technical assistant in the Flight Control Division he worked with scientists and astronauts on experiments for Skylab. He was a Flight Operations Manager for Experiment Operations in the MCC during flights.

After moving to NASA Headquarters in 1980, Hicks would often return to Houston to man the Headquarters console in Mission Control. As the concept for a space station developed, he joined the Headquarters Space Station Task Force that was headed by John Hodge.

Clay retired in 1984 after 31 years with NACA/NASA. His awards were numerous. He also spent another 15 years working with various aerospace companies, then finally retired in 1999 after 41 years in the space program.

Christopher C. Kraft, Jr.
Member of the STG Core Team

Born: February 28, 1924
Age at STG: 34
Spouse: Betty Anne Turnbull
Children: 2

BACKGROUND

"Chris" Kraft was born in Phoebus, Virginia. He was named after his father, the son of Bavarian immigrants. As a young boy, he played the bugle and became the State champion. He also loved to play baseball. He went to the Virginia Polytechnic Institute in 1942 and became a member of the Corps of Cadets. Because of the war, Virginia Tech was on a twelve month schedule and so he completed his degree in two years, graduating in December 1944 with a B.S. in aeronautical engineering.

NACA

Kraft joined the NACA Langley Memorial Aeronautical Laboratory in 1945 and was assigned to the Flight Research Division, headed by Robert Gilruth. Most of his years between then and the formation of the STG in 1958 were spent in the Stability Control Branch in the Flight Research Division. His work involved the development of gust alleviation systems for aircraft that were flying in turbulent air; most aircraft of that period were propeller driven and flew at lower (and rougher) altitudes than modern jet aircraft. He worked on deflecting control surfaces that would improve stability and he also investigated wingtip vortices. In addition, he worked on the flight testing of X-1 aircraft models that led to changes in the elevator design for stability.

STG

Robert Gilruth made Kraft a member of the Core Team which would work on the first manned space missions. He became the Assistant Chief for Plans and Arrangements in the Operations Division that was headed by Charles Mathews. His first assignment was to develop a mission plan to place a man into space and safely back to Earth. This broad task (at a time when no one had yet flown in space) involved starting from scratch in the creation of flight plans, timelines, procedures, mission rules, tracking, telemetry, ground support, communications networks, and contingency management.

This major effort led to the conceptual design of the Mercury Control Center, which he saw through to completion and operation. He realized that during orbital flight, the astronaut would be out of direct contact with the control center. This required another major effort to construct a world-wide system of tracking stations to be manned by flight controllers who would assist the astronaut; especially during abnormal situations. In addition, the astronaut could land anywhere in the world, so a new organization would be required to work with the Air Force and the Navy in recovering the astronaut and capsule.

As these organizations began to evolve, Kraft took up the role of Flight Director because he knew that these efforts would need to be coordinated and managed by the operational leader of the vast team of experts in various fields. Kraft held both of these management and operational positions throughout his three years with the STG.

NASA MSC/JSC

After the move to the Manned Spacecraft Center in Houston in 1962, Kraft maintained both key management positions and the Flight Director position through to Gemini 7 in December 1965. A multiple shift operation was required in order to manage the longer flights, so Kraft appointed John Hodge, Gene Kranz, and Glynn Lunney as Flight Directors. Later, others were also named for Apollo and the Space Shuttle.

Kraft went on to become the MSC Director on January 14, 1972, replacing Gilruth. In 1973 the MSC was renamed the Johnson Space Center (JSC). Kraft retired in 1982, then worked as a consultant for Rockwell International and IBM. He served on many review committees. He is fondly remembered by his colleagues as the "Father of Flight Operations." His memoir *Flight: My Life in Mission Control* was published in 2001 and widely acclaimed. His numerous awards include the NASA Outstanding Leadership Medal, presented by President Kennedy and NASA Administrator James Webb in the White House Rose Garden.

John B. Lee
Member of the STG Core Team

Born: June 13, 1924
Died: September 19, 2012 (88)
Age at STG: 33
Spouse: Dorothy "DeDe"
Children: Two daughters

BACKGROUND
Lee grew up on farms in Texas and Virginia. He graduated from King George High School at the age of 17. He majored in mechanical engineering at the Virginia Polytechnic Institute and was in the Corps of Cadets. During WW-II he joined the Army Air Corp at age 18 and was assigned to the Gulf Coast Training Command. He received his commission and wings as a fighter pilot at age 19, then flew 52 combat missions over Europe with the 20th Fighter Group of the 8th Air Force flying the P-51 Mustang. He was awarded the Distinguished Flying Cross, 6 Air Medals and 3 Battle Stars; the Battle of France, Battle of the Ardennes (commonly called the Battle of the Bulge); and the Battle of Europe. After the war he resumed his studies and graduated from the Virginia Polytechnic Institute in 1948 with a B.S. in mechanical engineering.

NACA
Armed with his B.S., Lee joined the NACA Langley Memorial Aeronautics Laboratory. He was recruited by Robert Gilruth to the PARD and was assigned to the Propulsion and Aerodynamics Branch headed by Paul Hill, where he devised a system for drop testing scale models of atomic warheads from supersonic aircraft at the Wallops Island Station. He worked with Paul Purser of the Aerodynamics Branch, Joe Thibodeaux of the Propulsion Branch, and Caldwell Johnson's model group. Lee also worked on wind tunnel tests of jet engine inlets and ramjets. He became an expert in the aerodynamics of dropping bombs from aircraft, including the top secret atomic bombs.

STG

When Robert Gilruth formed the STG in November 1958, Lee was one of the 36 people assigned to work on Project Mercury. Lee headed the Mechanical Systems Section of the Flight Systems Division led by Max Faget. After working on the parachute system he turned his attention to the propulsion systems on the capsule and the escape tower, working with Joe Thibodeaux and with the McDonnell Aircraft contractors.

POST-STG

In February 1962 Lee moved to the Manned Spacecraft Center in Houston, where he continued to work for Faget's Engineering and Development Directorate in various positions. This work included various spacecraft design efforts and analysis from 1961 to 1967. During this period, Lee was asked by Robert Piland to work as the manager for the Martin study contract on how to reach the Moon. This study, and two others, were factored into the Apollo program. Eventually North American Aviation was selected as the prime contractor for that spacecraft. The Apollo program organized the spacecraft into 42 subsystems, each of which had its own manager. Lee became the manager of the Subsystems Office, and met with the Head of the Apollo Program Office and the MSC management to resolve problems. He also worked on the redesign of the Apollo spacecraft after the Apollo 1 fire. Lee then worked on future programs, including early space station studies. He continued in Faget's engineering and Development Division until his retirement in 1980.

His many honors and awards included the NASA Group Achievement Award (3); the MSC Certificate of Commendation (2); the Silver Snoopy Award; the Presidential Medal of Freedom; and the Virginia Tech "Wall of Fame" for Aviation and Aerospace.

Glynn S. Lunney
Member of the STG Core Team

Born: November 27, 1936
Age at STG: 22
Spouse: Former Marilyn Kurtz
Children: 4

BACKGROUND

Lunney was born in the mining town of Old Forge, Pennsylvania, and was the eldest son of William and Helen Lunney. On graduating from the Scranton Preparatory School in 1953 he attended the University of Scranton for two years, then transferred to the University of Detroit where he enrolled in the cooperative training program of the NACA Lewis Flight Propulsion Laboratory to obtain some aeronautics experience. He graduated in June 1958 with a B.S. in aeronautical engineering.

STG

From June 1958 to September 1959, Glynn worked for John Disher at Lewis in a Section that was under George Low's Special Projects Branch, where he researched the thermodynamics of high speed re-entry vehicles. He was a member of the initial group of 10 Lewis engineers that joined the STG. (Other Lewis engineers were transferred later.) Some of Glynn's colleagues at Lewis who moved to NASA Headquarters included Low, John Disher, and the Lewis Director himself Abe Silverstein.

Lunney started at the STG in the Mathematical Analysis Section of John Mayer's Mission Analysis Branch, part of the Operations Division of Charles Mathews. He also worked in Jack Cohen's Operations Analysis Section on flight simulations and with Tec Roberts in the Flight Control Branch. In his first year, Lunney worked on simulations to train the Flight Operations people. One aspect of these simulations dealt with how astronauts and flight controllers might respond to potentially dangerous abort situations. This work led Lunney into the area of flight dynamics. He then began training under Tecwyn Roberts, a former AVRO engineer who was defining the role of the Flight Dynamics Officer (FIDO) during the early Mercury flights. Of importance to Lunney as a future FIDO were the various launch abort scenarios as well as the vehicle trajectories. He continued working in this area after the move to Houston.

POST-STG

After the STG was dissolved in November 1961 and the gradual relocation to Houston started, Lunney supported the manned orbital flight by John Glenn in February 1962 by serving as the FIDO at the Bermuda tracking station. His first support of a flight from the Mercury Control Center at Cape Canaveral was that by Scott Carpenter. Between launches, Lunney worked with the flight control people to develop the mission rules that defined the Go/No-Go procedures for various flight conditions and specified what the astronaut and the flight controllers should do in each situation. He continued as Chief of the Flight Dynamics Branch through the Gemini flight program and was selected as a Flight Director in 1964. He served on Gemini 3, the unmanned Apollo AS-201, and Gemini 9 through Gemini 12. He worked on most of the Apollo flights up until his final Flight Director slot on Apollo 15. Of particular note was his work on Apollo 13, leading the emergency improvisational work that helped to save the crew. For that, he and his entire operations team were awarded the Presidential Medal of Freedom.

Lunney was the Program Manager of the Apollo vehicles that ferried crews to and from the Skylab space station. In the same period, he also directed the U.S. efforts for the Apollo-Soyuz Test Project which flew in 1975. He served as the NASA Space Shuttle Program Manager from 1981 to mid-1985, then retired from NASA. He then worked for

Rockwell International and the United Space Alliance. He retired from full-time employment in 1998 and still serves the space program in a consulting role.

Amongst Lunney's numerous honors and awards were the following from NASA: the Group Achievement Award for Mercury, Gemini, and Apollo; several Sustained Superior Performance and Outstanding Performance Awards; several Exceptional Service Medals; two Distinguished Service Medals (1971 and 1975); and the Distinguished Executive Rank for Senior Executive Service (1983). He also received the Arthur S. Flemming Award in 1975 and the W. Randolph Lovelace II Award (AAS) in 1983.

John P. Mayer
Member of the STG Core Team

Born: May 10, 1922
Died: March 28, 1992
Age at STG: 36
Spouse: Geraldine Couch
Children: one son and two daughters

BACKGROUND
Mayer was born in Binghampton, NY and graduated from the University of Michigan in 1944 with a B.S. in aeronautical engineering and mathematics.

NACA
On graduating, Mayer joined the NACA Langley Memorial Aeronautical Laboratory and was assigned to the Flight Research Division. His first assignment was supersonic aerodynamics, including the X-1 research aircraft. His other work over the next five years was in supersonic boundary layer control systems. In 1949 he and his wife Gerry, a mathematician, moved to the High Speed Flight Research Station in California, where he was the chief analytical advisor for supersonic aerodynamics and she worked on wingspan stress analysis. In 1952 they returned to Langley to work in the Flight Research Division. During the next few years, Mayer joined Bill Tindall and Ted Skopinski in developing orbital mechanics and flight dynamics. This included the development of trajectory tools and early concepts for the mission planning process.

STG
When the STG formed in November 1958, Robert Gilruth included Mayer in the Core Team of Langley people. Mayer was initially assigned to head up the Space Mechanics Group, in Max Faget's Flight Systems Division. When the STG organization was redefined in August 1959 he became the Head of the Mission Analysis Branch in the Operation Division headed by Charles Mathews. It was then that the three Sections making up the Branch were defined; the Trajectory Analysis Section, the Operations Analysis Section, and the Mathematical Analysis Section. In effect, this was Mayer's previous work at Langley being applied to Project Mercury. Mayer's group developed detailed launch, orbital, and re-entry trajectories for all of the Mercury flights, including hundreds of abort scenarios. This original and creative work was instrumental in the success of Project Mercury. See Section 15.7 Mission Analysis and Trajectory Planning.

NASA
After the move to Houston, Mayer's Mission Analysis Branch was consolidated to produce the Mission Planning and Analysis Division (MPAD) under the Flight Operations Directorate. This developed the sophisticated trajectory analysis techniques that would facilitate space rendezvous and docking maneuvers for the Gemini and Apollo missions, in particular putting astronauts on the Moon and returning them safely to the Earth.

OTHER
After John and his wife Gerry retired from NASA, they joined two friends and opened the first Computerland store in the Clear Lake Area, near the Johnson Space Center. Mayer received the NASA Exceptional Service Medal in 1969.

George Merritt Preston
Member of the STG Core Team

Born: 1916
Died: 2007

BACKGROUND

"Press" Preston was born in Athens, Ohio, where he went to grade school and high school. He attended Ohio University in Athens and then transferred to the Rensselaer Polytechnic Institute in Troy, NY in 1936, from which he graduated in 1939 with a B.S. in aeronautical engineering.

NACA

Preston joined the Langley Memorial Aeronautical Laboratory straight out of college and was assigned to the Full Scale Wind Tunnel with Abe Silverstein as his supervisor. Preston tested numerous aircraft in the tunnel. In 1942 he followed Silverstein to the Lewis Flight Propulsion Laboratory and during WW-II tested the P-59 and F-80. There he met Kelly Johnson, who was visiting from the Lockheed "Skunk Works" to coordinate flight tests. In 1945 he became Chief of the Flight Research Branch. He conducted research on icing and, flying over Wallops Island, launched vehicles from an F-82 Twin Mustang to study drag and ramjets. On one occasion, he was involved in testing a B-57 using hydrogen fuel. He also carried out crash tests to determine the cause of fires. His report "Accelerations in Transport Airplane Crashes" resulted in a Flight Safety Foundation Award.

STG

When the STG was formed in November 1958, its Core Team included 10 people from Lewis; others joined later. Preston was Assistant Chief to the Operations Division headed by Charles Mathews. He was also Acting Head of the Preflight Checkout Section that carried out the tests and checkout of the capsule prior to launch. His initial task was to support the checkout of the Big Joe capsule and to learn about the operation at the Cape. This involved interfacing with the Air Force and the Atlas contractor Convair. For a year, he was commuting from Langley to the Cape and back again. Eventually the checkout team grew to about 35 people and was allocated Hanger S at Cape Canaveral. They supported all the Mercury flights. With the establishment of the Kennedy Space Center, this team became KSC employees.

POST-STG

Preston was appointed Manager of the Florida Operations in 1964 and Director of the Launch Operations Directorate in 1965. He held other Directorship titles involving the Gemini, Apollo, and Space Shuttle until his retirement in 1973 after more than 30 years with NACA/NASA.

OTHER
Among Preston's many honors and awards are the following: the Laura Tabor Barbour Award (1956); the NASA Leadership and Achievement Award (1962, 1963 and 1965); the Presidential Medal for Outstanding Leadership (1963); the Spirit of St. Louis Award from American Society of Mechanical Engineers (1969); and the Lifetime Achievement Award by National Space Club (1977). In addition he was an Associate Fellow of the AIAA.

A.2.3   Other STG Members

Several hundred people were hired by or transferred into the STG between 1958 and 1961. In 1996 JSC created the Oral History Project but many people did not participate or had already left. Consequently, the history of their contributions to the space program may either be lost or difficult to trace. Such was the case with many STG employees.

In addition to the Core Team, the following people (in alphabetic order) contributed a great deal to Project Mercury. Their biographies follow:

- Arnold D. Aldrich
- Peter Armitage
- Dr. William S. Augerson
- Harold D. Beck
- Dr. Charles A. Berry
- James A. Chamberlin
- William L. Davidson
- Dr. Lt. Col. William K. Douglas
- R. Bryan Erb
- Dennis E. Fielder
- George Harris, Jr.
- John D. Hodge
- Carl R. Huss
- Caldwell C. Johnson
- Kenneth S. Klienknecht
- Eugene F. Kranz
- Howard C. Kyle
- Charles R. Lewis
- John S. Llewellyn
- C. Frederick Matthews
- Owen E. Maynard
- Harold G. Miller
- Billy W. Pratt
- Tecwyn Roberts
- Rodney G. Rose
- Sigurd A. Sjoberg
- Howard W. Tindall, Jr.

- Robert F. Thompson
- Dr. Robert Voas
- Manfred "Dutch" von Ehrenfried
- Dr. Stanley C. White.

Sufficient detail wasn't found on others to include their biographies, but their names may be found in the organization listings and charts in Appendix 1.

Arnold Deane Aldrich

Born: July 7, 1936
Age at STG: 23
Spouse: Eleanor Jean Harris
Children: 3, Grandchildren: 4

BACKGROUND

Aldrich was born in 1936 and grew up in Lexington, Massachusetts. During the early to mid-1940s he developed a strong interest in contemporary aircraft and was an avid model airplane builder. He subsequently developed an interest in the assembly of high fidelity music systems. By the time he was in high school he knew he wanted to become an electrical engineer and to work in the field of aviation. After graduating from Lexington High School he enrolled in the five year cooperative education program at Northeastern University, and his "co-op" job was with General Radio Company; an early pioneer in the development of electrical and electronic standards and the manufacture of measuring equipment. Five years later, he graduated with a degree in electrical engineering and the bonus equivalent of two years of valuable experience working in the electronics industry.

In the fall of 1958, which was Aldrich's senior year at Northeastern, he interviewed with a variety of aviation and electronics companies. One day it was announced that there would be interviews by a new organization called the "National Aeronautics and Space Administration." During his interview, Aldrich was surprised and pleased to find that the NASA interview team was made up of Langley engineers rather than the traditional personnel management teams he had met with other organizations. Although NASA's subsequent offer was lower than some of others that he received, he decided that working for NASA on aeronautics research at Langley was the most exciting opportunity.

STG

Aldrich reported to Langley in the first week in July 1959 and was given the choice of working on aeronautics projects at the Langley Research Center or joining a new organization called the Space Task Group that was developing a human spaceflight program. Although he hadn't heard much, if anything, about this new program he excitedly chose to join the STG. He was assigned to the Control Central and Flight Safety Section of the Flight Control Branch headed by Gerald W. Brewer, which was in the Operations Division headed by Charles Mathews. He remained in this organization throughout the STG's three year existence.

Aldrich's first assignments were focused on the development of the flight control facilities and operations for the world-wide tracking network for Project Mercury. This was to include a prototype at Wallops Island and thirteen operational sites at strategic sites around the globe. The proximity of Wallops Island to Langley enable him to visit the prototype regularly to check out and test the equipment and operational procedures as they were being developed and to interface with NASA's Bell Laboratory and Western Electric network development contractors. Early on, Aldrich was also tasked to develop the operational requirements for a Flight Monitoring Trailer at Cape Canaveral as a backup to support early Mercury flights, owing to the tight schedule for developing the Mercury Control Center there. This trailer was built by McDonnell Aircraft and delivered to the launch site along with several trailers for other purposes, but it proved not to be needed in support of the flight program.

As the Mercury flights approached, Aldrich was also trained to serve as a remote site Capsule Communicator (CAPCOM). The CAPCOM was always the lead flight controller at each remote site, although for manned Mercury flights an astronaut was also assigned to key sites to perform any direct voice communications with the astronaut in orbit. As the flight program commenced, Aldrich was the CAPCOM at the Guaymas site in Mexico for MA-3 and MA-4 and at the Point Arguello site in California for MA-5 and MA-6. During MA-5 flight of the chimpanzee Enos it was decided to terminate the flight after two orbits and Aldrich sent the command that initiated retrofire and brought the spacecraft down in the desired landing area. This is the only time thus far that an American spacecraft designed to carry humans has been returned to Earth by ground command.

POST-STG

As the STG transitioned into the Manned Spacecraft Center (later to become the Johnson Space Center) Aldrich became the Spacecraft Systems Monitor in the Mercury Control Center at Cape Canaveral for the final three Mercury flights. He later held key management positions in flight operations for Gemini and Apollo. Subsequently, he served as Deputy Manager for the Skylab Program, as Apollo Spacecraft Deputy Program Manager during the Apollo-Soyuz Test Project with the Soviet Union, as Space Shuttle Orbiter Project Manager (overseeing fifteen successful flights as well as the construction of the orbiters *Discovery* and *Atlantis*), and as Space Shuttle Program Manager.

In the wake of the *Challenger* disaster Aldrich was appointed Director of the National Space Transportation System at NASA Headquarters, where he managed the Shuttle's return-to-flight work. He approved and oversaw the implementation of over 400 improvements to Space Shuttle flight and support systems which contributed greatly to 88 consecutive successful flights during the ensuing 14.5 years.

In 1988 Aldrich became Associate Administrator for Aeronautics and Space Technology, in which role he oversaw the agency's involvement in the National Aerospace Plane and the High Speed Civil Transport and he was also responsible for program and institutional activities at the Langley, Lewis, Ames and Dryden research centers. Later he served as Associate Administrator for Space Systems Development, overseeing the Space Station Freedom program, development of the super-lightweight external tank for the Space Shuttle, and other space system technology initiatives including single-stage-to-orbit concepts and feasibility. Aldrich also led political and technical initiatives with Russia that led to the incorporation of the Soyuz spacecraft as the on-orbit emergency rescue vehicle for the International Space Station.

OTHER

After 35 years with NASA, Aldrich left to join the Lockheed Missiles and Space Company in California, initially as Vice President for Commercial Space Business Development and later Vice President for Strategic Technology Planning. With the merger of Lockheed and Martin Marietta, he joined Lockheed Martin corporate headquarters in Bethesda, Maryland, where he oversaw the X-33 "Venture Star" single-stage-to-orbit program. Later, as Director of Program Operations, he pursued a broad array of initiatives designed to enhance program management capabilities across the company. He retired from Lockheed Martin in 2007 and is currently an aerospace consultant.

Aldrich gained a great many honors and awards during his long career including the NASA Distinguished Service Medal (3) and the Presidential Rank of Distinguished Executive. He is a member of the International Academy of Astronautics and of the National Research Council's Aeronautics and Space Engineering Board, and also an Honorary Fellow of the AIAA.

Peter Armitage

Born: March 5, 1929
Age at STG: 29
Spouse: June Blackett
Children: 4 sons

BACKGROUND

Armitage was born in Leeds, Yorkshire, England to Jack and Vera Armitage. His father was a tool and die maker, but was unemployed for many years during the "Great Depression." Vera was a dress maker and often became the bread winner. Armitage went to Hamble Senior School and left at age 14; an option there during WW-II. He joined the Air Service Training company, which offered "co-op" employment in aircraft design whilst attending Southampton University. Continuing this cooperative program, he later worked for the Cierva Helicopter Company as a design engineer. On graduating at age 21, he joined the Royal Air Force and trained as a flight engineer and co-pilot on the AVRO Lincoln II heavy bomber, then served with 617 Squadron, which had gained fame during the war as "The Dam Busters."

In 1952 Armitage joined AVRO Aircraft in Toronto, Canada as a flight test engineer for the CF-100; a two-seat, twin-engine interceptor. After three years AVRO awarded him a two year scholarship to attend the Cranfield College of Aeronautics (nowadays Cranfield University) in the U.K., where he received his master's in aeronautical engineering. He returned to AVRO in 1957 and became a senior flight test engineer on the CF-105 Arrow supersonic fighter. When this program was canceled on "Black Friday" February 20, 1959, Armitage and 14,000 other AVRO employees found themselves out of work. After Robert Gilruth at NASA STG heard of this opportunity to acquire top notch aeronautical engineers, an interview trip was planned and Armitage was hired along with many other AVRO engineers. See Chapter 8.

In 1969, Armitage received a Sloan Fellowship to attend the Graduate School of Business at Stanford University.

STG

On April 27, 1959 Armitage, along with many other AVRO Canadians, was processed in at NASA Langley. This was the same time that the astronauts were being familiarized with the STG. Armitage was hired as a research engineer in the Operations Division headed by Charles Mathews, initially working for Jerry Hammack in the Mercury Capsule Coordination Branch. After the STG organization became more formalized Armitage was assigned to the Recovery Branch headed by Robert Thompson, where he remained for his nearly three years at the STG.

Armitage's responsibilities in this area were related to qualifying the capsule for landing on land during a launch pad abort, as well as the planned end-of-mission recovery at sea. His work also involved airdrop tests from both aircraft and helicopters to assess the landing bag for both water and land landings. He developed operational tests for search and recovery of the capsule. His RAF experience with search and rescue and his AVRO work on ejection seats and general familiarization with parachute testing, gave him the insight necessary to complete the Mercury spacecraft landing and recovery flight qualification.

In addition, Armitage was responsible for a group of STG engineers assigned to develop and qualify electronic location aides and provide astronaut survival training. Pre-mission testing was done in realistic ocean conditions using Air Force Air Rescue Service aircraft,

Marine helicopter units, and Navy ships. Armitage developed the designs to enable Navy ships to retrieve a capsule under heavy sea states and aided in the development of techniques to retrieve an astronaut at sea. He also had an influence on the capsule design, by showing McDonnell Aircraft that the margin between the center-of-gravity and buoyancy of their design would make the capsule unstable on landing in the water under certain wind and sea state conditions.

POST-STG

Armitage supported recovery operations of all the Mercury flights throughout the STG period. He moved to Houston in 1962, becoming Chief of the Operation Evaluation and Test Branch of the new Landing and Recovery Division. He continued his recovery engineering responsibilities for the Gemini and Apollo programs. Armitage was often asked to take on additional duties; one such being to ensure the operational readiness of the Lunar Receiving Laboratory. In addition he reviewed the flight procedures of the Lunar Landing Training Vehicle and, after the crash of the second vehicle, took over the flight qualification of the third and final vehicle; the one flown by Neil Armstrong shortly prior to Apollo 11. Armitage became Manager of the Lunar Receiving Laboratory for Apollo 14 and 15. He then rounded out his NASA career in the Space and Life Sciences Directorate. After retiring in 1986, he worked for various companies including ACR Electronics and Space Services, Inc. He finally retired in 1992 and still lives in the Clear Lake area of Texas.

His many awards include a Superior Achievement Award for his recovery work on Apollo and the NASA Exceptional Service Medal for work on the Apollo Program.

Harold D. Beck

Born: April 12, 1933
Age at STG: 27

BACKGROUND

Beck was born in Melbourne, Florida, and grew up in Sanford, North Carolina. On graduating from Sanford Central High School in 1951 he joined the Air Force, which trained him as an Air Traffic Controller and stationed him at Itami Air Base, near Osaka in Japan, during the Korean War. He left the Air Force in 1954 and enrolled in North Carolina State College, from which he graduated in 1959 with a B.S. in aeronautical engineering.

NACA

After graduation, Hal joined Langley's Flight Research Division, working in the Flight Loads Branch. His initial duties included the data reduction and analysis of wind tunnel test data. He soon went to work with Jack Eggleston on an Earth orbit rendezvous study. This included the formulation of equations of relative motion and an analysis of the trajectories involved when a spacecraft approaches a space station. Their report was published as a Langley Technical Note. This was one of the earliest space rendezvous studies. All of the computations were performed manually using a Friden calculator because there were no computers in the organization.

STG

On moving to the STG in the summer of 1960, Hal joined the Mission Analysis Branch under John Mayer. It was there that he first gained access to the IBM 1620 operated by Shirley Hunt Hinson and Mary Shep Burton. The Branch personnel also had access to the Langley IBM 704 for carrying out trajectory computations and developing software tools. While in the STG, Hal supported the mission planning and trajectory analysis work for the early Mercury flights.

NASA MSC

Hal moved to Houston in April 1962 and was based with the mission planning organization in the Houston Petroleum Center, with John Mayer as Assistant Chief for Mission Planning. He was appointed Head of the Lunar Trajectory Section under Morris Jenkins. The initial task for this Section was to initiate the development of a capability for lunar mission design. This work began in 1962 and it resulted several years later in the Apollo Reference Mission Program that was used in end-to-end simulation of the Apollo lunar landing missions; for both nominal and contingency mission planning and analysis. In programmatic planning of lunar missions, Hal's responsibilities included extensive coordination with the launch vehicle developers at MSFC.

Hal continued to work in the lunar mission planning area throughout the Apollo program. In the years after Apollo, he continued working in the Mission Planning and Analysis Division.

- He played a lead role in the development of an agency-wide Shuttle Utilization Planning concept to support an extremely high rate of Space Shuttle flights (i.e. greater than fifty flights/year, a capability that never materialized).
- He supported the JSC Payload Integration Office, taking into account DOD requirements definition, documentation, and configuration management.
- He served as JSC operations representative in "The NASA/DOD Space Transportation Architecture Study (STAS), a major two-year agency-wide study started by presidential directive to report upon strategic planning for the future Space Transportation System. It involved all NASA field centers and the Air Force, and was undertaken in the context of the Strategic Defense Initiative commonly referred to as the "Star Wars" program.

- For several years in the mid-1980s, Hal was the NASA representative to the Fédération Aéronautique Internationale in Paris, France; he followed Carl Huss and Rod Rose in that position. The organization was (and still is) the official keeper of international aircraft and space records. The record keeping of the organization dates back to the days of the hot air balloon flights. NASA held many prestigious space records, as indeed did Russia, another member of the committee.

OTHER

After he retired from NASA in 1990, Hal worked for Ford/Loral/Lockheed for about ten years supporting the development of the Integrated Planning System for JSC's Mission Operations Directorate. After several years in semi-retirement, he went to work for Booz-Allen-Hamilton supporting the Directorate in MCC systems integration. He finally retired in 2012 at the age of 79. Among his honors and awards were NASA Outstanding Performance Ratings (1982, 1985, and 1986) and Sustained Superior Performance Awards (1983, 1985).

Dr. Charles A. Berry

Born: 1924
Age at STG: 34

BACKGROUND

"Chuck" Berry graduated from the University of California, Berkeley, CA in 1945, then went to the University of California Medical School in San Francisco and obtained his M.D. in 1947. He received his Aviation Medicine Residency Training Program at Hamilton AFB in California and from 1952 to 1955 was stationed at Albrook AFB in the Panama Canal Zone. He returned to the Aviation Residency program at the Harvard School of Public Health in Boston in 1955 which he completed in 1956, then remained at Harvard for a master's in public health *cum laude* later that year. He was a flight surgeon at Randolph AFB from 1956 to 1958, and was with the School of Aviation Medicine at the time of his assignment to the Mercury Astronaut Selection Committee.

STG

In 1959, Dr. Lt. Col. Berry, now in the Office of the Surgeon General in Washington, DC, was assigned to the STG on a half time basis. During the next three years, he supported the STG by commuting from Washington. As a flight surgeon, he spent some time at the Bermuda Control Center, the Canary Island tracking station, and the MCC at the Cape as an aeromedical monitor. He supported MA-3 at Canary, MA-6 at Bermuda, and both MA-7 and MA-9 at the Cape.

POST-STG

When the STG relocated to Houston in 1962, Dr. Berry was given an accepted Civil Service position so he could leave the Air Force and become a NASA employee. He was Chief of the Center Medical Operations Office from 1962 to 1966. Other positions that he held at the MSC were Director of Medical Research and Operations and, eventually, Director of Life Sciences, which he held until 1974. He supported all the Gemini and Apollo missions.

OTHER

After leaving NASA, Dr. Berry was made President of the University of Texas Health Science Center and Professor and Chairman of the Department of Aerospace Medicine at the University of Texas Medical Branch in Galveston. He was also Clinical Professor of Aerospace Medicine at the University of Texas School of Public Heath in Houston.

He received over forty national and international awards and honors, and had membership of many professional and honorary societies and associations. NASA awarded him its Exceptional Service Medal (1965) and Distinguished Service Medal (1973). His Air Force awards included its Commendation Ribbon (1957), its Certificate of Achievement (1957, 1962), and the Hoyt S. Vandenberg Trophy of the Arnold Air Society (1966). His medical awards included his being a Certified Specialist on Aerospace Medicine from the American Board of Preventive Medicine (1957); the Arnold D. Tuttle Award of the Aerospace Medical Association (1961); the Special Aerospace Medicine Honor Citation from the AMA (1962); and the Louis H. Bauer Founders Award of the Aerospace Medical Association (1966).

James A. Chamberlin

Born: May 23, 1915
Died: March 8, 1981 (65)
Spouse: Ella
Children: One son and one daughter

BACKGROUND

"Jim" Chamberlin was born in Kamloops, British Columbia. He earned a B.S. in mechanical engineering in 1936 from the University of Toronto and a master's in the same field in 1939 from the Imperial College of Science and Technology in London. After a brief time with the Martin-Baker Company, he joined Federal Aircraft Ltd., in Montreal, Quebec, to work on the Canadian variant of the British AVRO Anson from 1940–1941. He then spent another year at Clarke Ruse Aircraft in Dartmouth, Nova Scotia as chief engineer on anti-submarine aircraft. The remainder of WW-II was spent at Noorduyn Aircraft in Montreal as a research engineer.

In 1946, Chamberlin joined AVRO Aircraft Ltd., in Toronto, as chief aerodynamicist on the C-102 Jetliner and the CF-100 Canuck jet interceptor. Later, he became the Chief of Technical Design for the delta-wing supersonic CF-105 Arrow. When that project was canceled in 1959, Chamberlin and other AVRO engineers were recruited by the STG to join NASA and work on Project Mercury.

STG

Chamberlin led the AVRO team to the STG and became Chief of the Engineering and Contract Administration Division. He oversaw the development of the capsule at McDonnell, setting up an STG liaison office at the company and personally heading the Capsule Coordination Office. He managed two separate Branches: the Contracts and Scheduling Branch and the Engineering Branch. His staff worked the problems and changes which cropped up during Project Mercury. Eventually there were over 200 positions for technical and administrative support. By 1961, it was obvious that this task was too large for a single Division, and Engineering became its own Division and the task of administering contracts was moved to Office of the Director.

POST-STG

In 1961 Chamberlin became the first Gemini Program Manager and he was instrumental in the design of the two man spacecraft. During this period, he became a champion of the Lunar Orbit Rendezvous method for Apollo. In 1963, he left Gemini for Apollo, focusing on the Command and Service Modules and the Lunar Module.

OTHER

In 1970 Chamberlin left NASA to work for McDonnell Douglas Astronautics, to help them bid on the Space Shuttle; they won a support contract, which Chamberlin supervised. He personally supervised the company's facility at JSC until his death in 1981.

Chamberlin gained many honors and awards. NASA awarded him its Exceptional Scientific Achievement Award; Exceptional Service Medal; and Exceptional Engineering Achievement Medal. He was also a Professional Engineer of the Province of Ontario; an Associate Fellow of the Canadian Aeronautical Institute; a Member of the Institute of Aeronautical Scientists; and an Inductee in Canada's Aviation Hall of Fame (2001).

William L. Davidson

Born: June 21, 1937
Age at STG: 24
Spouse: Former Marguerite Ann Whitty
Children at STG: 3

BACKGROUND

"Bill" was born in Houston, Texas. His parents were Sam Lafayette and Ruby Lee Davidson of Arkansas, and he grew up with an older sister and two brothers. He attended the University of Texas and gained a B.S. in electrical engineering in 1959 and an M.S. in the same field in 1961. Bill met Ann in Austin and they married after his third year at UT. While in graduate school he worked at the Defense Research Laboratory and published his thesis on missile guidance, then worked for Texaco where cutting edge seismic research was underway to explore for oil.

STG

When Bill learned that the STG was sending a recruiting team to Houston, he interviewed, and accepted an offer of employment made by Chris Kraft. He moved his family to Langley and on December 4, 1961 reported for duty in the Flight Control Facilities Section of the Flight Control Branch in the Flight Operations Division. At that time the Section was planning ground mission control systems under the direction of Tec Roberts and Howard Kyle; in particular the Mercury Control Center at Cape Canaveral and the world-wide tracking network.

Bill became one of the team tasked with ensuring that the flight controllers got the telemetry data that they needed to meet the mission objectives and to keep the astronaut safe. This meant the operational requirements of the program had to be accurately determined, documented, and expeditiously issued to the MSFN for each mission. A Program Instrumentation Requirements Document (PIRD) included all of the STG's program mission requirements to the MSFN. Bill was accountable for assembling and assuring the completeness and accuracy of these telemetry requirements as he updated the PIRD telemetry section for each mission. He accomplished this responsibility by interacting with, and integrating the activities of each of the functional flight controllers (flight dynamics, booster, electrical & communications, etc.) who would use these data to execute critical mission operational decisions.

The MSFN also included two instrumented ships to cover gaps between land remote sites. When one of the ships was in the Baltimore Harbor for maintenance, Bill went to evaluate its telemetry readiness. The flight controllers were developing system flow diagrams and asking "what if" questions about what might fail and what recovery decisions and actions should be taken in each situation. Bill participated with the flight control teams' mission rules activities with regard to telemetry data requirements for systems and crew mission status.

NASA MSC

Bill moved to Houston in the late spring/early summer 1962. In April 1963 he became Special Assistant to John Hodge's Flight Control Division, with responsibility for advanced planning. After his Apollo duties he worked in the Advanced Missions Program Office on an Integrated Space Program that ultimately envisaged manned Mars expeditions. This work included flight operations analyses support to the Apollo Applications Program which led to the Skylab space station.

Bill also worked in the Earth Observations Aircraft Program Office, and flew as a Mission Manager in the high altitude WB-57F aircraft. And he worked in the Earth Resources Program Office to determine the rationale and data needed for a manned orbital space platform. Bill left NASA in 1975, but rejoined JSC in 1994 to work on risk management in its International Space Station Program Office. He moved to the Space Operations Program Office and retired in 2004 from the EVA Project Office.

OTHER

In 1975, Bill joined Brown & Root as a Project Coordinator in Alaska to build an oil gathering center that was to feed oil into the Alaskan pipeline then under construction. That was followed by Brown & Root Project Coordination in Mexico City for offshore oil platforms in the Bay of Campeche. He and family then returned to Houston, where he became a Systems Engineering Consultant for Eagle Engineering, with various Aerospace Corporations and NASA as clients. He also attended the University of Houston Clear Lake in the evenings and graduated in 1984 with a B.S. in computer science. After retiring from NASA, Bill continued to actively support various organizations, including the NASA Alumni League, the AIAA, The Mars Society, and the National Space Society.

Dr. Lt. Col. William K. Douglas

Born: September 5, 1922
Died: November 15, 1998 (Age: 76)
Spouse: Mariwade McIlroy
Children: One son

BACKGROUND

Douglas was borne in Estancia, New Mexico. After Phoenix High School he obtained a B.S. in science in 1942 from the Texas School of Mines and Metallurgy (now the University of Texas, El Paso). He received his M.D. from the University of Texas Medical Branch at Galveston in 1948. He then joined the Air Force but continued his medical studies. He earned a master's in public health from the Johns Hopkins School of Hygiene and Public Health in 1954, then was board-certified in aviation medicine in 1956. He became a flight surgeon at the U.S. Air Force Hospital at Langley Air Force Base.

STG

On April 1, 1959, Dr. Lt. Col. Douglas was selected as the personal physician for the Mercury astronauts and assigned to the Office of the Director of the STG, Robert Gilruth. He worked out of the Langley offices but when the flights began at the Cape he moved to Hanger S. As he was on leave from the Air Force, he still reported to the Office of the Assistant for Bioastronautics at Patrick Air Force Base; as did the astronauts' nurse 1st. Lt. Dee O'Hara. Dr. Douglas was in the Astronauts and Training Group headed up by Col. Keith G. Lindell. Also in the Group were Lt. Robert B. Voas, and Warren North who was formally with NASA Headquarters but worked on astronaut training. Dr. Douglas was able to get support from the Life Systems Branch that was headed by Dr. Lt. Col. Stanley C. White. In addition to his medical activities, Dr. Douglas also worked with bioinstru-mentation, pressure suits, physical stress tests, and flight simulations. At the time of a flight, he served in the Mercury Control Center. He supported the flights of Alan Shepard, Gus Grissom, John Glenn, and Scott Carpenter. Douglas often endured the same tests that he gave to the astronauts.

POST-STG

In 1962 Dr. Douglas became Assistant Deputy Director for Bioastronautics at the Air Force's Missile Test Center, with responsibility for all medical support for NASA manned flights into space. From 1966 to 1968 he was the Assistant Deputy Chief of Staff for

Bioastronautics and Medicine at the Air Force Systems Command located at Andrews Air Force Base in Maryland. He retired from the Air Force in January 1977 with the rank of a full colonel.

Dr. Douglas was a medical consultant from 1977 to 1986. During that time he was named a McDonnell Douglas Senior Fellow and authored many significant publications in the field of aerospace medicine. Also during this period, he joined the surviving members of the Mercury group of astronauts in establishing the Mercury Seven Foundation (now called the Astronaut Scholarship Foundation).

On November 15, 1998 he died from complications from a viral infection and pneumonia at the age of 76. Among his many honors and awards were: the Air Force Commendation Medal; the Air Force Legion of Merit with one Oak Leaf Cluster; the Air Force Association Citation of Honor; the Special American Medical Association Honor Citation; the W. Randolph Lovelace Award; and the Society of NASA Flight Surgeons.

R. Bryan Erb

Born: April 12, 1931
Age at STG: 28
Spouse: Former Donna Marie German
Children: 2

BACKGROUND

Bryan was born and raised in Calgary in Alberta, Canada. He was always interested in science and flying, and earned his B.S. in civil engineering at the University of Alberta, Edmonton, as well as a private pilot's license. He earned an M.S. in aerodynamics at Cranfield University in the U.K. as an Athlone Fellow, an M.S. in fluid mechanics at the University of Alberta, and an M.S. in management at the Massachusetts Institute of Technology as a Sloan Fellow. Later he was awarded an honorary doctorate by the University of Alberta in 1990.

Bryan began his aeronautical career as an aerodynamicist at A. V. Roe Aircraft in Toronto, Ontario, doing thermal analysis on the CF-105 Arrow. When that aircraft was canceled by the Canadian government, NASA sent a team led by Robert Gilruth to recruit company talent for Project Mercury, recently started at Langley, and Bryan was among a small group of AVRO engineers selected to join the STG, which he did in May 1959.

STG

One of Erb's first tasks at the STG was to analyze the performance of the heat shield that was planned for Mercury. His mathematical model for the performance of that ablative heat shield was the first such model for the expected re-entry heating of a manned spacecraft. However, in the spring of 1960 there was still a question about to how this re-entry protection was likely to work. Therefore a flight test named Big Joe was devised. That was the first test of the ablative heat shield, and Bryan used the data to calibrate his mathematical model.

He monitored the testing of a model of the Mercury capsule in the Unitary Wind Tunnel at Langley and analyzed the results. These tests, carried out in the Mach 4 to 5 range, explored how the heating rates varied over the front of the capsule and along the afterbody. One of the things observed was that the heating rate on the model was very high on the afterbody; often almost as high as on the front face. After the Big Joe capsule had been flown and recovered, it was discovered that the afterbody had buckled and been seriously damaged. It was obvious the heating rates on the afterbody were a lot higher than anticipated in the design. This required a redesign of the afterbody materials; in particular the provision of quarter-inch-thick beryllium "shingles" on the cylindrical portion of the capsule. This worked well on all subsequent flights. Bryan continued to follow the heat shield performance throughout the Mercury missions, with post-flight analyses.

By the spring of 1960, Bryan joined the Advanced Vehicles Team headed by Bob Piland as the thermal specialist. This little group of eight engineers laid the foundations for what became Apollo, and spearheaded three study contracts with industry to flesh out the options and details. A re-entry at lunar return speed brought new challenges in the thermal area, since not only was the convective heating much higher but there was also heating by radiation from the hot shock layer in front of the spacecraft. Little was known about such radiative heating at that time, so establishing its importance was a major task.

POST-STG

After the STG became the Manned Spacecraft Center, Bryan moved to Houston and continued doing the Apollo thermal studies. He became Subsystem Manager for the heat shield, and also Assistant Chief of the Structures and Mechanics Division. After his Sloan fellowship at MIT he returned to Houston and was made Manager of the Lunar Receiving Laboratory, a position that he held for the early missions when concern about biological contamination was at its greatest, with responsibility for quarantining the lunar astronauts and the examination and distribution of the lunar material by the principal investigators.

Subsequently, Bryan planned and managed important remote sensing experiments, including the first global inventory of wheat using satellite data. He was awarded the NASA Exceptional Service Medal for this accomplishment. He became Chief of the Earth Observations Division of NASA's Life Sciences Directorate in 1979, and later Manager of the Earth Resources Program Management Office. Immediately after retiring from NASA in 1985 he spent a year consulting with Eagle Engineering.

OTHER

In 1986 Bryan was recruited by the Canadian Space Agency as Assistant Director of Canadian Space Station Program, representing Canadian interests at the Johnson Space Center in Houston. In addition to his Space Station duties, Dr. Erb had, by 1991, become interested in the potential of solar power from space and he instigated and managed the Canadian Space Agency's Space Power Initiative. During his career Bryan authored or presented over 100 technical publications and major presentations, including invited papers for meetings in Canada, Brazil, France, Italy, Mali, Ghana, Australia, and the United States. He was also a lecturer at the International Space University Summer Sessions in Toronto, Huntsville, Houston, and Barcelona. After 26 years of service with NASA and another 16 with the CSA he finally retired in 2002.

Dennis E. Fielder

Born: 1930
Age at STG: 29

BACKGROUND

Fielder was born in Crouch End, North London and attended Wanstead Count High School. He won a national scholarship to enter the Royal Aircraft Establishment (RAE) as an engineering apprentice. After spending the next five years in all of the engineering workshops and technical departments of the RAE he gained a Higher National Certificate (B.S. equivalent) in electrical engineering in 1952. Fielder also earned a Glider Pilot License with a Silver-C Certificate. He remained with the RAE as an assistant experimental officer at the Cardigan Bay test range and missile recovery area until 1954.

PRE-NASA

In 1955, Fielder moved to Canada. While waiting for an opening at AVRO, he worked for the Canadian Electric Company, Ltd., on the weapons fire control systems for the CF-100. In 1956 he became a flight test engineer at AVRO working on instrumentation for the CF-100 weapons system. He also spent a year on temporary duty at the U.S. Navy R&D Center at Point Magu in California, where he worked on qualification tests of a CF-100 weapon system being considered for the CF-105 Arrow. In 1958 he returned to AVRO in

Canada and worked on flight testing the Hughes Falcon air-to-air missile for that aircraft. After the Arrow was canceled by the Canadian government in 1959, he was one of the many company engineers to be interviewed by the STG team.

STG
In the spring of 1959 Fielder was hired by the STG as an aeronautical research engineer in the Operations Division's Flight Control Branch headed by Gerald W. Brewer. His first job was to study the communications interfaces between the MCC, the BCC, the Goddard control center, and the Mercury Space Flight Network. He established requirements for communications, both voice and TTY, and coordinated them with the Langley Instrumentation Division that was the initial designer and contract manager for the system and the interface to the Goddard network team. He traveled to some of the sites, including the Bermuda site which included the Control Center and tracking station. Fielder was also involved with the simulation exercises that were related to the training of flight controllers. When the STG was dissolved in 1961, he moved to the Manned Spacecraft Center in Houston.

POST-STG
After Project Mercury, Fielder became involved with the communications requirements for the new Mission Control Center, in particular the interface between it and the Goddard voice, data and TTY. As technology was improving in the mid-1960s, the forms of communications were changing and Fielder worked to integrate new technology into the new control center's design, including using communication satellites to support future programs.

In 1966 Fielder became involved with Apollo and follow-on missions, including Skylab and studies of missions that could utilized the network of stations. He was Manager of the Program Planning Office until 1983, when he relocated to NASA Headquarters in the new Space Station Task Force. He worked there for a year, then transferred to the Space Station Program Office in Houston until retiring in 1985.

Fielder then created an independent company, DEF Enterprises, and for the next 15 years applied his experience to an array of future conceptual programs including advanced manned orbiting programs, large communities in orbit, solar power satellites, and Mars programs. He finally retired in 2000.

George Harris, Jr.

Born: July 5, 1929
Age at STG: 30
Current Spouse: Martha (née) McGowan
Children: Two boys and a girl

BACKGROUND

Harris was born in Willenhall, England, completed the British Electrical Engineers Apprentice Program with the Midlands Electricity Board, and in 1951 worked at its Television, Radio & Communications facility.

Harris left the U.K. in 1954, and the following year joined AVRO Canada where he was a member of the Experimental Flight Test Group and ran the technical side of their High Speed Flight Center. He also flew in the back seat of the CF-100 to test magnetic tape and telemetry systems that were intended to be installed in the CF-105 Arrow. After the cancelation of the Arrow in 1959, Harris performed similar work for North American Aircraft on the Navy A3J Vigilante.

STG

In January 1960 Harris was hired into the Flight Control Facilities Section of the STG. At that time his wife was Mary (née) Powell and they had one son, Kelly. After the move to Goddard they had a second son, Robert and a daughter, Sandy. The marriage ended in divorce owing to the time spent traveling in his job. He married Martha (née) McGowan in 1968. Harris worked with Dennis Fielder and Lyn Packham on telemetry systems, data routing and displays. He also worked with Paul Vavra and Bill Boyer of the Langley Instrumentation Research Division and Buck Heller from Goddard on the new Tracking and Ground Instrumentation Group (TAGIU). They established requirements for the MSFN. To test this network's capabilities and establish procedures, Harris equipped a DC-3 with Mercury communications, command, and telemetry systems. When the STG moved to Houston, Harris went to the Goddard Space Flight Center.

NASA GSFC

The DC-3 grew into a fleet of instrumented NASA aircraft, and Harris flew to all of the MSFN stations to conduct tests and to check out and train the ground station personnel. These aircraft evaluated the telemetry, voice, radar, and data gathering capabilities of the stations. He was also involved in the training of tracking station teams and spent about half the year visiting all of the sites around the world.

OTHER

In 1968 Harris left Goddard and became Operations Director of the Engineering and Operations Department of the European Space Research Organization (ESRO), applying his expertise to the design of the Darmstadt, Germany control center (ESOC) for European satellites and assisting in the training of the flight control team. Whilst there, he was the Flight Director for all the ESA flights.

In 1975 Harris returned to the U.S. as Chief of Systems Development for the USGS Center for Earth Resources Observation and Science at Sioux Falls, SD, where he led the team which developed the Earth Resources Image Processing and Enhancement Systems

(EDIPS) for the LANDSAT program. In 1979 he was made Assistant Vice President of Spacecom and Deputy Manager of the Tracking and Data Relay Satellite System (TDRSS) at White Sands, NM. The Inertial Upper Stage that was to place the first such satellite in geosynchronous orbit suffered a malfunction; Harris worked on the recovery effort that slowly maneuvered it into its operating position. In 1985 Harris worked briefly for the U.S. Information Agency's "Voice of America" on an automated control center for a White House interface. In 1987 he received a master's in engineering management from the American University.

Other positions held by Harris include: INTELSAT Liaison Officer at Arianespace in Evrey, France; Aerospace consultant to MCI Corporation for the Ariane 4 launch from French Guyana; Aerospace consultant for the HERMES manned spaceplane program at Alcatel, Paris; Executive Director, Commercial Space Development Office for the State of New Mexico; Manager/Flight Director, Joint CSA/NASA RADARSAT Antarctic Mapping Mission; and International Space Consultant for various aerospace clients.

His many awards include: the NASA Public Service Medal for his leadership of the TDRS-1 recovery (1984); the CSA/NASA Achievement Award as Flight Director for the RADARSAT Antarctic Mapping Mission (2000); the Institute of Electrical and Electronics Engineers' Life Time Senior Member Award (2001); and the Canadian Space Society/National Space Society Award-Member of the Space Task Group (1994).

John D. Hodge

Born: February 10, 1929
Age at STG: 30
Spouse: Audry
Children: Two boys and two girls

BACKGROUND
Hodge was born in Leigh-on-Sea in Essex, England and attended Minchenden Grammar School in Southgate, London. After WW-II, he went to the Northhampton Engineering College of the University of London and graduated with the equivalent of a B.S. in engineering in 1949. His first position was as an aerodynamics engineer with Vickers-Armstrong, Ltd., in Weybridge. In 1952 he and his wife Audry moved to Toronto, Canada, where he was hired by AVRO as Head of the Air Loads Section. He worked on the engine inlet designs of supersonic aircraft. After the cancellation of the CF-105 Arrow aircraft, he was interviewed by NASA and joined the STG at Langley.

STG

Hodge arrived at Langley in the spring of 1959; about the same time as the Mercury astronauts came onboard. He was assigned as Technical Assistant to Charles Mathews in the Operations Division. He became involved with the Bermuda Control Center, which was built in 1960 and operational in September 1961, and was the Bermuda Flight Director for the MA-4, MA-5 and MA-6 missions; the latter being when John Glenn made America's first orbital flight.

NASA MSC

Although the STG was dissolved at the end of 1961, the Mercury flights continued and Hodge was the second-shift Flight Director in the Houston MCC for the long duration MA-9 flight of Gordon Cooper in May 1963 that wrapped up Project Mercury.

Hodge was also a Flight Director for Gemini and was on duty during the Apollo 1 accident. He remained through the Apollo lunar landings of 1969 and then left in 1970 to work for the Transportation Systems Center in Cambridge, Massachusetts. He also worked for the Urban Transportation Development Corporation in Toronto, Canada. Hodge returned to the United States to work for the Department of Transportation, and in 1982 led the Space Station Task Force at NASA Headquarters until his retirement in 1987. He then formed J. D. Hodge and Company to work in international management and aerospace consulting.

Hodge's honors and awards included the following from NASA: the Group Achievement Award, Project Mercury (1962); the Quality Award (1963); the Superior Performance Award (1964); the Exception Service Medal (1967 and 1969); the Group Achievement Award, Gemini Program (1967); the Group Achievement Award, Apollo Program (1969); and the Presidential Rank Award of Meritorious Executive, NASA (1985). He received the Special Achievement Award (1979) from the Department of Transportation. In addition, he received the Arthur S. Fleming Award (1968).

Carl R. Huss

Born: July 11, 1925
Died: August 8, 1996 (Age 71)
Spouse: Margaret "Marge" (née) Kuskey (died in 1974)
Children: one son

BACKGROUND
Huss attended Wheeling High School and graduated in 1943. After serving in the Air Force from 1944 to 1946, he attended the University of West Virginia and graduated in 1949 with a B.S. in aeronautical engineering.

NACA
Huss joined the Langley Memorial Aeronautical Laboratory in August, 1949 and was assigned to a variety of tasks until transferring to the STG.

STG
In 1959 Huss joined the Mission Analysis Branch, Trajectory Analysis Section headed by John Mayer. His work involved trajectory analysis for the launch, orbit, and re-entry of the Mercury capsule, taking into account the evolving mission rules and constraints. This Section developed nominal and contingency trajectories and numerous abort conditions for different missions and determined the logic and equations for the computers.

As lead engineer for the Mercury Atlas orbital flights, Huss worked closely with John Mayer, Clay Hicks, Charlie Allen, Ted Skopinski and John Maynard on the preliminary definition of the mission planning process. Having specialized in planning the de-orbit maneuver, he became the first Retrofire Officer (RETRO) in the Mercury Control Center.

NASA MSC
Huss moved with the Mission Analysis Branch to Houston and continued supporting the flights out of the Cape Mercury Control Center. A mild heart attack shortly after Project Mercury was concluded with the MA-9 flight in May 1963, prompted Huss to move into management of the newly named Mission Analysis and Planning Division. In 1967 he became the Assistant Chief for Mission Design; a position that he held through the duration of his distinguished career. He received the NASA Outstanding Leadership Medal in 1981.

Caldwell C. Johnson

Born: 1919
Died: May 26, 2008 (Age 89)
Age at STG: 39
Spouse: Former Kathryn Lancaster
Children: 2

BACKROUND

Johnson was born in Wythe, Virginia, which isn't far from Langley. He went to Hampton High School, graduating in 1937. As a young man, he was interested in model airplanes and was very good at building them; so good in fact that Robert Gilruth hired the teen-ager out of high school to work in the Design Group of the Langley Memorial Aeronautical Laboratory as an artist and model builder. Caldwell attended the University of Virginia and studied engineering for a short while, but opted to return to Langley.

NACA

Johnson worked for the Pilotless Aircraft Research Division (PARD) during WW-II, launching aircraft models on various Army rockets from the NACA Wallops Island Station on the Eastern Shore of Virginia. These models were used in research applied to re-entry heating, aerodynamic drag, and the study of various aircraft/spacecraft shapes. Then his supervisors and many of his colleagues transferred to the STG and became some of its key managers.

STG

As the STG was forming in November 1958, Johnson was asked by Max Faget, his mentor and previous supervisor, to join the STG in the Engineering Branch of the Engineering and Contract Administration Division. When the STG was reorganized in 1960 Johnson became Head of the Systems Engineering Branch, a position that he held until the transfer to Houston in November 1961.

Caldwell's work at PARD directly related to the design concepts for a single-man spacecraft. He and his group, directed by Faget, reviewed the results of the previous studies that led to the "blunt end" designs for re-entry vehicles and then they performed conceptual spacecraft layouts. Faget would supply primitive drawings, and Caldwell's team would improve on them and pass them to the Langley model-makers. This process evolved into full scale mockups. In arranging the major systems within the capsule Caldwell had to take into account the sizes of the recently selected astronauts. He also considered the size and shape of the Redstone and Atlas, and their interfaces to the capsule. He and his team were given all due credit for the final design of the Mercury capsule.

POST-STG

Johnson was always comfortable with the up-front phase of design, so once Mercury was well underway he transferred to the new Manned Spacecraft Center in Houston. Initially he focused on Gemini design concepts, and then on various early Apollo concepts. From 1962 to 1963 he was Manager of the Apollo CSM Engineering Office, producing many conceptual drawings of various configurations for the Command and Service Module and also the Lunar Module with which it would have to operate during a lunar flight. He was

Assistant Chief of the Spacecraft Design Division from 1964 to 1968 and its Chief from 1968 until 1974.

When the idea came to cooperate with the Russians on a joint mission, Johnson traveled to Moscow to discuss joint docking systems. He promoted the merits of an androgynous system over the probe-and-drogue system of Apollo. Such a system was negotiated with the Russians and successfully used on the Apollo-Soyuz docking. Johnson was the principal investigator for habitability and crew quarters on the Skylab space station. Many of the interior changes on the workshop were due to recommendations by Johnson and his team. The Skylab experience had a great influence on the design of the American portion of the International Space Station.

OTHER

After retiring from NASA, Johnson became Chief Engineer for Space Industries Inc., founded by Max Faget with the aim of developing a space platform for microgravity experiments that would be serviced by a visiting Space Shuttle.

He has received numerous awards during his NACA/NASA career spanning 37 years. Along with M. A. Faget, A. J. Meyer, R. G. Chilton, W. S. Blanchard, A. B. Kehlet, J. B. Hammack, C. C. Johnson, and D. J. Bergeron, he shares the following patents: Mercury Capsule (1963); Space Capsule (1966); Spacecraft Operable in Two Alternative Flight Modes (1988); Spacecraft with Articulated Solar Array (1988); and Modular Spacecraft System (1989).

Kenneth S. Kleinknecht
Manager of Project Mercury

Born: 1919
Died: November 20, 2007 (age 88)
Age at STG: 40
Spouse: Patricia

BACKGROUND

"Kenny" Kleinknecht earned a B.S. in mechanical engineering from Purdue University in 1942, and during WW-II was a private in the U.S. Army Air Corp Reserve.

NACA

Kleinknecht joined the Lewis Flight Propulsion Laboratory right out of college. He worked on aircraft engine cooling and engine icing on many military aircraft of the war years, including the B-24, B-25, B-29, P-51, and P-59. Over his nine years at Lewis, he rose from junior mechanical engineer to aeronautical research scientist. In 1952 he went to the High Speed Flight Research Station at Edwards to work on the X-Series of aircraft, including the X-15, becoming a member of the Advanced Projects Management Officer under Walter Williams. In 1958 he took a course in space technology from the University of California. When the STG was formed in November 1958 he and others from the HSFS moved to Langley.

STG

At the request of Hugh Dryden, former Director of the NACA and now Deputy Administrator of NASA, Kleinknecht moved to the STG, along with Williams and Martin A. Byrnes. He was the Technical Assistant to Robert Gilruth as an aeronautical research scientist. His initial work was to learn the systems of the Mercury capsule and deliberate with McDonnell engineers on status and engineering issues. His focus was on safety and reliability. He also looked into the test and checkout that was going on at McDonnell, and worked with his former Lewis colleagues in the same activities in Hanger S at the Cape. He monitored this process and was on the Flight Safety Review Board for the capsule.

On January 15, 1962 Kleinknecht became Manager of the newly established Mercury Project Office. His primary responsibility was the technical direction of the McDonnell contract and all other industrial contractors hired to work on Project Mercury. There were initially 42 people in the Project Office responsible for scheduling, procurement, and technical monitoring tasks. The Office remained at Langley until after the MA-7 flight by Scott Carpenter and then move to the Manned Spacecraft Center in Houston. Kleinknecht remained on Project Mercury until it was wrapped up by MA-9 in May 1963.

POST-STG

In Houston, Kleinknecht became Deputy Manager of the Gemini Program Office, a position he held through to the completion of that program. In 1967 he became the Manager for the Apollo Command and Service Modules in the newly established Apollo Program Office. As plans were being made for the first American space station, Kleinknecht became the Manager of the Skylab Program Office in 1970. He served as the Johnson Space Center's Director of Flight Operations from 1974 to 1976 and was Assistant Manager of the Space Shuttle Orbiter Project Office from 1976 to 1977. He then went to NASA Headquarters as the Deputy Associate Administrator for Space Flight, Space Transportation Systems, ESA, and Spacelab. In 1979 he became Assistant Manager of the Space Shuttle Orbiter Project Office.

Kleinknecht retired from NASA in 1981 and joined the Martin Marietta Aerospace/Lockheed Martin Astronautics Denver, Colorado office, holding a number of senior management positions on spaceflight programs until his retirement in 1990. His honors and awards are too numerous to list, but they included NASA Group Achievement Awards for Mercury, Gemini, Apollo, Skylab, and Space Shuttle, as well as the agency's Distinguished Service Medal and Exceptional Service Medal.

Eugene F. Kranz

Born: August 17, 1933
Age at STG: 27
Spouse: Former Marta Cadena
Children: 6

BACKGROUND

"Gene" Kranz was born in Toledo, Ohio. His father was Leo Peter Kranz, the son of a German immigrant and local farmer. Gene was the third child, with two older sisters Louis and Helen. He graduated from Central Catholic High School in 1951, then went to St. Louis University's Parks College of Engineering, Aviation and Technology and graduated in 1954. He joined the Air Force, and then on completing his pilot training at Lackland AFB in Texas in 1955 he was commissioned a second lieutenant. Shortly thereafter, he married Marta. He was sent to South Korea and flew the F-86 Sabre along the DMZ. In 1956, after he finished his tour in Korea, he resigned his commission to work for McDonnell Aircraft Corporation. He was assigned as a flight test engineer to carry out research and testing of surface-to-air missiles at the Research Center at Holloman AFB in New Mexico.

STG

Kranz joined the STG at Langley in October 1960, working in the Flight Control Operations Section under the Flight Control Branch headed by Gerald Brewer. The Branch was under the Flight Operations Division headed by Charles Mathews and his assistant Christopher Kraft. At that time, Kraft was also the Mercury Flight Director and he selected Kranz as the Procedures Officer for the first Mercury Redstone flight, which occurred on November 21, 1960. Kranz remained in this organization throughout the STG period. Even after the move to the MSC in Houston he remained in Flight Operations.

In his early work on unmanned Mercury Redstone flights, Kranz developed the procedures that defined the "Go/No-Go" rules for launch and abort conditions. This work led to the formal process known as Mission Rules. He also developed the operational procedures for interfacing with the launch controllers. This required timely interface responses between the test conductor in the Launch Control Center and the Flight Director in the Mercury Control Center. Kranz was also responsible for developing the communications interface between the MCC and the MSFN whose stations and ships were positioned all over the world and required communications with, and direction from, the Mercury Control Center; usually via the Procedures Officer. By the time of Shepard and Glenn's flights, these operational procedures were in place and well developed. After Glenn's MA-6 flight, Kranz was promoted to Assistant Flight Director for the subsequent manned orbital Mercury flights.

POST-STG

In November 1961, the STG was folded into the new Manned Spacecraft Center and personnel gradually migrated to Houston. Kranz remained in Flight Operations in both mission planning and flight operations roles and became a Flight Director for Gemini and Apollo. After Apollo 7 and Apollo 9, he had the honor of manning that console for the Apollo 11 lunar landing of the *Eagle*. He was one of the Flight Directors on Apollo 13. In the 1995 movie *Apollo 13*, his role was played by Ed Harris, who received an Oscar nomination for Best Performance by an Actor in a Supporting Role. The movie also made Kranz an international celebrity. He continued as a Flight Director through to the final Apollo lunar flight. In 1974 he became Deputy Director of NASA Mission Operations and in 1983, with the Space Shuttle flying, he became the Director. He retired in 1994 after 34 years of spaceflight service.

He wrote his memoir *Failure is Not an Option* in 2000, and it was adapted for cable TV by the History Channel in 2004. He also appeared in the 2008 Discovery Channel documentary series *When We Left Earth*. His awards are too numerous to list, but include: the Presidential Medal of Freedom; the NASA Exceptional Service Medal; the NASA Distinguished Service Medal; the NASA Exceptional Service Medal; and the Robert R. Gilruth Award.

Howard C. Kyle

Born: January 1, 1921
Died: January 6, 1994 (Age: 73)
Age at STG: 37
Spouse: Josephine

BACKGROUND
No information was found on Kyle's early life.

NACA
As early as September 1958, shortly before NASA was created, Kyle's name appeared on a list of people considered for the STG. Whilst not on the initial Core Team, he was transferred from Langley's Instrument Research Division to the STG in early 1959 as the interface between the Flight Control Branch which needed a communications and tracking network and the Tracking System Study Group (TSSG) that was tasked with designing it. See Section 11.1.2.

STG
Kyle joined the Operations Division, Flight Control Branch, Control Central and Flight Safety Section as Assistant to Gerald W. Brewer. There he worked on the early design of the remote tracking stations from the viewpoint of the flight controllers that would operate them. He also worked on the design of the Mercury Control Center, from the standpoint of communications and tracking. When contracts were let for the network systems, Kyle ensured that they met the communications and tracking needs of the flight controllers. He assisted Flight Director Chris Kraft on the early Mercury Redstone flights and other flight controllers with their operational procedures in the new Mercury Control Center. During some early simulations Kyle actually doubled for Kraft as Flight Director and as CAPCOM.

POST-STG
No information was found on what Kyle did after Project Mercury.

Charles R. Lewis

Born: December 24, 1937
Age at STG: 24
Spouse: Carolyn
Children: one son and one daughter

BACKGROUND

"Chuck" Lewis was born in Lawton, Oklahoma. On graduating from high school and Cameron Junior College in Lawton he attended New Mexico State University, from which he received a B.S. in electrical engineering in 1961.

STG

After graduation Lewis began his career at the NASA Goddard Space Flight Center, working in the Operations Control Branch for the Minitrack network. On April 1, 1962 he transferred to the STG at Langley and began Flight Control Class 101 training as a remote site flight controller. In May he went to the Zanzibar (ZZB) tracking station to support Scott Carpenter's MA-7 flight as CAPCOM.

POST-STG

Back from Zanzibar, Lewis moved to the Manned Spacecraft Center in Houston. He continued supporting Project Mercury as a CAPCOM, first at the Canary Island tracking station for Wally Schirra's MA-8 flight and then at the Muchea station in Australia for Gordon Cooper's MA-9.

During the Gemini Program he was a remote site CAPCOM for GT-2 on the CSQ tracking ship; for GT-3 he was at the Guaymas (GYM) station in Mexico; for GT-4 he was on the CSQ; for GT-5 he was at the Carnarvon (CRO) station in Australia; for the first attempt for GT-6 he deployed to the Hawaii (HAW) station; and for the GT-7/6 rendezvous mission he was back on the CSQ. During the remainder of the Gemini Program, Lewis was in charge of developing the remote site conceptual definition for telemetry and command processing and control for Apollo and the generic requirements for hardware and software development.

During the Apollo program, Lewis was an Assistant Flight Director in the Mission Control Center for Apollo 5 (the first test of the Lunar Module, unmanned), Apollo 9, 10, 11, 12 and 13. Between missions, he was the Assistant Chief of the Flight Control Branch and later the Flight Operations and Recovery Branch. He became a Flight Director for Apollo 16 and 17, and then the flights to the Skylab space station.

In between serving as Flight Director, Lewis held various management positions including: Chief of the Communications and Data Systems Branch from 1974–1976, Chief of the Flight Training Branch from 1976–1978, and Chief of the Flight Operations Integration Office from 1984 through to his retirement in 1994. During the Apollo-Soyuz mission, he led the American flight control team in the Soviet Control Center. He was the on-orbit Flight Director on STS-1 and then the lead Flight Director on STS-2, STS-4 and STS-9.

OTHER

Chuck Lewis's many honors and awards from NASA included: the Exceptional Service Medal, (Skylab, 1973); the Exceptional Service Medal, (STS-1, 1981); the Outstanding Leadership Medal (STS-9/Spacelab 1, 1984); and the Outstanding Leadership Medal (Space Station, 1993).

John S. Llewellyn

Born: March 30, 1931
Died: May 8, 2012 (age 81)
Age at STG: 28
Spouse: Olga
Children: 2

BACKGROUND

Llewellyn was born in Dare, Virginia, in what is known as the Tidewater Area. He was the son of John Stanley Llewellyn and May Parker. After high school he joined the Marines at age 19 and was deployed to Korea with the 1st Marine Division. He survived the Battle of the Chosin Reservoir and was awarded two Purple Hearts and the Bronze Star. On leaving the Marines in 1954 he attended Randolph Macon College in Ashland, Virginia from 1954–1956. After a year of work he went to William and Mary College in Williamsburg, Virginia and graduated with a B.S. in physics in 1958.

NACA

In 1957, Llewellyn was hired by the NACA Langley Memorial Aeronautical Laboratory as a mathematics aide in the Pilotless Aircraft Research Division, gaining experience with re-entry heating experiments at the Wallops Island Station. Then he went back to college to finish his degree. After graduation in 1958 he returned to Langley as a research engineer in the Analysis and Computation Branch of the Structures Research Division.

STG

After the creation of the STG in late 1958, Llewellyn worked for the Flight Systems Division managed by Max Faget. He worked in the Performance Branch managed by Aleck Bond and in the Heat Transfer Section headed by Leonard Rabb. Others in that section included: Eugene L. Duret, R. Bryan Erb, Joanna M. Evans, Archie L. Fitzkee, Steven Jacobs, Robert O'Neal, Emily W. Stephens, and Kenneth Weston.

Sometimes the Operations Division would borrow people from Flight Systems to man various positions at the remote tracking stations. In 1961 Llewellyn was on the *Coastal Sentry Quebec* in U.S. waters, serving as a CAPCOM during training for the MR-3 flight of Alan Shepard. He then supported MA-4 at Zanzibar and, as his final such assignment as a member of the STG, he was at Kano in Nigeria for the flight of the chimpanzee Enos on MA-5.

POST-STG
After transferring to Houston, Llewellyn continued as a CAPCOM for John Glenn's flight on MA-6 at the Canary Islands. With his knowledge of re-entry heating and the intricate workings of the capsule clock, Llewellyn was persuaded by Tec Roberts to join Carl Huss in the Mission Analysis Branch in the Operations Division, working as a RETRO. His first flight in this role was in the Mercury Control Center for Scott Carpenter's MA-7 flight, which landed far down range and Llewellyn experienced an abnormal re-entry. He spent many years as a RETRO on Gemini and Apollo missions working as a member of the MSC Flight Dynamics Branch of the Flight Control Division. In the early 1970s he worked on various Earth and lunar experiments, and then in the later part of that decade he worked on Space Shuttle Payload Integration. In the early 1980s he worked in Aircraft Operations.

After retiring from NASA, Llewellyn became a rancher with a sideline as an engineer with SkyComm International.

C. Frederick Matthews

Born: November 28, 1922
Age at STG: 38
Spouse: Frances E. Hood
Children: 3

BACKGROUND
Matthews was born in Guelph, Ontario, Canada. He saw his first barn-storming plane at age 6 and while growing up he built model airplanes, read magazines about flying, and attended air shows. He graduated from Runnymede Collegiate Institute with honors and then started at the University of Toronto in 1942, but after one year he enlisted in the Royal Canadian Air Force (RCAF), becoming a pilot and ferrying and testing aircraft for two years. During that time, he married his high school sweetheart Frances Hood.

After the war he returned to the University of Toronto, but in the interim worked at Victory Aircraft milling Lancaster bomber wing spars. While in college he visited other major aircraft manufacturers, including Bell Aircraft and saw demonstrations of the P-59 Airacomet and the RP-63 Kingcobra. Other college trips included the National Research Council's wind tunnels, and demonstrations of Sikorsky's new helicopter. His summer jobs included the AVRO Stress Office working on the C-102 Jetliner. After graduating with a B.S. in aeronautical engineering, he returned to AVRO and continued work on the C-102 as well as the CF-100 twin-engine jet fighter. He was the first flight test engineer on the CF-100, along with test pilot Bruce Warren.

When the Jetliner was completed, Matthews worked with Howard Hughes on the test and demonstration flights out of Culver City, California airport. He flew on many pioneering test flights in the AVRO Jetliner; one of which was the first inter-city jet air mail from Toronto to New York. Some of these flights were the first jet transports ever to land at Midway and at La Guardia. Other AVRO flight test activity included the Lancaster-Orenda, icing issues and their solution, and redesigning the CF-100 canopy. Matthews was also involved with evaluating the ejection seats of the CF-105 Arrow. He worked on a real-time flight monitoring facility for the Arrow. This experience was applicable to the design of the Mercury Control Center, as was his close association with crews and the dangers of flight test. On February 20, 1959 the Canadian government canceled the Arrow and Matthews was suddenly one of 14,000 engineers out of a job.

STG

As one of the many AVRO engineers hired by NASA, Matthews joined the STG in the Control Center and Flight Safety Section within the Flight Control Branch headed by Gerald W. Brewer. Later this became the Flight Control Operations Section. They were in the Operations Division headed by Charles Mathews. Matthews would often travel to Cape Canaveral, Wallops Island Station, and NASA Headquarters for flight test and safety meetings. During space missions he would backup Chris Kraft and John Hodge in the Mercury Control Center. This led to his being put in charge of managing flight operations of the remote sites. NASA augmented Matthews's staff with 19 Philco engineers who assisted with both the flight controller training program and manning the remote sites along with the NASA people.

He worked with Dr. Capt. Bill Augerson, one of the astronauts' flight surgeons, regarding in-flight monitoring, telemetry displays, and the role of the doctors during flight operations both in the control center and at the remote sites. Matthews also worked with the Range Safety Officers and others at the Cape in pre-flight operations, regarding safety and contingency situations. Prior to the completion of the Mercury Control Center, Matthews ordered an interim Mission Control Trailer built but the MCC was completed before the trailer was needed; it was used later at the Navy's Johnsville centrifuge and other locations.

Matthews's group prepared some of the first operational flight control documents, including the Mission Rules, Flight Controller Handbooks, and related materials such as schematics of the capsule systems. He attached McDonnell engineers to the group to create these materials and to train flight controllers. In one instance, this training and analysis actually required the company to add a safety guard over the JETT RETRO switch. Makeshift world-wide flight simulations at Langley provided training prior to the actual world-wide deployment of the flight control teams in 1961.

Matthews contributed to the STG from the Little Joe launch to John Glenn's manned orbital mission, then left the STG in March 1962.

OTHER
Matthews moved to Lexington, Massachusetts to work for the Automated Systems Division of RCA. He was a senior engineering scientist and consulting engineer for a variety of advanced systems engineering projects. These included the Saturn launch vehicle, the EC-135 Looking Glass Airborne Command Post, the Advanced Airborne Command Post, the Lunar Excursion Module, and assisting the Tactical Air Force's development of automated aids for intelligence activities. When RCA was acquired by General Electric after 25 years, he developed a formal Systems Engineering Procedures for the Division, then retired from the company four later in order to pursue several independent consulting contracts until he finally retired in 1999.

Owen E. Maynard

Born: October 27, 1924
Died: July 15, 2000 (age 75)
Age at STG: 34
Spouse: Helen
Children: Two daughters and one son

BACKGROUND
Maynard was born in Sarnia, Ontario, Canada. He enlisted in the Royal Canadian Air Force at age 18, trained as a Mosquito pilot, and served in England during WW-II as a Flying Officer. After the war he returned to Canada and worked for AVRO, taking time off to earn a B.A. in aeronautical engineering from the University of Toronto in 1951. He was awarded an honorary doctorate in engineering from his alma mater in 1996.

Pre-NASA
Maynard started with AVRO in 1946 as a craftsman and worked on the CF-100 and the C-102 Jetliner. After gaining his degree in 1951, he worked as a senior stress engineer on the CF-105 Arrow. He applied his piloting and flight test skills to focus on the crew station

and the escape systems. When the Arrow was canceled in 1959, he was interviewed by the STG and hired on the recommendation of Jim Chamberlin for his flight test experience and engineering work.

STG

Maynard reported to the Flight Systems Division headed by Max Faget and the Systems Test Branch headed by William M. Bland. He was in good company, as other ex-AVRO engineers were in this Branch, including H. Kurt Strass, and Rodney G. Rose. Given his experience with crew safety, he worked on the capsule landing bag and designed drop tests for models and the actual capsule.

In July 1960 Maynard was part of the recovery effort for the flight test of the MA-1 capsule, following the failure of its launch vehicle. He performed a 30 foot free-dive to recover a critical component which proved that the Atlas adapter required reinforcement. See Section 13.2.7. He also assisted in the effort to identify the remaining technical issues on the capsule and assigned people to work the problems. By 1961 the STG was more formally organized and Maynard was in the Systems Integration Section of the Systems Engineering Branch under Caldwell Johnson.

POST-STG

After Project Mercury, Maynard became involved with Gemini and Apollo. By 1963 he was Acting Manager of the Spacecraft Systems Office, and later on Chief of the Lunar Excursion Module (LEM) Engineering Office. Throughout the late 1960s he held several management positions in the Apollo Spacecraft Program Office, including Chief of the Systems Engineering Division and Chief of the Mission Operations Division during the redesign which followed the Apollo 1 fire.

Maynard left NASA in 1970 and became the Senior Engineering Manager at Raytheon in Sudbury, Massachusetts, working on several aerospace programs. His interests there were the concept and feasibility of solar powered satellites. When he retired from Raytheon in 1992 he and his wife Helen settled in Waterloo, Ontario. He died at age 75 on July 15, 2000. He twice received the NASA Exceptional Service Medal.

Harold G. Miller

Born: September 22, 1937
Age at STG: 22–25
Spouse: Susan (Sue) Lynn Miller
Children: one son and one daughter

BACKGROUND
"Hal" Miller was born in Vanleer, Tennessee to Eugene N. and Mildred E. Miller. He went to Charlotte High School, graduating in 1955, and then earned a degree in electrical engineering from the Tennessee Polytechnic Institute in 1959.

STG
Miller was offered a job at graduation with the NASA Langley Research Center and reported to the personnel office on July 7, 1959. That day the Director of Personnel interviewed nine young graduates (Miller being one of them) and he sent every other man to work at either at the LRC or the newly created STG which was officially attached to Goddard Space Center but physically at Langley. Miller was sent to see Tom Markley of the STG; it could so easily have gone the other way! He reported to the Operations Analysis Section headed by Jack Cohen within the Mission Analysis Branch headed by John Mayer of the Operations Division headed by Charles Mathews. After several months he was assigned to Jack Cohen's Simulation Task Group, where he stayed throughout his time with the STG.

Miller's first task was to figure out how to simulate the communications system for Project Mercury. He wrote interface specifications for communicating between the Mercury Procedures Trainers (one at Langley and one at the MCC) and the Mercury Control Center that was being designed and built at Cape Canaveral during this time. He worked on a simulation at Langley which created "mock up" remote site tracking stations in the Full Scale Wind Tunnel building. Exercises were developed to train new flight controllers how to communicate with astronauts and the control center. He worked with the Western Electric contractors that implemented the actual systems.

Miller often traveled to Cape Canaveral to work on the simulations with the astronauts, the technical contractors at the MCC, and the flight controllers. As a member of the team that was developing the simulations for training purposes, Hal would simulate potential spacecraft and ground failures to exercise the procedures developed by the flight controllers. These exercises were conducted in real time for two to three weeks prior to each Mercury mission, and Miller was the Simulation Supervisor for Scott Carpenter's MA-7 flight. He participated in all of the Redstone and manned Mercury Atlas missions.

POST-STG
When Miller became Head of the Simulation Design Section at the Manned Spacecraft Center, he developed the specifications for the design of the simulation systems to be incorporated into the new facility. The simulation systems now included Gemini and Apollo, and interfaces with other NASA centers and contractors that were providing spacecraft hardware and software. He argued for and obtained one of the first fully digital real-time simulations of a flying spacecraft; the Agena that would serve as a docking target for Gemini. Digital simulations were to follow for the Saturn boosters and the Apollo spacecraft. His final year in the Flight Control Division was managing the support contractors.

Miller moved to NASA Headquarters in 1983 and worked with the Space Station Task Force. He also worked in the Office of Manned Space Flight, tracking Space Shuttle performance. Hal retired in January 1999 after receiving numerous awards during his 17 years with NASA.

Billy Warren Pratt

Born: March 25, 1938
Age at STG: 22
Spouse: Former Dorothy Mae Wyatt
Children: 2

BACKGROUND

Billy was born at home on a farm in Knott County Kentucky. His parents were Wiley C. and Mary Alice Pratt. His father was a farmer and coal miner. Billy graduated from Scott County High School in 1956 and entered the University of Kentucky that fall. At the end of his first semester, lacking money, he tried in vain to find work to enable him to return to school. After learning of Billy's unsuccessfully attempts, his uncle, Jasper Pratt, Jr., invited him to come to Hampton, VA to live with his family while he worked to earn the funds he needed in order to return to school.

NACA

Billy joined the NACA Langley Memorial Aeronautical Center in April 1957 as an engineering aide. Later, he was able to join the engineering cooperative program. As an aide and a "co-op," his work assignments included participation in wind tunnel testing. As his familiarity with test procedures grew, so did his responsibilities, which eventually included scheduling wind tunnel tests, drafting (drawing) models to be tested, and overseeing activities related to the fabrication and preparation of the models to be tested, conducting the tests, reducing the data, and writing the preliminary test results reports.

From 1957 he worked at several NACA facilities at Langley, including 8 foot Subsonic and Supersonic wind tunnels, Full Scale Tunnel, 8 × 10 foot Wind Tunnel, and the Pilotless Aircraft Research Division (PARD) where he participated in the launch of a multi-stage sounding rocket fired from the Wallops Island Station and the reduction of the resulting data. These experiences created a burning interest in the space program, and after the STG was formed Billy transferred in 1960.

POST-STG

Billy and his family moved to Houston in February 1962, where he joined the Apollo Program Office. An early task was to organize and manage the Apollo Document and Control group. He participated in the development of the initial Apollo contract with North American Rockwell; undertaking fact finding and assisting in the negotiation. This activity initially took place at the Rice Hotel in Houston, but then moved to a newly refurbished building at the nearby Ellington AFB.

In February 1964 the Manned Spacecraft Center finally became available for occupancy and the Apollo Program Office moved into what is now Building 1. Billy's initial assignment was to assist in the development, negotiation, and management of contracts for the automated checkout of Apollo spacecraft systems and hardware. The Acceptance Checkout Equipment (ACE-S/C) automated the checkout and acceptance testing of hardware at the manufacturer's facilities and for further testing at the Marshall Space Flight Center and in the Thermal Vacuum Chambers in Houston. It was also used in pre-flight checkout at the Kennedy Space Center. Billy eventually became the ACE-S/C contract Project Manager, with technical direction authority for contracts with General Electric, Radiation Incorporated, and the Control Data Corporation. He remained Project Manager until completion of delivery, installation, and acceptance of the 12 ACE-S/C stations that were installed at Rockwell, Grumman, Marshall, Houston, and the Cape.

Billy volunteered to work in the SPAN room in Building 30 for most Apollo missions. This was the communications hub and interface between the flight controllers in the mission control room and the rest of the aerospace community. It was for his participation in SPAN operations during the Apollo 13 mission, and as a part of the Apollo 13 Mission Operations Team, that he became one of the recipients of the Presidential Medal of Freedom.

In June 1988 Billy was promoted to Manager of the Engineering Integration Budget Office for the National Space Transportation System. In this branch-level position he was responsible for managing and supervising the people who planned, developed and managed the budgets of the Space Shuttle integration contractor. In 1989 he transferred to the Space Station Freedom Contract Management Division in Reston, Virginia, where he managed an organization which was responsible for business management of Grumman's contract for Space Station Freedom Integration; management of the contract for the Reston office; and overseeing all other Space Station Freedom prime contracts, including those that were directly managed by other NASA centers. Billy held this position until the Space Station Freedom Office closed in early 1994.

Returning to Houston in May 1994, Billy became manager of the Space and Life Sciences Business Management Office, a division-level position he kept until shortly before retiring in 1999 after 42 years with NACA/NASA.

Billy Pratt cherishes having been given the opportunity to participate in the NASA space program, and in particular the landing of men on the Moon, which he regards as the greatest endeavor undertaken by the United States of America during his lifetime.

Tecwyn Roberts

Born: October 10, 1925
Died: December 27, 1988
Age at STG: 34
Spouse: Doris Sprake
Children: one son

BACKGROUND
"Tec" Roberts was born in "Trefnant Bach" cottage in Llanddaniel Fab, Anglesey, Wales. He went to primary school at Ysgol Parc y Bont, and then the Beaumaris Grammar School, from which he graduated in 1942. Roberts began an engineering apprenticeship with Saunders-Roe located near his home town. He served for two years in the RAF and in 1944 resumed his work with Saunders-Roe. He attended Southampton University and also the Isle of Wight Technical College, earning a Higher National Certificate in aeronautical engineering in 1948.

In December 1952 Roberts moved to Canada to work for AVRO on the CF-105 Arrow, and after that was canceled in 1959 he was recruited by NASA.

STG
His first assignment for the STG was in the Control Central and Flight Safety Section under the Operation Division of Charles Mathews. His task was to define the requirements for the tracking and communications network and the Mercury Control Center. This led to the definition of the data flow required to enable the Flight Dynamics Officer to make launch and abort decisions. It was then logical that he should become the first FIDO. He supported all of the Mercury flights from MR-1 to MA-6; the latter being John Glenn's orbital flight. When the STG was folded into the Manned Spacecraft Center, Roberts took on the job of defining the requirements for the new Mission Control Center.

NASA GSFC
Roberts moved to the Goddard Space Flight Center in May 1962 as Head of the Manned Flight Division. Throughout Gemini, Apollo, Skylab, and Apollo-Soyuz, Roberts was in charge of the Manned Space Flight Network. He was also involved in using the Deep Space Network for the Apollo missions to the Moon.

OTHER
After retiring from NASA in 1979, Roberts consulted with Bendix Field Engineering for several years. Among his many honors and awards were the following by NASA: the Exception Service Medal (1964 and 1969); the Outstanding Achievement Award (1967);

the Distinguished Service Medal (1980); and the Robert H. Goddard Award of Merit (1984) which is the highest degree of recognition the GSFC can bestow on its employees. He was also made a Fellow of the American Astronautical Society (1976).

Rodney G. Rose

Born: August 10, 1927
Died: January 8, 2014 (86)
Age at STG: 32
Spouse: Leila
Children: two sons

BACKGROUND
"Rod" Rose was born in Huntingdon, England, graduated from grammar school in 1939, then attended the College of Technology in Manchester, graduating with a B.S. in mechanical and aerospace engineering in 1949. He won a scholarship to the Cranfield Institute of Technology and obtained an M.S. in aeronautical engineering in 1951. While attending college he was an engineering apprentice at the A. V. Roe Company in Manchester, where he worked in all the departments and learned the aircraft manufacturing business. From 1951–1957 he worked for Vickers-Armstrong, Ltd., as Chief of Performance Aerodynamics. He then moved to AVRO Canada to work on the CF-100 and CF-105 Arrow as an aerodynamicist. When the company folded he was one of the engineers hired by NASA.

STG
Rose joined the STG in the spring of 1959, about the same time the astronauts came onboard. His first assignment was to the Systems Test Branch in the Flight Systems Division headed by Max Faget. He served as the systems engineer on the Little Joe project at Wallops Island for a series of six flights. This work involved many aspects of the task, including engineering, range safety, and recovery operations.

   In late 1960 Rose transferred to the Engineering Division under fellow AVRO engineer Jim Chamberlin and joined the Project Engineering Branch to work on the Mercury capsule systems. This included working at the McDonnell factory, overseeing production. He also worked on the recovery systems, in particular the landing bag and the heat shield and its honeycomb protection. Rose participated in many tests, including drop testing the capsule in various water wave states, calculating the center of gravity, and various other aspects of capsule recovery engineering and operations.

MSC
When the STG was dissolved and the teams transferred to Houston, Rose became involved with the concept of a land landing for the Gemini Program. He spent two years evaluating and testing the Rogallo Wing; a paraglider system that was not actually used. He also evaluated the ejection seats and managed the abort and recovery systems for Gemini missions. He supported Apollo as the Technical Assistant for Flight Operations, serving as a troubleshooter for Chris Kraft. Later he became the Technical Assistant for the Space Shuttle and eventually its Assistant Director.

OTHER
After retiring from NASA in 1984 Rose worked for Rockwell International in various capacities until 1988. He also spent a year working for the prospective European manned space program in Paris. Amongst his honors and awards were the following by NASA: the Superior Performance Award (1968 and 1976); the Exceptional Service Medal (1969 and 1981); the Certificate of Commendations (1980); and the Superior Achievement Award (1981). He was an Associate Fellow of the Royal Aeronautical Society of England; an Associate Fellow of Manchester College of Technology; and an Associate Fellow of the AIAA.

Sigurd A. Sjoberg

Born: September 2, 1919
Died: March 26, 2000 (81)
Age at STG: 40
Spouse: Elisabeth Jane Ludwig
Children: Three boys

BACKGROUND
"Sig" Sjoberg was born in Minneapolis, Minnesota in 1919. He graduated from the University of Minnesota 1942 with a B.S. in aeronautical engineering, then joined NACA. He was awarded an honorary doctorate from DePauw University in Greencastle, Indiana in 1973.

NACA
Sig was hired as an aeronautical engineer at the Langley Memorial Aeronautical Laboratory in 1942. During WW-II he worked on stability control problems for most of the military aircraft of that era. In 1946 he spent a year at the NACA Muroc Flight Test Unit, working

on the Douglas D-558-II. By 1956 he was Head of the Airborne Analysis Section at Langley, and remained in that position until moving to the STG in 1960. He and his colleagues published many technical papers on aircraft directional stability, dynamic response, and various types of control systems.

STG

In 1960, Sig was assigned to the Flight Operations Division as Technical Assistant to Charles Mathews for Operations Coordination. During that time, both the Mercury Control Center at the Cape and the world-wide tracking network were being built, and he was involved with all issues regarding operational implementation.

POST-STG

When the STG was dissolved in November 1961, Sjoberg moved to the new Manned Spacecraft Center in Houston as Assistant to the Chief of the Flight Operations Division. In 1963 his focus was on Gemini flight operations and he was Manager of Operations, Planning and Development within the Flight Operations Directorate. After Gemini he turned his attention to Apollo, and by 1965 was Deputy Director of the Directorate. Then from 1969 to 1972 he was Director of Flight Operations.

In 1972, with the Apollo lunar missions winding down, he became the Deputy Director of the Manned Spacecraft Center and was involved with the Skylab space station and the Apollo-Soyuz Test Project. On his retirement from NASA in 1979, he worked with the Orbiting Astronomical Observatory (OAO) Corporation for a short period.

Amongst Sjoberg's many honors, awards and professional affiliations were the following by NASA: the Superior Performance Award (1966); the Exceptional Service Medal (1967 and 1969); the Certificate of Commendation (1970); and the Distinguished Service Medal (1971). He was also awarded: the Presidential Medal of Freedom (1970); the Space Flight Award of the American Astronomical Society (1977); the W. Randolph Lovelace II Award; and the National Space Award by the Veterans of Foreign Wars (1978).

John C. Stonesifer

Born: April 15, 1929
Age at STG: 32
Spouse: Marguerite (née) Vigneron
Children: two sons, two daughters

BACKGROUND
Stonesifer was born in Hanover, Pennsylvania to Robert and Edna Stonesifer, and was one of nine children. After graduating from Eichelberger High School in Hanover in 1947 he attended Gettysburg College in Gettysburg, Pennsylvania. The Korean War was in progress, and young men of draft age were being called into the services. Stonesifer joined the Navy in his senior collage year, and spent four years as an aerographer. Two of those years were spent at a Fleet Weather Center in North Africa forecasting the weather for ships in the Mediterranean Sea and aircraft operating in southern Europe and across Africa. Following those two years he served in the Fleet Weather Center in Miami, Florida that was responsible for tracking hurricanes in the Atlantic in coordination with the U.S. Weather Service. This center also directed the Hurricane Hunters, and he served as an observer on one flight into a hurricane. After separation from the Navy he completed his college degree at the University of Miami with a B.S. in physics.

NACA
Stonesifer joined the NACA Langley Aeronautical Laboratory in 1957 and was assigned to the Transonic Dynamics Branch, assisting in the conduct of aeronautical research, both analytical and wind tunnel model testing, to determine transonic flutter and aeroelastic characteristics of wing and tail surfaces on high performance military aircraft, including the F-105, F-104, P6M, B-58, and early model wing planforms for the F-111. Amongst these tests, he conducted a re-entry stability test of a small model of the Mercury capsule; but only after he had managed to have the closed down Vertical Blow Down Tunnel reopened.

STG
Given his Navy experience, in December 1961 Stonesifer interviewed by the Recovery Branch of the STG Operations Division and was processed into that organization. He was asked by his current organization to delay reporting to Recovery for a month while assisting with a model in the Transonic Wind Tunnel. Participating with the Recovery personnel, he assisted in planning and preparing documentation requirements for positioning ships and aircraft for the imminent (although much delayed) orbital flight by John Glenn. During that mission Stonesifer assisted Robert Thompson, who coordinated the landing and recovery operations, by working with the Department of Defense officials who were commanding their units from the Mercury Control Center.

POST-STG
For Mercury, Gemini, and early Apollo flights Stonesifer assisted and/or directed all phases of spacecraft operational planning, unique hardware development and testing, and the training of Department of Defense world-wide ship and aircraft forces. During this time, he served as the NASA Team Leader and technical advisor aboard a number of the primary recovery ships.

Stonesifer led the challenge of developing the guidelines, training, and procedures for use in the recovery of the Apollo 11 capsule and crew on their return from the Moon, to protect Earth from possible lunar pathogens. The requirements for isolation/quarantine of crew and apparatus were passed to NASA by the Interagency Committee for Back Contamination (ICBC) that was established by the National Academy of Sciences to formalize the viewpoints of many federal agencies. Stonesifer worked with this committee for the approval of the developed procedures. He also worked with the Secret Service in preparing for the visit of President Nixon and several cabinet members to the prime recovery ship, the aircraft carrier *Hornet*, to witness the helicopter arrival of the astronauts under the unusual conditions of quarantine/isolation.

The same procedures were followed for Apollo 12. After that mission, Stonesifer was made Chief of the Bioengineering Systems Division in the Life Sciences Directorate at the Manned Spacecraft Center, primarily to implement and manage a program to support the Skylab space station by the development, qualification, and integration of life sciences experiments into the spacecraft medical and biomedical systems. Related to this development of hardware for use in space was the related program to develop similar ground based equipment for comparison with the data obtained by crews conducting the experiments in space.

At the conclusion of the Skylab program, Stonesifer became Assistant to the Director of Life Sciences for Shuttle and Space Station Support, with responsibilities including the coordination and implementation of all proposed change activity emanating from the Directorate associated with the Space Transportation System in areas involving engineering, operations, and medical investigations. Once experiments and in-flight activities were approved, he would monitor the development progress of experiment equipment and crew procedures.

OTHER

After 31 years of service, Stonesifer retired from NASA in 1988 and became project manager of Krug Life Sciences; a company building biomedical equipment. They also performed training of crews for the conduct of the in-flight medical experiments and supported flight operations in the NASA control centers. While at NASA he received the Certificate of Commendation (1969) and

the Exceptional Service Medal (1974), as well as numerous Group Achievement Awards for the Mercury, Gemini, Apollo, Skylab, and Space Shuttle.

Robert F. Thompson

Born: May 16, 1925
Age at STG: 34
Spouse: Dot (née) Pritchett
Children: one son and one daughter

BACKGROUND
"Bob" Thompson was born in Bluefield, Virginia. On completing Graham High School in 1941 he attended the Virginia Polytechnic Institute in Blacksburg, Virginia, from which he graduated in 1944 with a B.S. in aeronautical engineering. Thompson joined the Navy and spent time at sea on destroyers in both the Atlantic and Pacific. He did not see active combat because it was near the end of hostilities. He was involved in decommissioning, and was released from the Navy in September 1946.

NACA
Thompson joined the NACA Langley Memorial Aeronautical Laboratory in early 1947 as an aeronautical engineer and was assigned to the Stability Research Division. He worked in wind tunnels on transonic aircraft programs, obtaining basic data for reports. Over the years, he was progressively promoted in the Division as project engineer, aerodynamicist, and stability and control research engineer. This work involved many different aircraft and included studies of control surface flutter, and various wing planform configurations, including those of the X-1 research aircraft. Much of his work was classified, and included coupling fighters to bomber wings. He also worked on supersonic bombers, and studied the stability and control issues of vertical takeoff.

STG
When Thompson was identified as a potential member of the Core Team of the STG, he was away on a two week assignment with the Navy and couldn't commit until he returned. Charles Mathews, Chief of the Operations Division, wanted him to head up the landing and recovery operations. With his Navy background and understanding of the magnitude of a world-wide recovery operation necessary for Project Mercury, Thompson coordinated the Department of Defense support for the landing and recovery operations. He developed the requirements and coordinated the support provided by the Air Force and the Navy. He made arrangements for NASA to reimburse the services for their work which, by requiring aircraft, ships, and many thousands of personnel, was expensive.

POST-STG
After the move to the Manned Spacecraft Center in Houston, Thompson became the Assistant Chief of Operational Support and Chief of the Landing and Recovery Division for the Gemini Program. He spent a year at NASA Headquarters as Mission Director, and then returned to the MSC in 1965. During Apollo, he was appointed Assistant Manager of the Apollo Applications Program and, later, Manager of the Apollo Program Office and the Skylab Program Office. He was the Manager of the Space Shuttle Program Office from 1970 to 1981.

OTHER
After retiring from NASA in 1981, Thompson became Manager of the McDonnell Douglas Space Operations in Houston. His honors and awards from NASA included:

the Outstanding Leadership Medal (1966); the Exceptional Service Medal (1969); the Distinguished Service Medal (1974); and the JSC Special Achievement Award (1976 and 1979).

Howard W. Tindall

Born: February 20, 1925
Died: November 20, 1995 (Age 70)
Age at STG: 36
Spouse: Jane

BACKGROUND

"Bill" was born in New York but grew up in Scituate, Massachusetts, where he went to high school and graduated in 1943. He then joined the Navy and served on destroyers in the Pacific. After the war, he went to Brown University and gained his B.S. in mechanical engineering in 1948.

NACA

After graduation, Tindall joined the Langley Memorial Aeronautical Laboratory and worked on wind tunnel instrumentation in the General Research Instrumentation Branch of the Instrument Research Division. He also worked on Project Echo (an inflatable balloon to reflect radio waves for communications) through to its launch on August 12, 1960, then transferred to the STG.

STG

In late 1960, Tindall joined the Mission Analysis Branch under John Mayer. He initiated early work to develop rendezvous software and also worked on orbit determination with the Goddard Space Flight Center. By June 1961 he was assistant Head of the Mission Analysis Division and went with it during the move to the Manned Spacecraft Center.

POST-STG

It was after moving to Houston that Tindall became more deeply involved in orbital mechanics. As part of the Mission Planning and Analysis Division he was involved in planning all 10 of the Gemini mission, including the test of the rendezvous concept that was necessary for Apollo. He was instrumental in coordinating the work at the MSC with the work at the MIT Instrumentation Laboratory on Apollo software. Tindall is best known for his work, beginning in 1967, as Chief of Apollo Data Priority Coordination. He held meetings that could include a hundred people that had very strong points of view. He was able to obtain everyone's input on software, procedures, and constraints for Apollo, and still arrive at a decision. His work was documented in what are now famously called "Tindallgrams" and more formally documented in a series of books that bear the title "Mission Techniques."

Tindall was appointed Deputy Director of the Flight Operations Directorate in 1970 and then Director in 1972. He later worked on the Skylab space station and the Space Shuttle. He retired in 1979 after 31 years with NACA/NASA. Among Tindall's many honors and awards were the NASA Exceptional Service Medal (1969) and the NASA Distinguished Service Medal (1973).

Dr. Robert B. Voas

Born: 1928
Age at STG: 30
Spouse: Carolyn M. Voas
Children: David W. Voas and Jeanette M. Voas

BACKGROUND
Voas was born in Evanston Illinois, attended Evanston Township High School, and obtained a B.Phil. from the University of Chicago in 1946. After relocating to California, he attended the University of California at Los Angeles and graduated with a B.A. in psychology in 1948, an M.A. in 1951, and a Ph.D. in 1953.

NAVY
In 1953, Dr. Voas began his career as a human factors scientist at the U.S. Navy Electronics Laboratory in San Diego, California. In 1954 he was commissioned as a lieutenant and sent to the School of Aviation Medicine in Pensacola, Florida where he participated in research on the selection and training of Navy aviators. In 1957 he transferred to the Naval Medicine Research Institute in Bethesda, Maryland, where he was involved in monitoring the pilots who undertook high altitude balloon flights and also the joint Army/Navy program that mounted the first space flights involving primates (squirrel monkeys).

STG
In September 1958, a few weeks before NASA was officially created, Dr. Voas was detailed to the STG as the first of three aeromedical consultants supplied by the armed

services. The others were Dr. Stanley White for the Air Force and Dr. Bill Augerson for the Army. Together with Dr. Alan Gamble from NASA headquarters, Dr. Voas was involved in developing the requirements for the Mercury astronauts. He was also on the committee that managed the astronaut selection program. His later responsibilities included the development of the astronaut training program, and when the astronauts joined in April 1959 he was appointed Astronaut Training Officer. An Astronauts and Training Group was formed under Col. Keith G. Lindell, who was also the Air Force's liaison officer to the STG. This group was large enough to have been a whole Division but Dr. Gilruth wanted to keep the astronauts, physicians, and trainers on his staff. But that was just on paper, because they had their own building and simulators at different locations at both Langley and Cape Canaveral. Air Force Dr. Lt. Col. William K. Douglas served as their flight surgeon. As the training officer, Dr. Voas had two assistants: George C. Guthrie and Raymond G. Zedekar, who were also assigned to the Astronaut Training Office. The Astronaut Training Office was responsible for supporting astronaut flight preparation activities at the Cape and for helping to develop the astronauts' flight plans. Dr. Voas was a member of the debriefing team which went to the landing site to meet an astronaut the morning after his flight. Following the move to Houston, Dr. Voas became Assistant for Human Factors in the Office of the Director.

POST-STG

In 1964 Dr. Voas left NASA to become campaign manager for John Glenn's first campaign to become the Senator for the State of Ohio. Following the end of that campaign, he was asked by Sargent Shriver to head up the Peace Corps Office of Field Selection. This was the department that determined the fitness of volunteers for overseas duty. He managed an organization with over 300 part-time Psy.D. psychologists traveling to the Peace Corps training sites in order to select the volunteers.

In November 1968, Dr. Voas joined the National Highway Traffic Safety Administration (NHTSA) Safety Bureau of the newly established Department of Transportation. In charge of driver safety research programs, he developed plans for and evaluated the effectiveness of the federal government's largest community highway safety program: the Alcohol Safety Action Projects (ASAPs) which were implemented in 35 localities across the U.S. In 1982 he retired from federal service and joined the Pacific Institute for Research and Evaluation (PIRE) as a senior research scientist. Since that time, he has been the principal investigator on numerous studies for the National Institutes of Health and for NHTSA. While at PIRE he produced over 200 publications and gave presentations at scientific conferences. Dr. Voas has received many awards including: the Lifetime Achievement Award of the Research Society on Alcoholism; the Widmark Award for Lifetime Achievement from the International Council on Alcohol, Drugs, and Traffic Safety; the NHTSA Public Service Award; and the James J. Howard Traffic Safety Trail Blazer Award from U.S. Governor's Highway Safety Association.

Manfred "Dutch" von Ehrenfried, II

Born: March 30, 1936
Age at the STG: 25
Spouse: Alice Jane Edmonds
Children: Two sons and a daughter

BACKGROUND

"Dutch" was born in Dayton, Ohio and raised in Lancaster County, Pennsylvania. He graduated from East Lampeter High School in 1954, then attended the University of Richmond, Virginia, from which he graduated in 1960 with a B.S. in physics and a minor in mathematics. He taught physics, mathematics and science at Colonial Heights High School for a year while applying for a job in physics.

STG

On the day prior to Alan Shepard's Mercury Redstone flight, Dutch interviewed with the STG. On finishing his teaching job he reported to the Flight Control Operations Section, with Gene Kranz as his immediate supervisor. His first year involved helping Kranz with the mission rules and going to flight controller classes on the Mercury capsule, the Atlas launch vehicle, and the tracking systems. Part of his training was to learn the communications for what would become the Manned Space Flight Network.

Dutch spent MA-4 and MA-5 at the Goddard Space Flight Center learning how voice and teletype was used for communications to the tracking stations. His first mission in the Mercury Control Center was John Glenn's MA-6. Before the eventual launch, he supported the mission rules effort and the communications between the control center and flight controllers at remote sites. Part of the training for the Operations and Procedures (PROCEDURES) position was the interface between the MCC and the various elements of the launch countdown and the Atlantic Missile Range. The call sign with the missile range was "DEVIL FOX BRASS ONE" and with the tracking stations it was "CAPE PROCEDURES." This work involved coordination with the RCA contractors on assignment to the MCC, most prominently John Hatcher, Andy Anderson, and George Metcalf.

In response to the infamous "Segment 51" signal on Glenn's flight, Dutch and Kranz sent TTY messages to all the tracking stations to determine whether they saw this signal indicating that the landing bag had been inadvertently deployed. Although the mission ended in success, this incident drove home the risks of flying in space and the need for training, simulations, and mission rules. Dutch continued to work on all the manned orbital Mercury missions in terms of countdowns, mission rules, and remote site communications and coordination.

POST-STG

After Project Mercury, Dutch was involved in upgrading the Mercury Control Center displays for the early Gemini missions and gathering and collating the flight controller requirements for the new "Integrated Mission Control Center" to be built in Houston. He was involved with the console layout and their displays, the communications panels, and the pneumatic tube system. Between missions he familiarized with the Gemini spacecraft and the Titan launch vehicle.

In 1964 Dutch was part of a team that went to the Carnarvon tracking station in Australia to check it out for the new Gemini missions. The team was headed by Dan Hunter and in addition to Dutch it included three of the Philco contractors assigned to the STG: Stu Davis, Bill Garvin, and Jim Moser. The team produced the first Flight Controller Handbook for the Gemini remote sites.

Dutch was a PROCEDURES officer on Gemini 3, but for Gemini 4, which saw Ed White's spacewalk, Kranz was named Flight Director and Dutch was Assistant Flight Director (AFD). Dutch continued to support Gemini through GT-4, GT-5, GT-6 and GT-7.

In 1966, after five months at the Martin Company in Denver, Dutch became one of the first GUIDOs to train for Apollo 1. This involved training with astronauts Grissom, Chaffee, White and people at the MIT Lincoln Laboratories to learn inertial guidance and navigation from Dr. Charles Stark Draper and his postgraduate students. Dutch also joined the Apollo astronauts at the Griffith Observatory in Los Angeles to study the stars that were to be used by Apollo as a backup mode of navigation. His position required the development of new console procedures and displays for the guidance officers. For the Apollo 1 "plugs out" test on January 27, 1967, Dutch and fellow GUIDO Will Pressley were conducting data flow tests, computer program checkouts, and command tests in the MCC when the spacecraft caught fire, killing the three astronauts.

After the "stand down" in the program to redesign the spacecraft, Dutch was assigned to the Apollo Program Office as the mission staff engineer for Apollo 7. This involved working with the crew of Walter Schirra, Donn Eisele, and Walt Cunningham. His supervisors for those two years were Cal Perrine, Owen Maynard, and George Low. Dutch was the lead engineer for the center-wide planning effort which set flight test objectives. The mission was 100% successful.

Once the decision was made to assign the Apollo 8 mission to a circumlunar flight, Dutch became the backup mission staff engineer to John Zarcaro. This work on flight test objectives involved working with the crew of Frank Borman, Jim Lovell, and Bill Anders. Dutch had the pleasure of being on duty in the MCC for the crew's famous readings from *Genesis* from lunar orbit on that Christmas Eve of 1968.

From 1967 to 1969 Dutch was a part-time Apollo pressure suit test subject, working for Jack Mays in the Crew Systems Division. This work included tests in various pressure suits, making treadmill runs, a ride in the "zero g" aircraft, making "g" runs in the centrifuge, and tests in the vacuum chamber. One test involved using one of Neil Armstrong's suits (he was issued three) to attempt to saturate the lithium hydroxide canister in the life support back pack to determine how long an astronaut could walk on the Moon without a carbon dioxide problem. The test was at an equivalent altitude of 400,000 feet. In 1969, Dutch was fitted for his own Apollo A7LB Skylab suit.

After Apollo 8, Dutch transferred to the Earth Resources Aircraft Program in 1969 to become the first Sensor Equipment Operator and Mission Manager for the RB-57F. This aircraft was on loan from the USAF 58th Reconnaissance Squadron and had been modified by NASA to carry a 4,000 lb. payload to extreme altitudes. This work required extensive coordination with scientists and the pilots to fly the missions in such a way as to achieve the objectives. Most flights were in the range 60,000–65,000 feet but one flight was to 70,000 feet.

During 1970 and 1971 Dutch was Chief of the Science Requirements and Operations Branch that was responsible for the definition, coordination, and documentation of science experiments assigned to Apollo and Skylab. These included the Apollo Lunar Surface Experiments Package (ALSEP), several versions of which were left on the Moon, and the experiments in orbit around the Earth and the Moon. This work also defined the procedures by which the astronauts were to deploy the packages and conduct experiments on the lunar surface.

OTHER

Dutch left NASA JSC to work with Wolf Research; a Goddard Space Flight Center contractor involved with the Earth Resources Technology Satellite that was later named LANDSAT. He worked there with former NASA flight controller Richard Holt. However, this contract lasted only a year.

From 1972 to 1975, Dutch worked for John Bryant, a former NASA JSC employee, at TRW as Manager of the Systems Engineering and Development Division on the World-Wide Military Command and Control System (WWMCCS) for the Secretary of Defense for Intelligence. After contract award, Dutch spent two years as the TRW representative to the Commander in Chief of the Pacific Fleet (CINCPACFLT) in Honolulu, Hawaii.

When the Nuclear Regulatory Commission was established in 1975, another former NASA flight controller, Tom Carter, encouraged Dutch to interview with the new agency. He became Chief of the Test & Evaluation Branch, with the responsibility for evaluating the safeguards of nuclear fuel cycle facilities against terrorism. Dutch acquired six Green Berets on loan from the Army to function as calibrated terrorists. Dutch and the team evaluated uranium and plutonium facilities across the U.S. As independent consultants, Dutch and former NASA flight controller Dick Sutton wrote contingency plans for the nuclear industry.

After graduating from the Federal Executive Institute, in 1976 Dutch joined International Energy Associates Ltd. (IEAL), to continue working with the nuclear industry. This involved defining security and safeguards for the Barnwell Plutonium Reprocessing Facility. Its purpose was to recycle spent reactor fuel for subsequent reuse as fuel. However, pressure from the anti-nuclear environmentalists convinced President Carter to close the facility.

In 1982 Dutch married the IAEL office manager Dayle Thompson, and worked in the FAA Aviation Safety Office for a year and then as a support contractor for the NASA Space Station Task Force. Dutch and his wife formed the Technical & Administrative Services Corporation (TADCORPS) and thereby continued to support the Space Station and other offices at NASA Headquarters until 2000. During those years, TADCORPS won the NASA-sponsored Small Business Administration (SBA) Administrator's Award for Excellence and also its Region V Award for Prime Contractors in 1995.

Dutch received Chartered Financial Consultant (ChFC) certification in 2004 and Chartered Life Underwriter (CLU) in 2007 from American College. He is now an independent Financial Advisor associated with the Raymond James Financial Services Corporation.

Among Dutch's awards and honors are: the Certificate of Participation First Manned Orbital Flight (1962); the NASA MSC Outstanding Performance Rating (1965); the NASA MSC Sustained Superior Performance Award (1966); the NASA Apollo Group Achievement Award (1969); the NASA Apollo Science Team Award (1969); the NRC Sustained Superior Performance Award (1977); the NASA Space Station Task Force (1984); the Marquis Who's Who in the East (1985, 1986); the Marquis Who's Who in the World (1989, 1990); the Eagle Scout Role of Honor; and the Chairmanship of the Washington Space Business Roundtable. In addition he has authored the following publications: *Adventures on Space Station Freedom*; an educational comic book (1989); *Adventures on Santa Maria*; an educational comic book (1991); *Nuclear Terrorism – A Primer* (2012); *Stratonauts: Pioneers Venturing Into the Stratosphere* (2014); *Birth of NASA: The Work of the Space Task Group* (2016); and has contributed his flight control experiences to *From the Trench of Mission Control to the Craters of the Moon* (2012).

Dr. Stanley C. White

Born: January 13, 1926
Died: September 10, 2011 (Age: 85)
Age at STG: 32
Spouse: Helene Rae (née) Ross
Children: Four sons and one daughter

BACKGROUND
White was born in Lebanon, Ohio. He graduated from Mason Village School in 1943 and later attended Miami (Ohio) University to earn an A.B. degree. He obtained his M.D. in 1949 from the University of Cincinnati, College of Medicine. He got his master's in public health in 1953 from Johns Hopkins University, Public Health & Hygiene. He entered residency training at the Headquarters Tactical Air Command Surgeon's Office at Langley Field, Virginia. He received Board Certification in Aerospace Medicine from the American Board of Preventive Medicine in 1956.

USAF
In 1954, Dr. White was assigned for two years as Chief of the Respiration Section, Physiology Branch, Aero-Medical Laboratory at Wright-Patterson AFB. While there, he was assigned to the planning group for the Air Force's Man-In-Space-Soonest (MISS) program. No Air Force-wide medical program had yet been developed in support of the MISS concept but he was assigned to work on life support hardware and crew selection. After the signing of the Space Act in 1958, a joint NASA-ARPA Manned Satellite Panel was formed to draft specific plans for a program of research leading to manned spaceflight. When White went to Washington to brief officials on the status of biomedical support in the proposed MISS concept, he was tapped for early service with NASA.

STG
In October 1958, Lt. Col. Dr. White was placed on temporary duty to the STG as the senior member of the aeromedical team assigned the task of establishing criteria for selection of the Mercury astronauts and then assisting with that selection; a process which took many months. Dr. White was made the Head of the Life Systems Branch within the Flight Systems Division headed by Max Faget. There were five Branches and a Computing Group in the Division. Dr. White's responsibilities were the Mercury capsule systems which supported the astronaut. Of initial concern was to man-rate the environmental control system. His group worked with the astronauts' physician Air Force Lt. Col. Dr. William Douglas, and Navy Lt. Dr. (Psychology) Robert Voas of the Astronaut Training Group. Several people in the Branch were also military officers assigned to Project Mercury, including Dr. Lt. Col. James P. Henry and Air Force Lt. Col. Dr. Capt. William S. Augerson.

Dr. White's team also wrote the requirements for aeromedical monitors that would support the flight surgeons in the MCC and the teams at the remote sites. It was involved in all the life systems including feeding and waste management, space suits, emergency survival apparatus, and oxygen supply and carbon dioxide removal. It coordinated with the other astronaut support personnel as well as with McDonnell Aircraft, the manufacturer of the capsule. These systems were related to all phases of flight, including launch, on-orbit, re-entry, and recovery.

POST-STG
As well as continuing to head the Life Systems Division at the Manned Spacecraft Center in Houston, Dr. White was involved with the selection of Gemini, Apollo, and Skylab astronauts. After retiring from the Air Force, he became the senior scientist at the Bionetics Corporation in support of the Kennedy Space Center in Florida. Among his many honors and awards were the Melbourne W. Boynton Award for Space Medicine Research, and the Louis G. Bauer Founders Award from the Aero-Space Medical Association. He was President of the Aerospace Medical Association and of the International Academy of Aviation and Space Medicine. In addition, he was a member of the American Rocket Society, the American College of Preventive Medicine, and the Association of Military Surgeons.

Appendix 3

STG Technology

In the 1958–1961 timeframe, the STG was using what now seems like "old school" technology. Almost every NASA/STG engineer and scientist had a slide rule; sometimes these were carried on their belts.

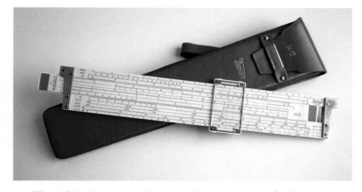

Fig. A.3.1 The K&E slide rule. (Photo courtesy of Wikipedia)

© Springer International Publishing Switzerland 2016
M. von Ehrenfried, *The Birth of NASA*, Springer Praxis Books,
DOI 10.1007/978-3-319-28428-6

Fig. A.3.2   The Friden calculator. (Photo courtesy of Wikipedia)

Fig.A.3.3   An IBM 704 computer at NASA in 1957. (Photo courtesy of Wikipedia)

The IBM 704 introduced in 1954 was the first mass-produced computer with floating-point arithmetic hardware. It had a control unit and several peripherals such as a punched card reader, alphabetic printer, magnetic tape units, magnetic core storage units, and magnetic drum readers. It used vacuum tube logic circuitry. The 704 was regarded as "pretty much the only computer that could handle complex math." The programming languages FORTRAN and LISP were first developed for the 704. It was used by the NASA laboratories and by all the military services, as well as by many aerospace companies. There were about 140 systems in the country in the late 1950s. The Navy used one for its Vanguard program, the Army used one for its Jupiter program and the Smithsonian Astrophysical Observatory used one to track Sputnik. The 704 installed at Langley was shared with the STG.

The computer on Bermuda was an IBM 709 but it still used vacuum tubes. The Goddard and Ames computers were transistorized 709s known as IBM 7090s.

Fig. A.3.4 The NASA Goddard computer room in 1962 with IBM 7090s. (Photo courtesy of NASA)

Fig. A.3.5 An IBM 1620 like that used by the STG. (Photo courtesy of courtesy of Sourisseau Academy for State and Local History, San José State University)

Fig. A.3.6 The Bendix G-15 computer. (Photo courtesy of Wikipedia)

The Bendix G-15 computer used a magnetic drum for storage whose rotational latency was too great to have each instruction follow on another in memory. Harry Huskey, the machine's designer, therefore required each instruction to include the address of the next one so that the instructions could be arranged optimally on the drum. The G-15 was fast for its time. Double-length registers simplified the programming of double-precision arithmetic operations. A single instruction could operate on either one word or a group of words. Furthermore, several G-15s could compute simultaneously and communicate with one other. A compiler was provided for the ALGO language. Bendix sold over 400 G-15s before its computer business was taken over by the Control Data Corporation. It was used by the STG Mathematical Analysis Section in the 1959 timeframe, in addition to the Langley IBM 704.

Fig. A.3.7 The GE/Burroughs Guidance Computer. This is the computer that sent commands to the Atlas booster for steering commands. (Photo courtesy of the Air Force Space Museum)

Fig. A.3.8 Goodyear Electronic Differential Analyzer. (Photo courtesy of vintchip.com)

In 1954, the High Speed Flight Station of NACA utilized an Air Force Goodyear Electronic Differential Analyzer (GEDA) to conduct flight simulations. NACA engineers Richard E. Day, Joe Weil, Donald Reisert, and Wendell H. Stillwell studied the ability of thrusters in a reaction control system to control aircraft in the near vacuum of extreme altitudes. This research began with a simulation of the Bell X-1B and led to the study of the RCS of the Iron Cross simulator, the JF-104, and the X-15, and ultimately translated directly into the Mercury capsule's RCS.

Appendix 4

Some Photos

The following photos are related to Project Mercury and have historical significance. They are not presented in any particular order

Figure A.4.1 is of the NACA Special Committee on Space Technology, known as the Stever Committee after its chairman, Guyford Stever. It is significant because after Sputnik NACA's Committee on Aerodynamics consisting of industry, the military, and academic representatives came to the conclusion that NACA should act promptly to avoid being ruled out of spaceflight research, and that increased emphasis should be placed on the problems of true spaceflight over extended periods of time. The committee then adopted a resolution that called for an aggressive program. NACA in turn voted to establish the special committee and named Stever, a physicist from MIT, as chairman. They met at the Lewis laboratory on May 26, 1958.

"Behind every great man there is an even greater woman." Before there were what we now call computers, there were ladies whose job description was "computer." Over the years, there were hundreds of them at Langley. The following are some excerpts from a Langley history article entitled, "When the Computer Wore a Skirt."

© Springer International Publishing Switzerland 2016
M. von Ehrenfried, *The Birth of NASA*, Springer Praxis Books,
DOI 10.1007/978-3-319-28428-6

Fig. A.4.1 The NACA Special Committee on Space Technology. Going clockwise from the left are: Edward R. Sharp, Director of Lewis Laboratory; Colonel Norman C. Appold, USAF; Abraham Hyatt, Department of the Navy; Hendrik W. Bode, Bell Laboratories; W. Randolph Lovelace II, Lovelace Foundation; S. K. Hoffman, Rocketdyne Division of North American Aviation; Milton U. Clauser, the Ramo-Wooldridge Corporation; H. Julian Allen, NACA Ames; Robert R. Gilruth, NACA Langley; J. R. Dempsey, Convair-Astronautics; Carl B. Palmer, Secretary of the Committee NACA Headquarters; H. Guyford Stever, Chairman and Dean of Engineering at MIT; Hugh L. Dryden, Director of NACA; Dale R. Corson, Department of Physics at Cornell University; Abe Silverstein, Associate Director NACA Lewis; and Wernher von Braun, Army Ballistic Missile Agency. (Photo courtesy of NASA)

Fig. A.4.2 The CF-105 Arrow. As related in Chapter 8, its cancellation in 1959 enabled NASA to hire AVRO engineers. (Photo courtesy of Don Rogers and Wikipedia)

Fig. A.4.3 A Langley "Computer." (Photo courtesy of NASA)

One of the first five NACA "computers' was Virginia Tucker, who reported to work on the day after Labor Day in 1935. Four other ladies were already there in the "Computer Pool." The job title designated someone who performed mathematical equations and calculations by hand. By the end of WW-II, Virginia headed a department of over 400 women assigned all across the Langley Memorial Aeronautical Laboratory. A good percentage were former school teachers, and most held at least a bachelor's degree, usually in mathematics.

Despite its sub-professional status, this job paid much better than the majority of jobs which were available to women in the 1940s and 1950s. It also afforded them an entry into the field of aeronautical research; an option other than teaching. Some of the job functions included reading film, running calculations, and plotting data. During wind tunnel tests, manometer boards would measure pressure changes and the "computers" read the photographic films and recorded the data on worksheets. They then ran different types of calculations, and analyzed and plotted the results on graph paper. All of this work was done by hand using slide rules, curves, magnifying glasses, and basic calculation machines such as the Marchant or the Frieden.

Many of Langley's ladies went on to become engineers, and with the dawn of the "modern" computer era they would become computer programmers and analysts. Some wrote technical reports and books, and some gained advanced degrees. Generally overlooked in the history of technology and even in the history of human computing, these women played a critical role in research at Langley and at the STG.

Fig. A.4.4 The Langley West End Model Shop. These are the men behind the scenes who built all the models for Langley, Wallops, and Project Mercury. (Photo courtesy of NASA)

Ever since the beginning of research operations in the early 1920s, Langley benefited from unique, highly professional design, fabrication, and maintenance efforts from several historical shops, both on the East Side of Langley AFB and on the West Side, namely the Dynamic Model Shop, the Machine Shop, the Fabrication Ship, the Composite Model Development Shop, and a Propeller Shop. The 1950s brought new challenges in fabrication techniques and a wide variety of vital research support functions to the West Model Shop. The wind tunnel research required extensive instrumentation, such as pressure ports and remotely actuated controls for efficiently operating the tunnels. The shop production peaked as more high speed wind tunnels became available.

The second research community was the Pilotless Aircraft Research Division (PARD) that had initiated major rocket-related research programs and founded the Wallops Flight Station to conduct rocket and missile flight testing as well as rocket-boosted flight tests to determine the performance, stability and control of emerging supersonic aircraft designs. They also supported the series of Little Joe tests for Project Mercury. The contributions of the West Model Shop to the space programs are some of the most historic events in Langley's history. In April 1958 it build the first test couches for astronauts. It built many models of the Mercury capsule. It also supported other space projects including Echo, Viking, and fabricated the parawings that were considered for the Gemini capsule. One member became well known; Beulah "Boots" Barger. She was the seamstress for these wings and the early spacesuits made for the astronauts. This shop also fabricated very large models of aircraft, including a Boeing 747 airliner. In addition, both the F-5 and F-16 fighter aircraft used in the Langley Full Scale Wind Tunnel.

Fig. A.4.5 The Missile Range Instrumentation Ship *Rose Knot Victor* used on Project Mercury. (Photo courtesy of Navsource.org)

Two tracking ships were used for Project Mercury. The *Rose Knot Victor* (the Victor added for its call sign RKV) and the *Coastal Sentry Quebec* (the Quebec added for its call sign CSQ) were owned by the Air Force and leased to the Pan American Guided Missile Range Division (GMRD). They were positioned in those locations that were not covered by the main tracking stations of the first three orbits. As the ground tracks moved to the

Fig. A.4.6 The Mercury launch vehicles. The six manned Mercury flights (l-r): MR-3, MR-4, MA-6, MA-7, MA-8 and MA-9. (Photo courtesy of Wikipedia)

Fig. A.4.7 The Mercury Control Center being modified for Gemini. The author is in the center in the blue suit and Gene Kranz is in the vest.

Fig. A.4.8 A post-flight celebration after Alan Shepard's MR-3 flight (l-r): Dr. Voas, Dr. Douglas, Scott Carpenter, John Glenn, Dr. Robert Gilruth, Alan Shepard, Walt Williams, Gus Grissom, Deke Slayton, "Shorty" Powers, Gordon Cooper and Wally Schirra. (Photo courtesy of Dr. Robert B. Voas)

west for the longer manned orbital flights, the ships were position in the Atlantic, Pacific, and Indian Oceans as required to provide critical functions.

As the MCC flight controllers gained experience with successive Mercury flights, the need for additional group displays became apparent. Each flight control group said what they needed, and then requirements were developed for the displays and the contractors would install them. Soon, more sophisticated software was developed to drive the displays. In one early 1958 concept for a control center, the contractor suggested that a desk with three phones would be adequate! In one early flight, the day/night terminator on the wall map broke. The RCA Support Coordinator John Hatcher rigged up a motorized roll of paper (à la toilet paper style), half painted black and half white, and it moved across the wall map to inform the team whether the capsule was in the dark or daylight. Even Fig. A.4.7 shows how the MCC evolved in three years. Look at what control centers today have for group displays.

Appendix 5

Quotes

Before providing illuminating quotes by NACA, NASA, and STG people, I think it is important to begin with the speech given by General James H. Doolittle, at that time Chairman of NACA, before the Senate Special Committee on Space and Astronautics on May 6, 1958. His powerful and timely presentation was critical in making Congress understand the rationale for creating a civilian space program.

Dr. James H. Doolittle, Chairman
National Advisory Committee for Aeronautics

Before
Special Committee on Space and Astronautics
United States Senate
May 6, 1958

Mr. Chair and members of the Committee, and counsel:

On January 31, 1958, the first of our satellites was put into orbit. The chairman of this Committee, the Honorable Lyndon B. Johnson, Senator from Texas, had noted that, and I quote, "The Explorer is a triumph of persistence against great odds... Our satellite is very aptly named. It is truly an explorer – a representative of the free people searching for the facts of a totally new dimension into which men and women will soon step. And, in its search for facts, it brings us face to face with a sharp reality which we cannot ignore..."

The previous month, on the 54th Anniversary of the historic first flights by Wilbur and Orville Wright, I said, "There has been exploration since the beginning of mankind, since the beginning of curiosity. The airplane has made well known most of the remote spots on this globe, but exploration will continue. The new exploration will be in science and space. We, as a Nation, must have vision and must work hard if we are to be leaders in this new type of exploration... We, the United States of America, can be first. If we do not expend the thought, the effort, and the money required, then another and more progressive Nation

© Springer International Publishing Switzerland 2016
M. von Ehrenfried, *The Birth of NASA*, Springer Praxis Books,
DOI 10.1007/978-3-319-28428-6

will. They will dominate space, and they will dominate the world. There is a Nation with this ambition. We must not let it prevail."

It is with these thoughts in mind that I speak in support of Senate Bill 3609, cited as the "National Aeronautics and Space Act of 1958." On April 2, the day the President transmitted to the Congress a special message recommending the establishment of a National Aeronautics and Space Agency, he sent a memorandum of instructions to the Secretary of Defense and to me, as Chairman of the National Advisory Committee for Aeronautics. I should like now to present to this Committee a copy of this communication. The President's instructions included preparation and presentation by the NACA to the appropriate committees of the Congress of a full explanation of the proposed NASA legislation and its objectives. I am in complete agreement with the proposed legislation and am, of course, happy to respond to the President's directive. Even if the President had not so instructed, I would have welcomed this opportunity to support the Bill.

For the record, I should like to make one further introductory comment. The absence at this hearing today of Dr. Hugh L. Dryden, Director of the NACA, is because, several months ago, the scientific community of Free Germany paid him the signal honor of inviting him to deliver tomorrow in Munich, the second annual lecture in honor of the late Ludwig Prandtl, the eminent Germany physicist and aeronautical scientist. The fact that Dr. Dryden was invited to be the first non-German to deliver the Prandtl lecture testifies to the great international reputation that he so rightly enjoys. Except for this compelling reason, Dr. Dryden most certain would be here today.

It was just fifty years ago today – May 6, 1908 – that the Wright brothers went back to Kitty Hawk to sharpen their piloting skill before they made the world's first demonstration and sale of a military airplane to the U.S. By 1908, the controversy had begun as to whether the airplane could ever be a militarily decisive weapon. It was not until the Second World War that the argument was finally settled, beyond all question. In the thirty some years between, the airplane was transformed from a contraption of wood and cloth and wire, driven by unreliable, low power engines, into all-metal craft that could and did accomplish the military tasks required. The very great improvement in those years, and since, was due to unceasing research and development effort in which the NACA has played a leading role.

The point I wish to make is that today, so far as space technology is concerned, we are at the early state of development comparable to that of the airplane a half-century ago. On the basis of the present state of the art, it may be difficult to see clearly the full range of future civil and military uses of space vehicles. But, already enough is known for us to be sure that we must lead, not lag, in space technology.

My comments about the uses of satellites and space craft are confined to the non-military aspects. They are only examples of the more obvious uses.

What makes the space vehicle of such tremendous interest to the scientists is the fact that they can use it as a tool to study the universe, including the Earth, in a way that previously was completely impossible. We live at the bottom of a deep, atmospheric ocean. As we look skyward through the miles of our atmosphere, even when we use the finest optical and radio telescopes that man has been able to devise, we get only burred images of the wonders of the heavens. Dr. Joseph Kaplan, chairman of the U.S. National Committee for the International Geophysical Year, says that the astronomers have a lot to put up with: "Looking at the stars through our atmosphere is like looking through a dirty window." And Dr. Lloyd V. Berkner, President of Associated Universities, Inc., has pointed out that the differences between radio observations of the heavenly bodies made from Earth and from a satellite will be the difference between a picture in black and white and in full color. Stated another way, the atmosphere filters out all but a single octave of light and, similarly, all but a narrow band of the broad spectrum of radio waves that travel Earthward from the Sun and distant stars.

If we can furnish the scientists with a steady platform out in space beyond our atmosphere, it will provide priceless opportunities to study the phenomena associated with the physical behavior of the Sun and the stars. It is possible, for example, to construct a telescope of useful size, together with the television and transmitting equipment needed to send its observations back to Earth that could be carried in an orbiting satellite that weighs only a ton or so. A first step in developing such a satellite is provision of accurate automatic stabilization and control, a task that admittedly will be difficult. It is however, a job that is essentially straightforward. It can be accomplished with reasonable promptness if sufficient scientific and technical manpower is assigned to the work.

Your committee, Mr. Chairman, has performed a very real public service by including in your #1 "Compilation of Materials on Space and Aeronautics," the "Basic Objectives of a Continuing Program of Scientific Research in Outer Space," as drafted by scientists participating in the IGY. This document will enjoy wide readership as a consequence of your action. The many data-gathering projects it proposes should help our citizens to understand the importance of space vehicles for scientific purposes.

In this connection, the question will inevitably and properly arise: What good will all this new information be to the people of the United States. After all, they have to pay the bill, and it is a bill that annually, for years to come, will probably be counted in the hundreds of millions of dollars. I don't know all the good it will do and I doubt if any man alive today can give specific answers. But, in this connection, I am reminded of the story they tell about Michael Faraday, the English physicist, whose pioneering work in electromagnetics had a profound effect upon our later understanding of electro-dynamics leading to useful electric power. About a hundred years ago, Mr. Faraday is supposed to have been asked, in the British Parliament, about the value of his electro-magnetic experiments. His answer, so the story goes, was, "I can't tell you what it'll be good for. But I'll tell you this: One of these days you'll be taxing it."

I can't tell you precisely what of great value will come out of our moving into space to probe the secrets of the universe. However, I have the conviction, and in this I find myself in the company of some very wise men, that a century from now, perhaps much sooner, people will say that this venturing into space that we're planning now was one of the most practical, intelligent investments of our national wealth to be found in history. If we, in the United States, take the wisely bold action necessary to lead in exploiting the possibilities of space technology for science, all mankind will benefit. If Russia wins dominance in this completely new area; well, I think the consequences are fairly plain – probable Soviet world domination.

Now, fortunately, there are peaceful uses of satellites and space platforms that offer a more immediate pay-off. One of the most obvious, and immediately valuable, has to do with meteorology. The weather people calculate that if they can forecast long-range and local weather with greater accuracy, this will, each year, be worth billions of dollars to the citizens of the United States. Over the years, great progress has been made in this direction, but there are definite limits to what we can hope to accomplish in weather prediction so long as we operate entirely within the Earth's atmosphere. For example: The behavior of the air masses over the oceans often determines the weather over the inhabited land masses. These great areas of water, as every schoolboy is taught, cover more than two-thirds of the Earth's surface. We know so little about how the world's weather is generated over these vast ocean masses and over the polar areas that we are unable to forecast the resulting weather accurately. This is particularly important in the case of devastating typhoons and tornadoes. We have made some progress in our aerial study of hurricanes that form in the Caribbean, but the cost to expand this kind of effort around the world would be great and the information obtained insufficient.

With properly instrumented satellites, the meteorologists can watch storms form and move and disappear, all around the world on a 24-hour basis. They can also study the physical processes that make our weather, how the Earth's surface absorbs heat energy from the Sun in varying amounts, and how the heat circulates unevenly between the equatorial and polar regions. By observing, measuring and then understanding these complicated heat-transfer processes, the meteorologists expect to be able to predict normal and abnormal weather including the onset of destructive droughts, catastrophic windstorms, and flood-producing rains. Beyond all this, they are thinking and planning for the day when, finally, they will have fully comprehended the meaning of their new knowledge and thus be able to control the weather. The worth of such a possibility is literally incalculable.

I am not an electronics expert but the use of satellites in communications also offers great advantages and economics.

Today, we are at the very edge of one of the great frontiers of history. The frontiersmen who explore the vast reaches of this unknown region will be the space scientists. They will undertake their explorations, because they are compelled by the drive that motivates every true scientist to seek new knowledge, valuing it for itself. For us to attempt to gauge each of these adventures into space on the basis of the value of its expected immediate results would, I fear, defeat the very purposes of the essential master plan for the peaceful conquest of space.

This thought brings me to where I wish to emphasize as strongly as possible my conviction that the direction of our national space program must be the responsibility of a civilian agency, as stipulated in Senate Bill 3609. I speak as a civilian who was for many years privileged to serve in our Armed Forces. I support the Department of Defense position that the Military Services should be permitted to undertake the projects that have military value. I also support the DOD position that the Military Services need freedom to conduct research, or to contract for such research, on space matters that may offer promise of future military value. A little later, I shall touch upon the essentiality of the civilian space program and the military space program being closely coordinated.

Around the world there are signs that, even in Russia, the peoples of this Earth are beginning to realize that the new weaponry man has invented cannot be fully used without upsetting our civilization. I wish to be clear on this point. I don't for a moment believe that the Millennium has at last arrived. I don't for a moment believe that the Russians have discarded their implacable drive for world domination. But I do believe that the Russians may finally have realized that, in the face of our armed might, any attack they launched against us would result in their receiving mortal wounds. It is quite possible that they have learned, as we have, that all-out nuclear war is no longer a satisfactory way to impose national desires. If so, we may expect Russia to attempt world domination by other means, such as infiltration, insurrection, and limited warfare. Over the long haul they will certainly attempt to achieve world domination through scientific, technological, industrial, and economic pre-eminence. This, the Free World must not permit. We must remain strong militarily and economically. Thus we must be able to make whatever military use is desirable and possible in the field of space technology. It means we must lead in the development of space technology for peaceful purposes. We and the Russians are competing to capture the minds of the peoples of the world. In this struggle, gaining and retaining world leadership in space science will provide us with tremendous opportunities.

The language of Senate Bill 3609 is broad. The authorities it provides are necessary to enable NASA to marshal American's resources as may be needed to carry forward our national space program. The Bill provides that the National Advisory Committee for Aeronautics will become the nucleus of the new agency. What needs to be done in space technology, however, cannot be accomplished merely by expanding the work load of NACA in the space field where it is already focusing nearly half of its research effort. Nor will a mere name change, for purposes of symbolism, be enough.

For example, the NASA must have authority to contract for a substantial part of its space research to be done by laboratories outside the Government. NACA's research contract authority is limited to $500,000. This is obviously inadequate for the NASA. NASA needs authority to develop, test, launch, and operate aeronautical and space vehicles. NASA needs to have its appropriations on a no-year basis, so that vital space programs can be undertaken and financed without regard to a single fiscal year. NACA does not have that authority, except for construction of laboratory facilities. NASA needs authority to pay salaries that are reasonable comparable with prevailing rates paid to non-federal employees for similar work. As we see it, about half of the work performed by NASA will soon be done by private agencies under contract and about half by the federal employees who make up the NASA organization. It would be extremely difficult, if not impossible to get, and keep on the NASA staff, the high caliber people we must have, if they know they can go across the street and do the same kind of work, also for NASA, at higher pay rates. The NACA for the past 12 years, has successfully applied the principle of reasonably competitive pay to 43 percent of its personnel, known as wage-board employees (trades and crafts). This method of establishing rates is thus being extended to the remaining employees.

The Bill calls for the new NASA to study problems of manned or unmanned flight within the Earth's atmosphere as well as out in space. Continued research on airplanes is important for two reasons. For some time to come, the quality of our overall military posture, our ability to deter enemy aggression, will depend upon our airplanes being able to accomplish their assigned tasks. The rate at which obsolescence occurs in military aviation has never been more rapid than today. It is essential that the NASA assume the vital role NACA now plays in improving the performance of our military aircraft. Similarly, commercial aviation depends to an important degree upon the results of NACA research. This research NASA must also continue to do.

To sum up, the NASA will use the 8,000 scientists and supporting workers of the NACA, and its aerodynamics, structures, and propulsion laboratory facilities, which cost nearly $400,000,000, as the foundation for the new organization and programs. The added authorities I have briefly mentioned are essential for it to carry out its new mission – the conduct of our national space program.

One of the questions most often asked about our national space program is: How there can be a clear-cut distinction between the space projects which should be under military control and those which should be under NASA. The answer is that on the one side there will be projects clearly and obviously military, and on the other side, projects clearly and obviously civilian. In between, there will be projects with both civilian and military interest. Here, and I expect this will include many projects, there needs to be the closest sort of consultation to determine whether NASA or the Department of Defense (ARPA) should do the work, or whether it should be done cooperatively. For 43 years, NACA has worked most effectively with the Military Services. I would expect this ability to be carried over into NASA. As a matter of fact, Dr. Dryden – whom I certainly feel is the man in the United States most capable for the job of NASA director – is already working out a good relationship, covering this broad region of mutual interest, with top officials of the DOD. Members of his senior staff are working closely with ARPA Director Roy W. Johnson and ARPA Chief Scientist Herbert York. The caliber of the men and their desire to get the overall job done in the national interest will, I believe, be even more important in handling this matter than the precise language in the Bill. The legislation is, however, necessary in order to provide the framework in which they will operate.

I have spoken about the close relationship required between NASA and the Department of Defense. A similar close, effective relationship must exist between the new agency and the Atomic Energy Commission. This is particularly important in the area of nuclear power plants. The same can be said respecting the relationship with the scientific community, largely represented by the National Science Foundation and the National Academy of Sciences.

Finally, I wish to say just a word about the most exciting, and perhaps, ultimately, the most important, aspect of our national program – the putting of man into space. This activity will be conducted as an integral part of NASA's effort, along with research on space technology problems, and development and use of unmanned vehicles to gather desired scientific data. Just as rapidly as research can provide the necessary data, we should employ this information in the development and use of manned vehicles to deepen our penetration into the far distances of our solar system, first to the Moon and then on to the planets.

The kind and magnitude of space program that our national interest requires will, as I said earlier, cost hundreds of millions of dollars each year for many years to come. As a Nation, we have the scientific and the technical competence that is needed. We have the resources to pay the bill for this new, great exploration. We can and must succeed in finding our destiny in space.

Quotes from Presidents

- *"I would like for you as Chairman of the Space Council to be in charge of making an overall survey of where we stand in space... Do we have a chance of beating the Soviets by putting a laboratory in space, or by a trip around the moon, or by a rocket to land on the moon, or by a rocket to go to the moon and back with a man. Is there any other space program which promises dramatic results in which we could win?"*
  President Kennedy to Vice President Johnson, one week after Yuri Gagarin's historic orbital flight. Correspondence dated April 20, 1961.

- *"Look, I want to be first, now do something."*
  President Kennedy to Robert Gilruth with regard to a lunar program in 1961.

C
O
P
Y

April 20, 1961

MEMORANDUM FOR THE VICE PRESIDENT

   In accordance with our conversation I would like for you as Chairman of the Space Council to be in charge of making an overall survey of where we stand in space.

   1. Do we have a chance of beating the Soviets by putting a laboratory in space, or by a trip around the moon, or by a rocket to land on the moon, or by a rocket to go to the moon and back with a man. Is there any other space program which promises dramatic results in which we could win?

   2. How much additional would it cost?

   3. Are we working 24 hours a day on existing programs. If not, why not? If not, will you make recommendations to me as to how work can be speeded up.

   4. In building large boosters should we put out emphasis on nuclear, chemical or liquid fuel, or a combination of these three?

   5. Are we making maximum effort? Are we achieving necessary results?

   I have asked Jim Webb, Dr. Wiesner, Secretary McNamara and other responsible officials to cooperate with you fully. I would appreciate a report on this at the earliest possible moment.

/s/ John F. Kennedy

Fig. A.5.1 President Kennedy sent this memorandum to Vice President Johnson eight days after Yuri Gagarin's historic orbital flight, expressing his frustration regarding America's state of space flight preparedness. He had only been in office a little more than three months and knew we were far behind the Soviets.

Quotes by NACA/NASA Administrators

- *"I'm reasonably sure that travel to the moon will not occur in my lifetime."*
  NACA Director Hugh L. Dryden in 1953. He died on December 2, 1965. He almost made it.

- *"Yet, the Mercury program was one of the best organized and managed of any I have been associated with."*
  NASA Administrator T. Keith Glennan.

Quotes by NACA/NASA Center Directors

- *"The scientific study of the problems of flight with a view of their practical solution."*
  The sign on Langley Director Floyd L. Thompson's desk to remind him of the primary role of NACA. Sometimes the military pressured NACA to do more than research.

- *"You designed this thing (attack submarine) as a surface ship but it spends most of the time under the water. Why don't you design it as something that spends most of its time under the water and it comes up to the surface every once in a while, sort of like let it fly through the water like an airplane."*
  Floyd L. Thompson chastising several Admirals in 1946 that they were not designing attack submarines properly. Soon, Langley was doing research on hulls for submarines in the Full Scale Wind Tunnel, employing the wind tunnel for hydrodynamics, as well as aerodynamics.

Quotes by or about Dr. Robert R. Gilruth

- *"You know that guy (NASA Administrator T. Keith Glennan) told me to go ahead (with the space program) but he didn't tell me how. I haven't got an organization. I don't know how I'm supposed to do this."*
  Robert R. Gilruth talking to Acting Langley Director Floyd L. Thompson in 1958, to which Thompson replied, "Well, why don't we just create the Space Task Group."

- *"We will have to move the program to Africa because there aren't enough chimpanzees in the United States to do this."*
  Robert R. Gilruth, Director of Project Mercury after one of the participating doctors recommended that we should not fly an astronaut until we had run fifty chimpanzees through a centrifuge program at Johnsville, 1959.

- *"... perhaps the most important single requirement in our programs [is] that designs, procedures, and schedule must have the flexibility to absorb a steady stream of change generated by a continually increasing understanding of space problems."*
  Robert R. Gilruth discussing his policy regarding the checkout of the capsule to the checkout crew at the Cape, 1960.

- *"During the life of the Mercury project, several million people have unselfishly contributed their knowledge, skill, and ingenuity to make this pioneering space project a successful national effort. The people came from all walks of life; they came from government agencies, from industrial organizations, and from many educational institutions. Also, valuable co-operation has been received from a number of people and governments of other nations to enable the accomplishment of the essential support activities around the Earth. Each person is to be commended for his untiring efforts and his unfaltering faith in the ultimate successful fulfillment of the objectives of the project."*
  Dr. Robert R. Gilruth, Director of the NASA Manned Spacecraft Center and of Project Mercury, 1963.

- *"It is fitting that the Manned Spacecraft Center expresses its sincere appreciation to the Langley Research Center for the invaluable contributions that the Center has played in our initial manned spaceflight program. The Manned Spacecraft Center owes much to Langley, since Langley really was its birthplace."*
  Dr. Robert R. Gilruth, Director of the MSC to Floyd L. Thompson Director of the LRC, 1962.

- *"I don't think you could live through many of these Mercury programs. It was something you do when you're young. It was the case of working all the time for the first year or so. But it was rewarding."*
  Robert R. Gilruth in an interview well after the completion of the Apollo program.

- *"There is no question that without Bob Gilruth there would not have been Mercury, Gemini or an Apollo program. He built in terms of what he felt was needed to run a manned spaceflight program. It is clear to all who have been associated with him that he has been the leader of all that is manned spaceflight in this country."*
  George M. Low upon the death of Robert R. Gilruth, 2000.

Quotes by Christopher C. Kraft Mercury Flight Director

- *"I have a very soft spot in my heart for the network and the people who operated it. These people were as much a part of the success of our efforts as were the flight controllers and the other people at Houston and the Cape. Whatever we did was in large measure dependent on the reliability of the world-wide tracking and communications network. This is truly one of the unsung accomplishments of the space program."*
  Christopher C. Kraft, 2001

- *I was part of the crowd, then part of the leadership that opened space travel to human beings. We threw a narrow flash of light across our Nation's history. I was there at the best of times."*
  Christopher C. Kraft, 2001 (from the final page of his memoir *Flight*).

Quotes by Gene Kranz – Mercury Cape Procedures

- *"To recognize that the greatest error is not to have tried and failed, but that in trying, we did not give it our best effort."*
  Gene Kranz (from *The Foundations of Mission Control*)

- *"My wish as I close this book is that one day soon, a new generation of Americans will find the national leadership, the spirit, and the courage to go boldly forward and complete what we started."*
  Gene Kranz, 2000 (from his book *Failure is Not an Option*)

Quotes by Other STG People

- *"You know, this place (STG) is not going to make it. I think you ought to think more than once about whether or not you want to go, because when it does fail, there isn't any job for you back here."*
  A manager at PARD advising Alan B. Kehlet not to join the STG, 1958.

Quotes by Others

- *"The launch of Sputnik is one of the greatest scientific moments in history."*
  Arthur C. Clarke, 1957

- *"Classmates, the future lies before us, the school behind us. There are a thousand diverging paths from which we must choose. Once started, we can never retrace our steps. Only a few times more shall we meet together, but let us, as members of the class of 1925, be one unit in that we shall never be satisfied to keep any position except the highest, to do anything except our utmost, and to give any service except our best."*
  From Ira Abbott's speech to his high school graduation class. He went on to spend 32 years with NACA and NASA. He chose his path to be a founding father of the space program.

- *"The network worked better than it could have in the most optimistic dreams... Langley can take a well-deserved bow."*
  Edmond C. Buckley, former Head of the Langley Instrument Research Division and manager of the Tracking and Ground Instrumentation Unit. He went on to become the Director of the Office of Tracking and Data Acquisition at the Goddard Space Flight Center.

- *"General familiarity with travel and maneuvering in space today can obscure just how uncertain almost everything was (in the mid-1950s) relating to the first vehicle designed to carry a human into suborbital flight."*
  Christian Gelzer and Curtis Peebles from their book marking the Centennial of NACA, scheduled for publication in 2016.

Appendix 6

Stories and Trivia

A.6.1 Birds

I would like to kick off this Appendix with a story from 1932. It isn't an STG or Mercury story but a story about a NACA engineer that says a lot about the engineers and scientists of NACA, NASA, and the Space Task Group. The story begins in the lunch room of a Langley laboratory. The conversation is about the aerodynamic characteristics of birds. There were a lot of birds in the vicinity; especially sea gulls because Langley lies on the shore. Of course people have been watching birds soar for millennia. So engineer Tom Collier, taking the question seriously, goes and shoots a buzzard and then proceeds to freeze it with the wings outstretched to enable him to test the bird in the NACA towing tank. The results indicated that the frozen bird was inherently unstable and therefore couldn't fly. He also tested a frozen sea gull. It turns out that birds are, in fact, unstable, but this has never prevented them from flying! It's no wonder that the community of Hampton, Virginia called the boys at Langley "Those crazy NACA nuts."

© Springer International Publishing Switzerland 2016
M. von Ehrenfried, *The Birth of NASA*, Springer Praxis Books,
DOI 10.1007/978-3-319-28428-6

Fig. A.6.1 NACA engineer Tom Collier with his frozen buzzard. (Photo courtesy of NASA Langley)

A.6.2   Cape Canaveral

The Cape Canaveral Air Force Station was originally called the Naval Air Station Banana River. On June 1, 1948, the Navy transferred it to the Air Force, which named it the Joint Long Range Proving Ground. Three months later, it became Patrick Air Force Base. The first rocket launch was an American manufactured V-2 fitted with a WAC Corporal upper stage in a configuration called Bumper.

Fig. A.6.2 First Cape Canaveral firing. (Photo courtesy of Wikipedia)

A.6.3 Wallops Island PARD

The crews from the Pilotless Aircraft Research Division (PARD) did some crazy things during the 1940s and 1950s. These were the guys that built aircraft and capsule models and then built rockets to launch them out over the Atlantic.

Alan B. Kehlet tells the story of how one man had his hand cut off. Derwood A. Dereng was working on a model that was about to be launched on top of a rocket. They had disconnected its external power, so the model was now on internal power ready for the launch. When the rocket didn't fire, Derwood went to reinsert the external power plug to save the batteries in the model. The rocket sat on what is known as a "zero length" launch platform, meaning that there was no launch stand or platform; the rocket literally sat on the ground. As Derwood climbed up to put the ground power plug back in, the rocket fired and as it ascended the stabilizing fin struck his hand with such force that it severed his hand!

The remedy they devised for this problem was to put switches on the model which could be turned using a screwdriver. Alan tells the story of how he would stand on a platform about ten feet away and use a special ten-foot-long screwdriver to turn the switches on, then jump off the platform and take cover behind a concrete barrier before the rocket went. It reminds one of the old adage, "I wouldn't touch that with a ten-foot pole."

A.6.4   Johnson and Low

Caldwell Johnson was working on a Mercury capsule in the shop. He was in an area that was roped off to isolate the capsule work from other activities in the large building. A guy who did not have the proper badge stepped over the rope and started to look around. Johnson, irritated, walked over to the guy and asked, "What's your business?" The man apologized and said, "Oh, I'm sorry; I didn't mean to interfere," and left. About then Johnson saw someone else and said, "Who's that guy?" "Oh, that's George Low; he runs the whole program." Johnson half thought he'd get fired the next day.

A.6.5   Military Service

In doing research for this book, I found that I was working alongside men who served in combat during WW-II and Korea. At the time, I had no idea who some of my colleagues and managers really were, and what they did before the STG. Here are just a few examples:

- Max Faget was a submariner and was actually in (under) Tokyo Bay during WW-II.
- John B. Lee flew 52 combat missions in Europe and won the Distinguished Flying Cross.
- Joe "Guy" Thibodaux was in the Army Corps of Engineers and helped to build the Lido Road in the China-Burma Theater of Operations.
- John S. Llewellyn served as a Marine in the "Battle of the Chosin Reservoir" during the Korean War, and received a Bronze Star.
- Lt. Col. Dr. William S. Augerson won a Silver Star in Viet Nam for gallantry in action against a hostile force on January 8, 1968.

A.6.6   Recovery

Bob Thompson tells the story about the recovery of the Big Joe capsule in 1959. Thompson was on one of the many destroyers and ships in the Atlantic where the capsule was supposed to land. Since it was night, all the ships were told to look for a "shooting star" and provide the azimuth to Bob's ship. John Mayer was back at the Cape getting the computer data to give Bob a predicted impact point based upon the launch trajectory and cutoff conditions. At this time, the Navy used CW radio for communications.

The report was late; it was expected about 10 minutes after the Atlas engine cutoff. Bob was in the Combat Information Center with Captain Wright, the Task Force Commander, who was waiting for Bob to tell him where to pick up the capsule. The report finally came in and it said, "It left here but we don't know where it is!" As Bob's ship hadn't seen anything, he requested the Captain to ask the next ship up the line if they had seen it. They said they saw it, but it was up-range. So Bob asked the Captain to ask the next ship up the line whether they'd seen it. That ship reported they had seen it and it was down-range. Bob went to the map with all the recovery forces accurately positioned on it and placed his finger directly in between those two ships and told the Captain, "Let's go there!" The Navy sent a P-2V to look for the capsule, and spotted it about 2 miles from where Bob's finger had been on the map!

A.6.7 Recovering Sam

Little Sam was a male rhesus monkey who received training and conditioning at the School of Aviation Medicine; thus his name. They tried to launch him several times, but the weather was very stormy. Finally it cleared slightly, and Sam was launched on the Little Joe 2 rocket out of Wallops Island on December 4, 1959. It was still very bad weather and the sea states were too high for the helicopters to launch from the recovery carrier and try to retrieve the capsule. Bob Thompson was on the nearby destroyer to pick Sam up.

While the destroyer was going just fast enough to keep steerage, the sailors attempted to put a hook on the capsule in order to lift it aboard with a davit. The ship started to roll back and forth, with the motions approaching 45 degrees. The capsule with Sam inside banged against the side of the ship several times. Just as the capsule was swung onboard, one of the sailors got washed overboard. While he was waving frantically, they secured the capsule and turned the destroyer around to pick up the sailor.

The destroyer recovery team had expected the helicopter to pick up the capsule and deliver it to the carrier, where the Air Force veterinarians and technicians were. No one on Bob's ship had ever seen the monkey, and wasn't sure what to do. They called over to the carrier and were told to go alongside and transfer Sam, still in his container,

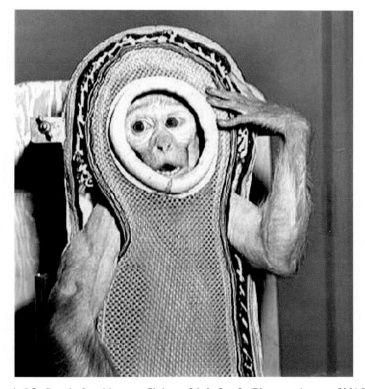

Fig. A.6.3 Sam before his spaceflight on Little Joe 2. (Photo cou\rtesy of NASA)

across on a line. The sea states were still too high and the line broke. That was the end of that! It was therefore decided to retain Sam on the destroyer. Bob, the Captain, and the pharmacist's mated took Sam down to sick bay to take him out of the aluminum container, complete with his little astronaut's couch and his flight suit. Sam reached out and grabbed Bob's hand. He happily accepted some apples and oranges. They put him on display for the whole crew, who strolled past the sick bay to catch a glimpse of their first "astronaut."

When the ship got to Norfolk, the veterinarians were waiting for Sam and put him in a cage and covered it up because no one was supposed to see him, then carried him away. Bob didn't have the heart to tell them that they'd been playing with him all along. Sam was returned to his training colony and, no worse for his experience, lived over twenty more years before dying in 1982.

A.6.8   Support for John Glenn's Flight

Glenn's MA-6 flight was supported by about 19,300 people, of which 15,600 were associated with the recovery effort. About 2,600 people were involved with the launch and 1,100 people manned the Mercury Space Flight Network. No parade since John Glenn's has ever generated that much "ticker tape."

A.6.9   Who the Hell is Kranz?

It was during MA-3 in 1961 that Gene Kranz was starting to feel comfortable in his Procedures role at the Cape. With a number of suborbital flights accomplished, he prepared for what was to be the first Mercury orbital flight. The flight control teams had deployed across the globe. Flight Director Kraft trusted Kranz to approve all the teletype traffic to the remote sites.

MCC Surgeon Dr. Stanley White was not only looking after the astronaut's health, but those of the military surgeons/aeromedical monitors deployed to the remote sites as part of the flight control teams. So Dr. White drafted a message to all the remote sites requesting their evaluation of the living, sanitary, and health conditions at the sites. After this message had been approved by Kranz it was broadcast to all sites.

The only flight controller sent to Kano, Nigeria was my neighbor, Larry Wafford, one of the experienced Philco contractors in the flight control organization. He was required to perform as CAPCOM, SYSTEMS and AEROMED. Larry immediately responded to the message, clearly describing the local food, housing, atrocious medical conditions, and finally wrapping up with some not-so-kind words about the work force. Kranz received the message and acknowledged Larry's observations, then put a copy into Dr. White's message tray in the MCC.

The next afternoon, NASA Headquarters received a message from the Department of State requesting copies of all radio traffic between Kano and the MCC. Unbeknownst to NASA, the Nigerian government had read Larry's message and was so incensed that it was threatening to eject NASA from the Kano site, and also the recently arrived Kennedy Peace Corps. The State Department message rippled through the NASA channels and finally arrived at Flight Director Kraft and Operations Director Williams' consoles.

On entering the MCC the next morning Kraft held up the message and asked, "Who the hell authorized this?" Kranz replied, "I did." Kraft grew angrier and said, "Dammit, don't you know enough not to trust any of those damned Surgeons? Who needs this crap anyway! Headquarters was really pissed when they called Williams. Everyone wanted to know who the hell Kranz was. Walt told them to kiss off!" Kraft looked at Gene and said, "Since everyone in the NASA chain of command now knows who Kranz is, don't screw up again!"

The U.S. apologized to the Nigerian government and the flight control teams stayed in place to support Glenn's mission. Kranz got his first lesson in international diplomacy! After that he was very careful in reviewing traffic to the remote sites. Half a century later, everybody knows who Kranz is!

A.6.10   Pulling one over on Kraft

Sometimes we spent way too much time in the MCC considering all the training, simulations, group discussions about mission rules, and preparing for the flights. Many a day we spent 16 hours in that room, which got very small over time. Usually, we were all very disciplined and serious but sometimes someone had to break the ice.

One of my favorite people was John Hatcher, the RCA Support Coordinator for the flight controllers in the control center. He would get the facility up to flight status, and would get us anything that we required. We were all set to launch Glenn several times, but each time there would be a hold which led to a scrub. This went on for two months. One night, Kranz had the idea to put one over on Flight Director Kraft. He discussed his harebrained idea with RETRO Carl Huss. Huss replied, "Yeah, it might work but your life will never be the same." The next day, Kranz discussed the plan with John Hatcher, whose help would be needed to pull off the prank. They both "hatched" the plot.

We were scheduled to run a pad test and a simulation in parallel the following day. Flight Director Kraft would lead the simulation while Kranz ran the pad test; two separate activities running simultaneously. Kranz knew that Kraft would watch the pad video on his console TV. The plan was to fake Kraft into thinking that he had inadvertently launched the Atlas! Hatcher substituted an old video of an Atlas launch and synchronized the liftoff to the moment when, during the simulation, Kraft would throw the "Firing Command" switch. This would start the simulation and the clocks.

As the simulation proceeded, Kraft would ask Kranz how the pad test was going and Kranz would give him a quick status check with a straight face and his head down. As the simulation got down to liftoff, at just the same moment that Kraft threw the switch, Hatcher started the old Atlas liftoff video on Kraft's console. Kraft's eyes bulged and his forehead wrinkled. He turned to Kranz and asked, "Did you see that?" Of course Kranz played dumb, "See what?" Without a pause, Kraft explained, "The damned thing lifted off!" Hatcher and Kranz tried to keep straight faces but couldn't hold back their laughter. Kraft demanded, "Who the hell did this?" He then realized he'd been "had" and gave a half-hearted laugh. Kranz and Hatcher pulled Superman's Cape and survived!

A.6.11   The Reluctant Astronaut

One of our entertainments at the Cape was listening to Bill Dana's TV skit as "Jose Jimenez, the reluctant astronaut." In addition, he and Al Shepard would entertain the crowd at the hotel. Jose was dubbed the "eighth astronaut."

Now it is May 5, 1961 and Shepard has been in the Mercury Redstone capsule for hours with many holds in the count. There was a brief hold at T-30 minutes, and then a hold for 15 minutes for weather at T-15 minutes. Then there was a problem with the Redstone that required a recycle to T-35 minutes and yet another hold. Suddenly, a high pitched voice came over the Air-Ground (A/G) loop, "My name is Jose Jimenez. I don't want to go to space without my Crayolas and coloring book."

Jose went on and on, and everybody listened and tried to figure out who was making the call and where it was coming from. Kranz started punching all the communications loops to figure it out. Kranz thought that if it was coming from the control center, Kraft and Williams were going to give him hell. The MCC communications guy told Kranz that it was not coming from there; it seemed to be coming from the blockhouse.

Moments later, the test conductor reported that astronaut Gordon Cooper and flight surgeon Dr. Bill Douglas had patched the Jose Jimenez recording into the A/G loop because they thought Shepard needed to relax a bit after spending so many hours in the capsule. As it turned out, there were further holds and Shepard spent over four hours in the capsule before finally lifting off.

A.6.12   1958, A Dangerous Year

The year that NASA was born was a very, very dangerous year. The Cold War was at its highest in relation to the nuclear threat and bomb testing had moved from underground to underwater, to the atmosphere, and into space. The testing of nuclear explosions involved the most sophisticated instrumentation and analysis of its blast, shock, and radiation effects to date. The launch of the first American satellite had revealed the existence of the van Allen radiation belts and scientists knew that even small nuclear explosions in the atmosphere or in space would cause radiation to be trapped and affect communications. This would affect our ability to communicate in order to retaliate in case of a nuclear attack.

There was so much nuclear testing going, even in previous years, that some 9,000 scientists from 43 nations petitioned the United Nations for a nuclear test ban treaty as early as January. That didn't slow down the U.S., USSR, and Great Britain, which were conducting nuclear tests almost every month and often more than once a month; like the Soviets in February. The U.S. launched six nuclear weapons into the atmosphere and into space that August. This was also the period of testing the ICBMs designed to carry our nuclear weapons; albeit experiencing many spectacular failures. Just three days after the Space Act initiating the civilian space program on July 29, a Redstone rocket launched the nuclear test "Teak" which was exploded in space at an altitude of 48 miles. On September 6, the U.S. launched "Argus III" to an altitude of 300 miles; the highest known launch of a nuclear weapon. The creation of a civilian space program didn't slow down the military space program.

And just to make 1958 even more dangerous, Nikita Khrushchev became the Soviet Premier in February and Fidel Castro took over Cuba. Historians and political scientists were discussing the dangers of the Cold War. For example, military historian David A. Rosenberg penned what was to become a seminal article "The Origins of Overkill;" nuclear strategist A. J. Wohlstetter wrote "The Delicate Balance of Terror;" political scientist Kenneth N. Walz wrote "The Spread of Nuclear Weapons: More Maybe Better;" and political scientist Morton A. Kaplan wrote "The Calculus of Nuclear Deterrence." Just to add to the worry of nuclear war, on March 11 a B-47E accidently dropped a Mark 6 nuclear bomb (with no fissionable material) near a farmhouse in Mars Bluff, South Carolina. But finally, in November the U.S, USSR, and Great Britain agreed on an informal moratorium on nuclear tests; it didn't last long though.

To put the year in which NASA was created into historical perspective, the following events are also associated with 1958:

- Sugar Ray Robinson beat Carmen Basilio for the heavyweight boxing championship.
- The *Bridge on the River Kwai* won best picture and its star Alec Guiness best actor.
- *At the Hop* was the number one song.
- The Yankees beat the Milwaukee Braves for the World Series.
- Van Cliburn won the Tchaikovsky competition.
- Tim Tam won the Kentucky Derby and the Preakness.
- Brazil won the World Soccer Championship with Pele's help.
- Congress approved statehood for Alaska, effective January 3, 1959.
- The Supreme Court ordered the integration of the Little Rock Central High School.
- The World's Fair was held in Brussels, Belgium and featured the cultural differences between the USSR's technology and the American way of life.
- Eisenhower's voice was the first to be relayed from space.
- Charles de Gaulle was elected President of France, effective January 8, 1959.
- The International Geophysical Year ended.
- The price of the 3 cent stamp was increased to 4 cents.
- The U.S. population was 174 million and the world's 2.945 billion.
- The U.S. debt was 280 billion; it is currently 18.5 trillion!

A.6.13   Remembrances

After over 50 years, you remember certain people in special ways. Why certain memories about them stand out, I don't know. Here are a few stories of my deceased flight controller friends. By now, there is a complete space launch and flight control team in heaven!

- One night Carl Huss, the Retrofire Officer, and I roomed together in Cocoa Beach and we rode out to the Cape to go to work at the MCC about 4 AM. It was about a 15–20 minute drive. This was during one of the manned Mercury Atlas launches and the flood lights of Pad 14 could be seen off in the distance. From several miles away, it's an awesome sight. The shiny Atlas just glistened in the bright lights. Carl was eleven years my senior, and a good friend. I liked him and his wife Marge very much. As we got closer he turned to me and said, "Someday you'll look back and remember this." That day comes often! Thanks, Carl.

- "Tec" Roberts, the first Flight Dynamics Officer, was a Welshman who had a distinctive accent. He was about 10 years older than most of us young flight controllers and was very much a respected gentleman. When President Kennedy came into the control center after John Glenn's flight to wish us all well, he would enter the mission operations room from my left and I would be the first person he would approach as he came up two steps to the level of the Flight Director. He often had back pain, and placed his hand on my shoulder for support as he came up the stairs. I sat at attention. He went past us to Chris Kraft who welcomed him to the Mercury Control Center. Kraft chatted with the President and began to introduce him around. "Tec" could see the President slowly approaching, pausing along the way to talk to the individual flight controllers. "Tec," the man from Great Britain, was becoming excited as the President drew near and he attempted to compose himself. When the President reached the Flight Dynamics console, "Tec" jumped to his feet and greeted him with the honorific "Your Highness!"
- No collection of Mercury stories would be complete without one about John S. Llewellyn. In fact, an entire book could be written about this very unusual man, who is a dear friend. John was a decorated combat Marine who survived the "Battle of the Chosin Reservoir." He looked like a Marine, acted like a Marine, and could cuss like a Marine. But there was another John inside; not necessarily eager to come out. When he was in the control center he was another man. He had a master's in physics and was a Retrofire Officer. He could compute a re-entry firing time and chew an astronaut out for not achieving the correct re-entry attitude. He was a father and husband, and loved to raise cattle and horses. He loved to go head-to-toe with you in a discussion but could cut you in half with a unique form of repartee. He would pull off the most unusual antics. One time he lost his parking pass for having been given too many tickets so he rode his horse to work and parked it on the lawn. Another time he was late for a simulation in the control center but couldn't find a parking spot, so he drove up on the grass in front of the control center and went in to work. I could go on and on. But just let it be known that whether you were in mission control or just out in life, when times were tough you wanted John Llewellyn on your side.

Appendix 7

Author's STG Experience

I graduated from the University of Richmond, Virginia in June 1960. I was applying for jobs as a physicist to the national laboratories such as Los Alamos and Lawrence Livermore but with only a B.S in physics I wasn't getting very far. I had been married since 1959 and had one child at the time, so I took a job at the local high school to make ends meet. The new Colonial Heights High School didn't have a physics teacher, therefore I was hired. I taught physics, mathematics, solid geometry, trigonometry and general science for that one year while I continued looking for a job in physics. I knew I didn't want to continue teaching. Since I was also in the Air Force ROTC in college, I had always thought about being a pilot. That was my backup plan.

On May 4, 1961, I took a pilot flight physical at Langley Air Force Base, which I flunked by checking the box about having asthma as a child. I walked out of the base with my tail between my legs and saw a sign that said "NASA." The AFB and NASA were, and still are, co-located. I thought, "While I'm here, I might as well see what they're doing." I had heard about the NASA Langley Research Center but knew very little about it. I'm convinced that there was some divine intervention going on as I walked off the base, feeling very disappointed. I could've just as well gotten in the car and gone home.

I walked into the NASA building and asked to speak with someone about employment. I was introduced to Chris Critzos, who told me all about Project Mercury and that they needed flight controllers. I thought he meant flight control surfaces on aircraft; I simply didn't have any idea of what he was talking about. Because I was a new physicist, I thought he must be referring to aerodynamics. I asked him about salary and he took out a chart and said, "With your degree in physics and one year of experience, you'd be a GS-7." I said, "What's that?" He said that was $6,345/year. I said, "I don't need that much; I'm only making $3,600/year teaching." He said, "That's what a starting GS-7 makes." Of course, I have learned a lot about salary negotiation since then!

© Springer International Publishing Switzerland 2016
M. von Ehrenfried, *The Birth of NASA*, Springer Praxis Books,
DOI 10.1007/978-3-319-28428-6

Mr. Critzos introduced me to several people who seemed quite busy but took the time to say hello. It was then that I learned they planned to launch Alan Shepard the next day. Critzos gave me an application form and told me to report to work as soon as I could. The next day, I turned on the radio in my class and we listened to the launch of MR-3. I couldn't wait until the school year was over. I sent my application in that day; the day that Shepard made his suborbital flight. Years later, shortly before Shepard flew to the Moon as commander of the Apollo 14 mission, I had the occasion to tell him that story.

On June 19, 1961 I packed up the car, my wife, and son, and drove to Langley. It was a hot day, as I recall. I went into the building that displayed the STG sign and told the receptionist I was reporting for work. She could not locate my name on the list. I told her that Chris Critzos had hired me, so she went to find him. After straightening things out, he introduced me to six people in the Flight Control Branch. That was where I first met John Hodge, Fred Matthews, Arnie Aldrich, Gerald Brewer, and Jack Koslosky. I became the 639th employee of the STG, and I was really happy to be there. I was 25 years old.

During the first several days, I met the group of 16 Philco technicians assigned to the Flight Operations Division. They were in a separate room because they were contractors but they were there to help us new, young NASA flight controllers with communications, electronics, and data systems. And they had experience from other defense tracking stations that would be helpful in augmenting our training and staffing at the remote tracking stations which made up the MSFN. See Section 10.2.2.

Eugene Kranz, a former F-86/F-100 fighter pilot, took me under his wing and pointed me to operational areas related to the MSFN and the MCC. Most of 1961 was spent attending classes about the Mercury spacecraft, the Redstone and Atlas launch vehicles, and the capabilities of the MSFN. There were no college courses to prepare you for supporting a space flight. This period of my training also kicked off many years working with mission rules. See Section 15.1.

My first actual supporting role was helping Kranz with the mission rules for Gus Grissom's MR-4 flight. Within one month of teaching kids in the classroom, I was involved in an historic spaceflight. Of course, my first tasks were at the "gofer" level, but I was involved and learning. The word "newbie" hadn't yet been invented. During my first six months, there was the MR-4 flight by Grissom and the unmanned MA-4 test. Meanwhile the Soviet Union sent one of their men into orbit for a day-long mission.

I spent MA-4 and MA-5 at the new Goddard Space Flight Center, learning how the flight controllers communicated with the Mercury Control Center. That placed me in line to be an Operations & Procedures flight controller, since that position was the prime interface with the remote site flight controllers and required detailed knowledge of how the MSFN functioned. I was trained by Robby Robertson (Goddard Voice) and Bob Plaumann. This involved the NASA Communications Network (NASCOM), consisting of voice and teletype (TTY). The command and telemetry was handled by the Satellite Tracking and Data Acquisition Network (STADAN). I could tour the IBM 7090 computer complex. I could "plug in" at the Goddard Voice console, and listen to mission simulations and to the actual flight communications between the MCC, the BCC, and the MSFN flight controllers. This gave me an understanding of how flight controllers communicate. It definitely was a different language. The communications are short, and convey what is important without extraneous words.

Communications were very primitive in the era immediately before satellites. Some stations were connected to underwater cables and land lines back to Goddard, and then on to the MCC and BCC. NASA paid AT&T extra to continuously monitor and tweak these lines for a mission. Very remote sites and tracking stations in the Indian, Atlantic, and Pacific oceans needed up to half an hour to establish contact. It was done with Single Side Band HF radio, much like "ham" radio. I had to learn that unique lingo. The role of CAPE PROCEDURES was to communicate with the remote sites at different times in the mission, to keep them informed of the countdown, synchronize their mission clocks, send them the latest mission rule updates, and send them TTY messages of critical mission information in between their overhead passes. This was what I had to learn during my first six months at the STG and at Goddard. It was invaluable experience and prepared me for my next assignment in the MCC, supporting the MA-6 orbital mission by John Glenn.

After MA-5, I spent the next few months helping Kranz with the mission rules for MA-6, the procedures and communications to the launch pad and the range people, and internal procedures within the MCC. This was in preparation for my job as CAPE PROCEDURES. I was getting a lot of help from John Hatcher (call sign SUPPORT) and George Metcalf (call sign COM TECH) who were full-time RCA technical support people for the MCC. If anybody needed something in the MCC, then John Hatcher was the man to consult. Sometimes he would give us lessons on the data flow into and out of the MCC from all the organizations that supported a launch at the Cape.

Most of the MCC contractors were very experienced with radar, data, voice and teletype as a result of having supported many military and NASA launches. My favorite TTY guy was Andy Anderson. He worked in a small room just off the main control room, and both he and the room reeked of teletype oil. When the door was opened, the sound that emerged resembled a bunch of loud sewing machines. TTY was the primary way to communicate with the remote sites. There was a strict protocol and format for its messages. They often contained critical information but had to be cryptic and precise because they were transmitted at low speed. My Goddard training was very helpful in this regard.

Beneath the VIP viewing room next to the operations control room, there was a very small room. It was in there that people convened to discuss what had gone right and what had gone wrong with a particular mission simulation. These meetings were often "testy" and quite frank. The outcomes and decisions had to be documented, reviewed, and approved. Kranz and I took copious notes in preparation for revising the mission rules and procedures. Nobody could take notes like Kranz. I'm convinced that in a previous life he was a scribe to the Pharaohs.

John Glenn's launch slipped many times. It appeared like we spent several months down at the Cape. Between challenging simulations and a lot of rewriting documentation, we did have some "down time." We had "scrub parties" at the motel, pool parties and played a lot of volley ball. It seemed very weird for me and the other "20-something" flight controllers to be playing with an Air Force General, a McDonnell VP, and a couple of flight surgeons.

Finally, Glenn was launched on February 20, 1962. Although it was my first real mission in the MCC, I was able to step outside and watch the liftoff, then run back inside. Jack Koslosky and I sat alongside to Kranz as assistants to PROCEDURES. Things went fine until the famous "Segment 51" signal. This situation is best described in Kranz's book

Failure is Not an Option. My small role in this situation was to assist Kranz prepare TTY messages that Andy Anderson and Eshleman would send out to the remote flight controllers to find out whether they also had the signal, and any other information that would shed light on the problem. Did the heat shield actually deploy? Some people in the MCC feared the worst; some were talking to the designers to come up with an analysis and a recommendation. I remember Kranz and I with McDonnell engineers John Yardley and Ed Nieman in the small area off the TTY room drafting messages for the remote sites. It was a very tense situation. Meanwhile, Chris Kraft, Walt Williams, Max Faget and other capsule experts strategized on the proper response to the telemetry and possible changes to the re-entry procedures in the time remaining to the retrofire maneuver. These new procedures were radioed to Glenn, who was still unaware of what was going on. The MCC was very quiet during the period of radio blackout during re-entry! There was a great sigh of relief when Glenn called the MCC, and an even greater relief when he was recovered safely. It could have turned out very differently. And this was my first mission in the MCC!

Some days later, President Kennedy came into the MCC to congratulate everybody. We were all dressed up in suits and ties. The PROCDEDURES console was the first one as you enter the main room of the control center. Kennedy's back often gave him some pain, so in going up the two steps to our floor level he placed his hand on my shoulder for some support. I just sat there at attention. He walked past Koslosky and Kranz to greet Flight Director Chris Kraft, who then gave him a tour. Meanwhile, General Curtis LeMay used the top of my console to check out the medal which he was going to present to John Glenn, who was already in the MCC. The General said, "Do you want to see Glenn's medal?" as he showed it to me. I just smiled and continued to sit at attention. A photo of that visit and "Certificate of Participation" for the team that launched the first American into orbit is still on my office wall. Glenn's autographed picture of MA-6 has faded somewhat after over half a century.

By this time, the STG had been folded into the new Manned Spacecraft Center, but many of us were still living in Hampton, Virginia. Kranz told me to go find the flight controllers a place to live in Houston. I found a flat field with some concrete slabs and a house or two set in a new development near the Hobby Airport called Sun Valley. It was just a few miles north of where the Manned Spacecraft Center was supposed to be built. I bought the lot on the corner of Regal and Welk. I reported back to Langley, and told Kranz that we could all get new three bedroom houses with air conditioning for $16,500. I thought I could afford it, since I was making almost $7,000/year! Many flight controllers and contractors bought houses in that development, which was later called "Flight Controller Alley." After wrapping up the MA-6 mission, people started leaving for Houston.

Prior to MA-7 with Scott Carpenter, I continued to support the remote site flight controllers and became involved with the test conductor's countdown and particularly how that drove the activities of the MCC and remote sites. It became clear that we needed to integrate the various countdowns. There were a lot of groups supporting a launch and they each had to take specific actions at certain times. So I started to work on an "integrated countdown" for the MCC.

Carpenter was launched on May 24, 1962; a little under a year since I was hired by the STG. My most vivid memory of this flight was the long re-entry owing to Carpenter not being in the proper attitude for retrofire and making that maneuver late. Not only that, he was almost out of fuel. RETRO John Llewellyn realized that Carpenter would be about 250 miles down range of the prime recovery ship. While the recovery team began to work that problem, I began calling down the Atlantic Missile Range to determine whether anybody had any radar or voice contact with the capsule. My training in military procedures for voice communications came in handy that day. I was "DEVIL FOX BRASS ONE" calling the range stations on SSB radio;, and they all had the word "TOWNSEND" as part of their call sign.

I saw Flight Director Chris Kraft in a heated discussion with the Air Force General running the Atlantic Missile Range, who wanted to take over the recovery operation. Kraft was backing up the Recovery Coordinator's recommendation to let the Navy recover Carpenter. The General learned that the Flight Director was in charge of the mission. This issue was ultimately passed to the Secretary of the Air Force. See Section 13.1. There were usually a couple of Generals and Admirals, and other military officers in the control center in monitoring roles. The only *active* position to have a military officer on loan to NASA was the Network Controller. However, a casual observer might gain the impression that the military were in charge. They were older and in uniform with their rank and medals. Then there were us 20-somethings at consoles. Kraft was 37 years old, and Kranz was 29. The average age for the flight control team was 27. I learned a lot about operations and the decision-making skills that were required of a Flight Director. After this flight my family moved into our newly completed home in Sun Valley. I'd been with NASA for a year and life was pretty good!

The next Mercury mission was MA-8 with Wally Schirra, launched on October 3, 1962. By this time, I had greater experience and was helping to coordinate the mission rules with various flight controllers. As you would expect, the astronauts had a big input. Once Schirra invited me to his beach hotel to coordinate his inputs to the latest draft. We sat on the beach and went over each rule. I would offer some insight into why a particular rule was written and he'd give me his input, which I would later review with Kranz and the other flight controllers. Then we'd get into Wally's 1957 Maserati and drive down to Cocoa Beach for lunch. Over the years, we developed a good relationship, and after the Apollo 1 fire, I worked with him, Walt Cunningham, and Donn Eisele on the flight test objectives for Apollo 7.

Try as you might, it is difficult to plan the birth of your children to coincide with the launch schedules. We were still going back and forth from Houston to Cape Canaveral. My son Kevin was due soon, but we were scheduled to start preparations for MA-9, the final Mercury flight. I knew that I'd have to leave soon, so I arranged for a maid to assist my wife, Jane. Sure enough, Kevin was born on May 1 and I left a day or two later. That really goes over well! To top it off, the maid was more of a hindrance than a help. Fortunately, my wife's aunt resided in Houston and she came to help.

Gordon Cooper's MA-9 mission launched on May 15, 1963. Because it was to last at least a day, two shifts of flight controllers were required. John Hodge would relieve Chris Kraft as the Flight Director. John had been the Bermuda Flight Director on earlier flights. The mission was fairly routine until the last couple of orbits, when things started to fail in the spacecraft; but all ended well. My role, both pre-flight and in-flight was fairly routine; mission rule development, remote site communications, keeping up with the countdown interactions and basically helping Kranz.

In summary, this was my experience with the STG and Project Mercury. Gemini was next with orbital rendezvous and spacewalking, and then Apollo and the Moonwalks. My first two years with NASA were the chance of a lifetime, with things getting better, more complex and more exciting. I feel enormously privileged to have worked with so many outstanding people during what was arguably the "Golden Age" of spaceflight.

References

Books by NASA (Available from the U.S. Government Printing Office-bookstore.gpo.gov)

NASA (No SP #): *Results of the Second U.S. Manned Suborbital Space Flight,* 1961
NASA SP-6: *Results of the Second United States Manned Orbital Space Flight,* 1962
NASA SP-12: *Results of the Third United States Manned Orbital Space Flight,* 1962
NASA SP-45 *Mercury Project Summary & Results of the Fourth Manned Orbital Flight,* 1963
NASA SP-4201: *This New Ocean: A History of Project Mercury,* 1966
NASA SP-4001: *Project Mercury: A Chronology,* 1963
NASA SP-4003: *Space Medicine in Project Mercury,* Mae Mills Link
NASA SP-4105: *The Birth of NASA: The Diary of T. Keith Glennan,* Introduction by Roger Launius, 1993
NASA SP-4407: *Exploring the Unknown,* Edited by John M. Logsdon, 1995
NASA (No SP #): *Hugh L. Dryden's Career in Aviation and Space,* Michael H. Gorn, 1996
NASA SP-2007-4232: *Read You Loud and Clear: The Story of NASA's Spaceflight Tracking and Data Network,* by Sunny Tsiao, 2008
NASA SP-2007-4409: *The Wind and Beyond: A Documentary Journey into the History of Aerodynamics,* Edited by James R. Hansen
NASA SP-2011-595: *Dressing for Altitude, U.S. Aviation Pressure Suits – From Wiley Post to Space Shuttle,* Dennis R. Jenkins
NASA SP-2010-4319: *Revolutionary Atmosphere,* Robert S. Arrighi, 2010
Manned Spacecraft: Engineering Design and Operation, Purser, Faget & Smith. Fairchild Publications, Inc., 1964 (Not published by NASA but authored by STG personnel)
Project Mercury Familiarization Manual, A joint NASA/McDonnell Aircraft book, 1962
Expanding the Envelope: Flight Research at NACA and NASA, Michael H. Gorn, 2001

Springer-Praxis Books

Friendship 7: The Epic Orbital Flight of John H. Glenn, Jr., Colin Burgess, 2015
Freedom 7: The Historic Flight of Alan B. Shepard, Jr., Colin Burgess, 2014
Liberty Bell 7: The Suborbital Flight of Virgil I. Grissom, Colin Burgess, 2014
Selecting the Mercury Seven: The Search for America's First Astronauts, Colin Burgess, 2011
Escaping the Bonds of Earth: The Fifties and Sixties, Ben Evans, 2010
Project Mercury NASA's First Manned Space Programme, John Catchpole, 1957 & 2003
Praxis Manned Spaceflight Log 1961–2006, Tim Furniss and David J. Shayler, 2007
Animals in Space: From Research Rockets to the Space Shuttle, Colin Burgess and Chris Dubbs, 2007.

© Springer International Publishing Switzerland 2016
M. von Ehrenfried, *The Birth of NASA,* Springer Praxis Books,
DOI 10.1007/978-3-319-28428-6

Books by NASA Flight Controllers

Flight: My Life in Mission Control, Christopher C. Kraft, Jr., 2001
Failure is Not an Option, Gene Kranz, 2000
Highways Into Space, Glynn Lunney, 2014
From the Trench of Mission Control to the Craters of the Moon, Third Edition, 2012. Written by the Flight Controllers in the Flight Dynamics Branch.
Go, Flight: The Unsung Heroes of Mission Control, 1965–1992, Rick Houston, J. Milt Heflin. Foreword by John Aaron, 2015

Books by Others

How NASA Learned to Fly in Space, David M. Harland, 2008
Deke: U.S. Manned Space from Mercury to the Shuttle, Deke Slayton and Michael Cassutt, 1994.
Arrows to the Moon: Avro's Engineers and the Space Race, Chris Gainor, 2001
Project Mercury, Ray Spangenburg and Kit Moser, 2001
American in Space, Steven Dick and Robert Jacobs, 2007
Suddenly, Tomorrow Came: The NASA History of the JSC, Henry C. Dethloff, 2013
Mission Control: Inventing the Groundwork of Spaceflight, Michael Peter Johnson, 2015

Internet Links

NASA Headquarters
http://www.nasa.gov/topics/history/index/html
http://history.nasa.gov/refcoll.html

Langley Research Center Sites
http://crgis.ndc.nasa.gov/historic/Space-Task-Group
http://crgis.ndc.nasa.gov/history/Project_Mercury
http://www.larcalumni.org/about-the-laa

Johnson Space Center Sites
http://www.jsc.nasa.gov/history/
http://www.jsc.nasa.gov/history/oral_histories/oral_histories.htm
Historian Rebecca Wright www.jsc.nasa.gov/history
https://www.nal.jsc.org/

Glenn Space Center Site
Historian Anne K. Mills anne.mills@nasa.gov

Armstrong Flight Research Center Site
Historian Christian Gelzer Christian.gelzer-1@nasa.gov

Ames Research Center
Historian April D. Gage april.d.gage@nasa.gov/ames

U.S. Navy Centrifuge
http://crgis.ndc.nasa.gov/historic/Naval_Air_Station_-_Johnsville,_PA

YouTube Videos

Creation of NASA: Message to employees of NACA from T. Keith Glennan
The Moment Sputnik Terrified and Thrilled Americans
The Story of the Sputnik Moment
Sputnik Declassified History Channel Documentary
Avro Canada-CF-105 Arrow History
The Arrow CF-105 Definitive Documentary

Credits

The following photographs were provided courtesy of NASA unless otherwise indicated:

Front Cover   "Mercury Memorial at Cape Canaveral Complex 14" by U.S. Air Force
Back Cover   Four STG photos and author's photo courtesy of NASA
Frontispiece   Sputnik Art created to mark the 50th anniversary of the launch of
                     Sputnik on October 4, 1957 by Gregory R. Todd by Wikimedia Commons

Chapter 3   All declassified documents courtesy of the Eisenhower Archives

Fig. 7.1   Langley Unitary Flow Wind Tunnel Building 1251
Fig. 7.2   The Original Langley Memorial Aeronautical Laboratory
Fig. 7.3   Entrance to the South Side of the Building with the STG sign
Fig. 7.4   The STG sign over the entrance to the building.
Fig. 7.5   Building 60 used by the Astronauts and Training people
Fig. 7.6   MASTIF at the Lewis Research Center
Fig. 7.7   Testing the posigrade rocket separation of the capsule
Fig. 7.8   Testing the retro package
Fig. 7.9   Testing the Mercury Escape Tower
Fig. 7.10   H. Julian "Harvey" Allen
Fig. 7.11   Alfred J. Eggers, Jr. beside the Atmospheric Entry Simulator
Fig. 7.12   Larger version of the Atmospheric Entry Simulator
Fig. 7.13   An ablation test of the ¼-inch scale plastic Mercury capsule
Fig. 7.14   Wallops Island in 1961
Fig. 7.15   The Wallops Flight Facility in 2010

Fig. 10.1   The Philco Team at the STG

Fig. 12.1   The building site for the Manned Spacecraft Center in 1962
Fig. 12.2   Just another view of where we're moving
Fig. 12.3   The new Mission Control Center in 1964
Fig. 12.4   The completed Manned Spacecraft Center in 1965

© Springer International Publishing Switzerland 2016
M. von Ehrenfried, *The Birth of NASA*, Springer Praxis Books,
DOI 10.1007/978-3-319-28428-6

Fig. 13.1   The first NASA Headquarters home from October 1958 until October 1961
Fig. 13.2   Gilruth and the Mercury Seven examining couches in 1959
Fig. 13.3   John Glenn's couch manufactured by McDonnell Aircraft
Fig. 13.4   Pilots testing the couches at the Johnsville centrifuge in 1960
Fig. 13.5   Liberty Bell 7 recovered by Curt Newport and the Oceaneering Expedition
Fig. 13.6   Liberty Bell 7 awaiting restoration. By Kansas Cosmosphere and Space Center
Fig. 13.7   Liberty Bell 7 being restored by Greg "Buck" Buckingham. Courtesy KCSC
Fig. 13.8   The fully restored Liberty Bell 7. By the Kansas Cosmosphere
Fig. 13.9   Gus Grissom inspecting his Redstone at Marshall Space Flight Center
Fig. 13.10   Gordon Cooper's Atlas being off loaded at the Cape
Fig. 13.11   NACA pilot Stanley Butchart flying the Iron Cross
Fig. 13.12   Mercury 1 lb. thruster. Photo courtesy of Historic Space Systems
Fig. 13.13   Mercury 24 lb. thruster. Photo courtesy of Historic Space Systems
Fig. 13.14   Thrusters installed on the Mercury capsule. Courtesy of Historic Space Systems
Fig. 13.15   Arnie Aldrich demonstrating how he threw the Retrofire switch

Fig. 14.1   The Mercury Control Center during MA-6
Fig. 14.2   The restored MCC flight control area at KSC
Fig. 14.3   The restored MCC front row of consoles
Fig. 14.4   The restored Environmental and Systems consoles
Fig. 14.5   The restored Flight Director's console
Fig. 14.6   The restored upper row of consoles
Fig. 14.7   The restored Retrofire Officer and Flight Dynamics Officer consoles
Fig. 14.8   The MCC in 2010 prior to demolition
Fig. 14.9   The end of an historical landmark
Fig. 14.10   The MCC historical marker
Fig. 14.11   The Bermuda Control Center
Fig. 14.12   The NASA C-Band radar at Cooper's Island
Fig. 14.13   The Manned Space Flight Network
Fig. 14.14   John Glenn in the Langley Mercury Procedures Trainer
Fig. 14.15   The Mercury simulation control room at the Cape

Fig. 16.1   The Space Launch System vehicles
Fig. 16.2   The Orion attached to the Service Module
Fig. 16.3   The ISS derived concept for a Deep Space Habitat Module
Fig. 16.4   John Glenn's parade down the Canyon of Hero's

Appendix 1, Organization Lists, Charts and Manning
             All organizational information was courtesy of NASA
             The Manning lists are courtesy of Gene Kranz and Arnold Aldrich

Appendix 2, Biographies
             Photos were from individuals, NASA, Google, or Wikipedia

Appendix 3, STG Technology
Fig. A.3.1   The K&E slide rule. Courtesy of Wikipedia
Fig. A.3.2   The Friden calculator. Courtesy of Wikipedia
Fig. A.3.3   An IBM 704 computer at NACA in 1957. Courtesy of Wikipedia
Fig. A.3.4   The NASA Goddard computer room in 1962
Fig. A.3.5   An IBM 1620 like that used by the STG. Courtesy of San José State University
Fig. A.3.6   The Bendix G-15 computer. Courtesy of A History of Modern Computing 1998
Fig. A.3.7   GE/Burroughs Guidance Computer. Courtesy of the Air Force Space Museum
Fig. A.3.8   The Goodyear Electronic Differential Analyzer. Courtesy of vintchip.com

Appendix 4, Some Photos
Fig. A.4.1   The NACA Special Committee on Space Technology
Fig. A.4.2   The CF-105 Arrow built by AVRO
Fig. A.4.3   A Langley "Computer."
Fig. A.4.4   The Langley West End Model Shop Employees
Fig. A.4.5   The Missile Instrumentation Ship Rose Knot. Courtesy of Navsource.org.
Fig. A.4.6   The Mercury manned launch vehicles. Courtesy of Wikipedia
Fig. A.4.7   The Mercury Control Center modified for Gemini.
Fig. A.4.8   A post-flight celebration of Alan Shepard's flight. Courtesy of Robert Voas.

Appendix 5, Quotes
Fig. A.5.1   Kennedy's memo to Johnson, April 1961

Appendix 6, Stories and Trivia
Fig. A.6.1   Tom Collier and his frozen buzzard.
Fig. A.6.2   V-2 Bumper, the first flight from Cape Canaveral
Fig. A.6.3   Sam before his flight

About the Author
Fig. AA.1   The author as a young STG flight controller, late 1961
Fig. AA.2   The author with two of his supervisors
Fig. AA.3   The new MCC Houston-White Team

Glossary

ABMA	Army Ballistic Missile Agency
ACE	Acceptance Checkout Equipment
AEDC	Arnold Engineering and Development Center
AF	Air Force
AFD	Assistant Flight Director in the MCC
AFMTC	Air Force Missile Test Center
AFRC	Armstrong Flight Research Center
A/G	Air Ground communications loop
AIAA	American Institute of Aeronautics and Astronautics
AMR	Atlantic Missile Range
AOMC	Army Ordinance Missile Command
ARC	Ames Research Center
ARDC	Air Research and Development Command
ARGMA	Army Rocket & Guided Missile Agency
ARPA	Advanced Research Projects Agency
ASIS	Abort Sensing and Implementation System
ATS	Atlantic Tracking Ship
AVRO	A. V. Roe and Company
BDA	Bermuda (Tracking Station)
BECO	Booster Engine Cut Off (Atlas)
BJ	Big Joe
BMD	Ballistic Missile Division
BOOSTER	Call sign for the Booster Monitor in the MCC
B.S.	Bachelor of Science
CAL	Point Arguello, California (Tracking Station)
CAPCOM	Call sign for the Capsule Communicator
CCAFS	Cape Canaveral Air Force Station
Cdr.	Commander

© Springer International Publishing Switzerland 2016
M. von Ehrenfried, *The Birth of NASA*, Springer Praxis Books,
DOI 10.1007/978-3-319-28428-6

ChFC	Chartered Financial Advisor
CIA	Central Intelligence Agency
CINCPACFLT	Commander in Chief of the Pacific Fleet
CLU	Chartered Life Underwriter
COMTECH	Communications Technician
CORONA	Code name for a series of reconnaissance satellites
CRO	Carnarvon, Australia (Tracking Station)
CSQ	Coastal Sentry Quebec (Tracking Ship)
CTN	Canton Island (Tracking Station)
CV/A	Convair/Astronautics Division
CYI	Canary Island (Tracking Station)
DOD	Department of Defense
DOS	Department of State
DFRC	Dryden Flight Research Center (now AFRC)
ECG	Electro Cardiogram
FAA	Federal Aviation Administration
FLIGHT	Call sign for the Flight Director in the MCC
FIDO	Call sign for the Flight Dynamics Officer in the MCC
FRD	Flight Research Division
FSD	Flight Systems Division of the STG
GD/A	General Dynamics/Astronautics
GEDA	General Electric Differential Analyzer
GMRD	Guided Missile Range Division of PanAm
GSFC	Goddard Space Flight Center
GUIDO	Call sign for the Guidance Officer in the MCC
GYM	Guaymas, Mexico (Tracking Station)
HAFB	Holloman Air Force Base
HAMC	Holloman Aerospace Medical Center
HAW	Hawaii (Tracking Station)
HSFS	High Speed Flight Station
IBM	International Business Machines
ICBM	Intercontinental Ballistic Missile
IEAL	International Energy Associates, Ltd.
IFEP	In-Flight Experiment Panel
IGY	International Geophysical Year
IOS	Indian Ocean Ship
IRBM	Intermediate Range Ballistic Missile
IRD	Instrument Research Division of Langley
JPL	Jet Propulsion Laboratory
JSC	Johnson Space Center
KNO	Kano, Nigeria (Tracking Station)
KSC	Kennedy Space Center
LeRC	Lewis Research Center
LJ	Little Joe
LLRV	Lunar Landing Research Vehicle

LLTV	Lunar Landing Training Vehicle
LOX	Liquid Oxygen
LRC	Langley Research Center
LRL	Lunar Receiving Laboratory
MA	Mercury Atlas
MAC	McDonnell Aircraft Company
MASTIF	Multiple Axis Space Test Inertial Facility
Max-Q	Maximum Dynamic Pressure
MCC	Mercury Control Center, Mission Control Center
MECO	Main Engine Cutoff (Atlas)
MIT	Massachusetts Institute of Technology
MPAD	Mission Planning and Analysis Division
MPT	Mercury Procedures Trainers
MR	Mercury Redstone
MR-BD	Mercury Redstone Booster Development
M.S.	Masters of Science
MSC	Manned Spacecraft Center
MSFC	Marshall Space Flight Center
MSFN	Manned Space Flight Network
MUC	Muchea, Australia (Tracking Station)
NACA	National Advisory Committee for Aeronautics
NASA	National Aeronautics and Space Administration
NASCOM	NASA Communications Network
NIE	National Intelligence Estimate (CIA)
NOAA	National Oceanic Atmospheric Agency
NRL	Naval Research Laboratory
NSC	National Security Council
NSTS	National Space Transportation System
PARD	Pilotless Aircraft Research Division
PIRD	Program Instrumentation Requirements Document
PSAC	President's Scientific Advisory Committee
PROCEDURES	Call sign for the MCC Operations and Procedures Officer
Q & A	Question and Answer
RAE	Royal Aircraft Establishment
RAF	Royal Air Force
RB-57F	Reconnaissance Bomber-57F (Air Force/NASA aircraft)
RCA	Radio Corporation of America
RCS	Reaction Control System
RETRO	Call sign for the Retrofire Officer in the MCC
RKV	Rose Knot Victor (Tracking Ship)
ROTC	Reserve Officers Training Corps
RSO	Range Safety Officer
SAM	San Antonio Medical Research (Rhesus monkey)
SECO	Sustainer Engine Cutoff (Atlas booster)
SLS	Space Launch System

SRD	Stability (or Structures) Research Division
STADAN	Satellite Tracking and Data Acquisition Network
STDN	Satellite Tracking and Data Network
STEM	Science, Technology, Engineering and Mathematics
STG	Space Task Group
STL	Space Technology Laboratories
STMD	Space Technical Mission Directorate
SURGEON	Call sign for the flight surgeon in the MCC
SYSTEMS	Call sign for the Systems Flight Controller
TADCORPS	Technical and Administrative Services Corporation
TAGIU	Tracking and Ground Instrumentation Unit (Langley)
TDRS	Tracking and Data Relay Satellite
TDRSS	Tracking and Data Relay Satellite System
TSSG	Tracking System Study Group (Langley)
TEX	Corpus Christi, Texas (Tracking Station)
TM	Telemetry
TTY	Teletype
U.K.	United Kingdom
U.N.	United Nations
U.S.	United States
USA	United States Army
USAF	United States Air Force
USGS	United States Geological Survey
USN	United States Navy
UT	University of Texas
V-1 and 2	Vengeance Weapons 1 and 2
VIP	Very Important Person/People (a room in the MCC)
VOA	Voice of America
WB-57F	Weather Bomber-57F (NASA aircraft)
WECO	Western Electric Company
WFS	Wallops Flight Facility
WIS	Wallops Island Station
WOM	Woomera, Australia (Tracking Station)
WSMR	White Sands Missile Range
WSPG	White Sands Proving Ground
WWMCCS	World-Wide Military Command & Control System
ZZB	Zanzibar (Tracking Station)

About the Author

Manfred "Dutch" von Ehrenfried II had the extremely good fortune to have interviewed for the NASA Space Task Group the day before Alan Shepard was launched on MR-3. At the time, he had very little knowledge of Project Mercury and naïvely expected that since his degree was in physics he would be assigned in that area. As fate would have it, he was assigned to the Flight Control Operations Section under Gene Kranz, who became his supervisor and mentor. Most of his activity on Project Mercury would be in the areas of mission rules, countdowns, operational procedures, and coordination with the remote tracking station flight controllers. During his first six months, Dutch was in training to be a flight controller, and he spent MA-4 and MA-5 at the Goddard Space Flight Center learning about the communications between the Mercury Control Center and the Mercury Space Flight Network. For John Glenn's orbital flight on MA-6 he was learning the Procedures flight control position in the Mercury Control Center under Gene Kranz. He went on to support the follow-on Mercury missions of Carpenter, Schirra, and Cooper in the Mercury Control Center.

Dutch supported all the Gemini missions up to and including the first EVA on Gemini 4 and the first rendezvous in space in December 1965 by Gemini 6 and Gemini 7. In 1966 he became a Guidance Officer on Apollo 1 and was on-console for that ill-fated pad test on January 27, 1967 when a fire took the lives of Gus Grissom, Ed White and Roger Chaffee.

As the program resumed, Dutch became the Mission Staff Engineer for Apollo 7 and backup for Apollo 8. During this period he was also an Apollo Pressure Suit Test Subject. This afforded the opportunity to test pressure suits in a vacuum chamber to an altitude exceeding 400,000 feet, including one test of Neil Armstrong's suit. He also experienced 9 "g" in the centrifuge and took part in trials aboard the "zero g" aircraft. As part of this program, he had his own Apollo A7LB Skylab suit. These experiences enabled Dutch to join the Earth Resources Aircraft Program and become the first Sensor Equipment Operator and Mission Manager on the high altitude RB-57F.

Dutch also worked in the nuclear industry for seven years and on the Space Station Program for ten years. He has written several books about his experiences.

© Springer International Publishing Switzerland 2016
M. von Ehrenfried, *The Birth of NASA*, Springer Praxis Books,
DOI 10.1007/978-3-319-28428-6

Fig. AA.1 The author as a young STG flight controller in late 1961. (Photo courtesy of NASA)

Fig. AA.2 The author on the right with two of his supervisors; Gene Kranz and George Low. (Photo courtesy of NASA)

Fig. AA.3 The new MCC in Houston. The White Team for the Gemini 5 mission in August 1965. Front row (l-r): Larry Armstrong, Stu Davis, Larry Keyser. Seated: the author and Gene Kranz. Back row (l-r): Neil Armstrong, Buzz Aldrin, Dr. Duane Catterson, Will Fenner, Dave Massaro, Don Bray, Al Chop, Henry Stephenson, John Aaron, Bill Johnson, Capt. Andy A. Piske, Ron Cagle, and Russ Nickerson. (Photo courtesy of NASA)

The Front Cover

The photograph on the cover is the Mercury Memorial at Cape Canaveral Launch Complex 14. Made of titanium, it combines the astronomical symbol of Mercury with the "7" for the seven Mercury astronauts. Buried beneath the sculpture is a time capsule of technical documents for Project Mercury. It is schedule to be opened in 2464; 500 years after the official conclusion of the program.

This site was first used by the Atlas launch vehicle in 1957 and then for the Mercury Atlas Big Joe test in 1959. All four of the manned Mercury Atlas missions, including Colonel John Glenn, Commander Scott Carpenter, Captain Wally Schirra, and Colonel Gordon Cooper were launched from here. It also launched seven Atlas Agena target vehicles for Gemini rendezvous missions. The site was decommissioned and dismantled in the 1970s.

In 1997 the Air Force 45th Space Wing embarked on a restoration using industry funds and non-military volunteers. The blockhouse and parking area was made into a conference facility which contains historical documents, photos and memorabilia from Project Mercury as well as photos of the blockhouse prior to and during the restoration.

© Springer International Publishing Switzerland 2016
M. von Ehrenfried, *The Birth of NASA*, Springer Praxis Books,
DOI 10.1007/978-3-319-28428-6

Index

A

Abbott, Ira H., 198, 204, 205, 318
Ablation, 35, 49, 102, 103
ABMA. *See* Army Ballistic Missile Agency
Abort Sensing and Implementation System
  (ASIS), 82, 115, 121, 123, 124
Adams, Gov., 8, 9
Advanced Research Project Agency (ARPA), 2,
  16, 27–28, 36, 113, 170, 197, 198, 201,
  208, 226, 294, 314
AEDC. *See* Arnold Engineering Development
  Center
Aeromedical Monitor (AEROMED), 67, 131,
  294, 324
Aerospace Corporation, 96, 227, 255
AFRC. *See* Armstrong Flight Research Center
Aikenhead, Bruce A., 57, 64, 75
Air Force Scientific Advisory Board, 30
Air Research and Development Command
  (ARDC), 27, 30
Aldrich, Arnold D., xi, 73, 74, 126, 127, 244, 330
Aldridge, Roy, 60
ALGO, 299
Algranti, Joe, 42
Allen, Charlie C., 70
Allen, H. Julian, 46, 47, 211, 302
Ames, Joseph S., Dr., 200
Ames Research Center (ARC), 89, 170, 200
Anders, Bill, 291
Anderson, Andy, 81, 290, 331, 332
Anderson, Melvin S., 39, 228
Apollo 1, 107, 122, 123, 135, 163, 164, 276,
  291, 333
Appold, Norman C., Col., 302
Arabian, Donald D., 60, 61, 77

ARC. *See* Ames Research Center
Armitage, Peter J., 56, 57
Armstrong Flight Research Center (AFRC), 50, 120
Armstrong, Neil A., 27, 51, 118, 163, 344
Armstrong Siddeley Holdings, Ltd., 54
Army Ballistic Missile Agency (ABMA), 28–29,
  53, 75, 79, 113, 170
Army Ordnance Missile Command (AOMC),
  29, 79
Arnold Engineering Development Center
  (AEDC), 53
Arnold, Henry "Hap," Gen., 53
ARPA. *See* Advanced Research Project Agency
Arrowhead Rubber Company, 165
ASIS. *See* Abort Sensing and Implementation
  System
Atlantic Ship (ATS), 155
Augerson, William S., Capt. Dr., 67, 79, 244,
  294, 322
Avco, 113
A. V. Roe & Co. (AVRO), 3, 54, 55, 281

B

Bailey, James W., 62
Bailey, John R., 78
Baillie, Richard F., 78
Ballinger, E.R., 103
Ballistic Missile Division (BMD), 63, 69, 82, 96,
  112, 114, 116
Barger, Beulah, 305
Barker, Al., 81
Barnard, Jack, 78
Barry, Bill, xi
Basilio, Carmen, 327

© Springer International Publishing Switzerland 2016
M. von Ehrenfried, *The Birth of NASA*, Springer Praxis Books,
DOI 10.1007/978-3-319-28428-6

Batdorf, Samuel, 36, 198
BCC. *See* Bermuda Control Center
Beatty, LaMar D., 62
Beck, Harold D., 244, 249
Beck, Janette H., 78
Beckman, David A., 71
Beeding, Eli L., 103
Behuncik, John A., 70
Bell Telephone, 24, 45, 48, 82
Belly Band, 111, 112, 124
Beltsville Space Center, 44
Bender, Joseph, 41
Bendix Corp., 46
Bendix G-15, 298, 299
Bennett, James A., 78
Benson, Harold, 68
Bergen, Jim, 66
Berkner, Lloyd V., Dr., 310
Bermuda, 3, 33, 45, 46, 69, 82, 124, 125, 132,
     151–155, 161, 240, 252, 260, 263, 297, 333
Bermuda Control Center (BCC), 3, 151–153, 161,
     252, 260, 263, 330, 331
Berry, Charles A., Dr., 67, 79, 244, 251
Beryllium, 66, 102, 103, 107, 208, 231, 258
Big Joe, 41, 42, 66, 84, 85, 102, 103, 109–111,
     164, 208, 229, 231, 243, 258, 322
Black Friday, 55, 248
Blanchard, Robert C., 69
Bland, William M., 39, 65, 104, 106, 228, 232, 276
Blume, Donald D., 60, 61
BMD. *See* Ballistic Missile Division
Boatman, Dana, 80
Bobola, Robert E., 68
Boeing, 113, 305
Boiler plate, 102, 103, 111, 122, 233
Boler, Leonard J. 77
Bond, Aleck C., 39, 65, 103, 228, 230
Borman, Frank, 291
Bost, James E., 78
Boswick, Guy W. Jr., 60, 85
Boyer, William J., 39, 43, 228
Brewer, Gerald W., 73, 74, 246, 260, 270, 274, 330
Bronk, Detlev W., Dr., 10, 14
Brown, B. Porter, 75, 103
Brumberg, Paul G., 71, 73
Brundin, Robert H., Lt. Col., 114
Brush Beryllium Company, 102
Bryant, John, 292
Buckingham, Greg "Buck," 108
Buckley, Edmond C., 84, 179, 318
Buglia, J., 101
Buller, Elmer H., 75
Burcher, Margaret, 60
Burns and Roe, Inc., 80, 143
Burton, Mary Shep, 72, 235, 250

Butchart, Stanley, 51, 117
Byrnes, Martin A., 89, 92, 226, 267

C
California Institute of Technology (CALTECH), 170
Campbell, Jack A., 41, 61
Canary Islands (CYI), 32, 151, 155, 252, 271, 273
Canning, Thomas, 46
Canton (CTN), 33, 155
Cape Canaveral (CNV), 154
Capsule Communicator (CAPCOM), 55, 73, 121,
     127–129, 138, 144, 152, 156, 246,
     270–273, 324
Carley, Richard R., 57, 69
Carlson-Ewell, 80, 143
Carnarvon, 271, 291
Carpenter, Edward A., 62
Carpenter, Malcolm Scott, Lt., 64, 128–130,
     134–137, 140, 240, 332
Carrillo, Antonio, Amb., 32
Carter, Nancy K., 72
Carter, Tom, 292
Catalong, Thomas M., 41
Catterson, Duane A., Dr., 67, 79
Central Intelligence Agency (CIA), 21–23
CF-105 Arrow, 55, 56, 96, 248, 253, 257, 259,
     261, 262, 274, 275, 280, 281, 303
CF-100 Canuck, 253
Chaffee, Roger B., 107, 123
Chalmers, Frank J., 57
Chamberlin, James A., 56, 57, 60, 76, 86, 87, 116,
     244, 252
Chambers, Thomas V., 57, 69
Champine, Robert A., 104, 106, 216
Chapman, Arthur C., 62
Chilton, Robert G., 39, 68, 69, 228
Chrysler, 82, 113
CIA. *See* Central Intelligence Agency
C-102 Jetliner, 55, 56, 253, 274, 275
Clark, Arthur C., 62, 318
Clark, George Q., 45
Clements, Henry, Capt., 145
Coastal Sentry Quebec (CSQ), 132, 155, 271,
     272, 306
Cochran, Jacqueline, 97
Cohen, Jack, 57, 64, 70, 71, 162, 240, 277
Cohn, Stanley H., 57, 72
Collier, Tom, 319, 320
Collins, Carter, Lt., 106
Compton, Harold R., 69
Control Data Corp., 279, 299
Convair, 63, 72, 82, 96, 103, 112–116, 123, 144,
     208, 219, 243, 302
Coons, Owen D., Dr., 67

Cooper, Leroy Gordon, Jr., Capt., 64, 115, 125, 131, 132, 136, 137, 141, 166, 171
Cooper's Island Wildlife Observation Tower, 153
Corning Glass, 106
CORONA, 23
Corson, Dale R., 302
Cour-Palais, Burton, G., 57
Covington, Ozro M., 45, 52, 53
Critzos, Chris C., 69, 87, 329, 330
Cross, Dick, 81
Crowley, John W., 99, 198, 205
CSQ. *See* Coastal Sentry Quebec
Culbertson, Phillip E., 115
Cunningham, Walter, 292, 333
Cutler, Gen., 9

D
Dabbs, John H., 74
David Clark Company, 165
Davidson, Robert E., 71
Davidson, William L., 74, 244, 254
Davis, Stu, 67, 291
Day, Richard E., 300
Deans, Philip M., 68
Debus, Kurt H., Dr., 150
Deep Space Network, 154
DeFrance, Smith J., 170
De Gaulle, Charles, 327
Deluca, Lou, 81
Dempsey, J. R., 302
Dereng, Derwood A., 321
Diefenbaker, John G., P.M., 55, 56, 96
Disher, John H., 41, 240
Dodson, Joe W., 77
Doll, Edward B., 114
Dolley Madison House, 2, 99, 198, 199, 208, 219
Donegan, James J., 84
Donlan, Charles J., 36, 39, 56, 198, 210, 212, 228
Doolittle, James H., Gen., 16, 24, 169, 197, 309
Douglas, William K., Lt. Col. Dr., 63, 67, 79, 138, 244, 256, 289
Drummond, William E., 62
Dryden, Hugh L., 5, 41, 95, 99, 117, 119, 169, 170, 198, 200, 202, 206, 226, 267, 302, 310, 316
Dulles, Allen, 21, 22
Dulles, John F, 21
Dunseith, Lynwood C., 73
Duret, Eugene L., 57, 67
Dyna-Soar, 5, 28, 30, 113, 220

E
Eddy, Bob, 64, 162
Edwards, Elwood S., 62
Eggers, Alfred J., 36, 46, 48, 101, 199, 211, 214
Eggleston, Jack, 250
Eglin AFB (EGL), 155
Eisele, Donn F., 135, 291, 333
Eisenhower, Dwight D., Pres, 2, 7, 98, 104, 169, 170
Enderson, Lawrence W. Jr., 65
Engel, Jerome N., 72, 235
Enos (chimpanzee), 30, 92, 126, 246, 273
Erb, Donna M., 57, 257
Erb, R. Bryan, 56, 57, 67, 244, 257, 272
Ernull, Robert E., 74
Eshelman, J., 81, 332
Evans, Joanna M., 67
Ewart, David D., 56, 57, 66
Explorer, 29, 31, 44, 80, 84, 309
Eziaslav, N. Harrin, 76

F
Faber, Stanley, 64, 74, 75, 162
Faget, Maxime A., 39, 65, 83, 101, 102, 104, 106, 198, 215, 228, 322
Farbridge, Joseph E., 57, 66
Farmer, Norman B., 57, 68
Ferguson, Paul O., 61
Ferrando, James A., 70, 235
Fiberglass-phenolic, 66, 102
FIDO. *See* Flight Dynamics Officer
Fielder, Dennis E., 58, 73, 74, 244, 259, 261
Fields, Edison M., 39, 65, 228
Fisher, Louis R., 65, 77
Fisk, James B., Dr., 16, 24, 169
Fitzgerald, Evelyn B., 77
Fitzkee, Archie L., 67, 272
Flight Director (FLIGHT), 58, 69, 73, 80, 82, 120, 121, 127–130, 135, 143–145, 148, 152, 158, 160, 161, 163, 164, 237, 240, 261–263, 268–271, 291, 317, 324, 325, 328, 332, 333
Flight Dynamics Officer (FIDO), 58, 71, 73, 75, 82, 121, 160, 162, 240, 280
Flight Operations Division (FOD), 254, 268, 283, 330
Flight Research Division (FRD), 40, 74, 218, 225, 229, 232, 237, 241, 250
Folwell, Paul A. II, 62
FORTRAN, 72, 297
Fountain, Jo Ann S., 60, 85
Fraleigh, William, 32
Franklin, Arthur E., 64, 75
Franklin, Marion R. Jr., 60

Freedman, Gilbert M., 76
French, John C., 76, 116
Friden, 250, 296
Frutkin, Arnold W., 32
Funk, Jack, 69

G

Gagarin, Yuri, 124, 138, 314, 315
Gage, April, xi
Gainer, Mary E., xi
Gainor, Chris, 54
Galezowski, Stanley, 58, 69
Gallagher, Thomas F., 61
Gallup, Benjamin, 152
Gamble, Allen O., 213, 289
Gardner, Benson B., 62
Garland, Benjamin J., 101
Garvin, Bill, 291
GE/Burroughs Guidance Computer, 161, 299
GEDA. *See* Goodyear Electronic Differential
     Analyzer
Gelzer, Christian, xi, 120, 318
Gemini, 2, 3, 32, 46, 53, 54, 57, 71, 93, 103, 111,
     131–133, 136, 141, 143, 145, 151, 155,
     161–164, 168, 208, 212, 216, 219, 222,
     232, 236, 237, 240–243, 246, 249, 252,
     253, 263, 265, 267–269, 271, 273, 276,
     277, 280, 282–287, 291, 294, 305, 307,
     317, 334
General Dynamics, 58, 113, 208
Gibson, Thomas F. Jr., 69
Gilkey, John E., 41, 229
Gilruth, Robert R., Dr., 3, 36, 39, 59, 79, 83, 86,
     95, 102, 104, 123, 161, 165, 170, 179, 198,
     216, 228, 269, 302, 318, 319
Glennan, T. Keith, Dr., 28, 32, 36, 43, 57, 59, 90,
     98, 170, 179, 198, 199, 208, 210, 316
Glenn, John H., Lt. Col., 64, 92, 93, 101, 123,
     133, 138, 157, 176
Glynn, Francis I., 62
Goddard Space Flight Center (GSFC), 31–33,
     44–46, 50, 71, 72, 81, 82, 85, 87, 89, 99,
     133, 136, 137, 143, 153, 170, 201, 261,
     271, 280, 281, 287, 290, 292, 318, 330
Goett, Harry J., 45, 100, 170
Goodpaster, Andrew J., Gen., 7
Goodrich, B. F., 63, 79, 165, 166
Goodwin, Burney H., 60, 61, 85
Goodyear Electronic Differential Analyzer
     (GEDA), 51, 117, 118, 300
Gorman, John, 81
Goslee, John B., 78
Gough, Melvin, 103

Graham, John B. Jr., 76, 286
Grana, David C., 77
Grand Bahamas, 33, 155
Grand Turk, 33, 155
Granger, Harold E., 76
Graves, George B., 84
Gray, Wilbur H., 80
Gregory, Donald T., 77
Grissom, Virgil I., Capt., 64, 103, 106, 107, 114,
     122, 123, 133, 138, 163, 166, 256, 291.
     307, 330
Guaymas, 32, 155, 246, 271
Guidance Officer (GUIDO), 107, 160, 291
Guiness, Alec, 327
Gulfgate Shopping Center, 92, 93
Guthrie, George C., 63, 64, 74, 289
Guy, Walter W., 68

H

Hagen, Mr., 9
Hagerty, James, 10
Haise, Fred, 163
Ham (chimpanzee), 106, 121
Hamby, William H., 66
Hammack, Jerome B., 39, 65, 76, 77, 113, 228, 231
Hand, Arthur A., 64, 71, 162
Haney, Paul P., 60
Hare, Linda J., 60
Harrin, Eziaslav N., 76
Harrington, Robert D., 69, 82, 115
Harris, George Jr., 58, 244, 260–262
Harrison, Albertus S., Gov., 93
Hartung, Jack B., 70
Hasselblad, 136, 137
Hasson, Dennis F., 66
Hatcher, John, 81, 87, 290, 308, 325, 331
Hatley, Shirley J., 40, 70, 228
Havenstein, Paul L., Cdr., 59, 69, 79, 143, 159
Hawaii, 155, 271, 292
Hawker Siddeley, 54, 55
Hayes, William C., 76
Heat sink, 35, 66, 102, 230
Heberlig, Jack C., 40, 60, 65, 104, 228, 235
Henry, James P., Lt. Col. Dr., 67, 79, 165, 294
Hermann, Robert A., 65
Hibbert, John, 82
Hicks, Claiborne R., 40, 70, 228, 234
Higgins, Rodney F., 62, 64, 75
High Speed Flight Station (HSFS), 35, 50–52, 97,
     117, 120, 199, 217, 226, 267, 300
Hill, Ann W., 60, 61, 85, 204
Hinson, James, K., 68
Hinson, Shirley Hunt, 250

Historic Space Systems, 118, 119
Hodge, John D., 56, 58, 66, 69, 152, 236, 237, 244, 255, 262, 263, 274, 330, 333
Hodge, Leon B., 76
Hoffman, S. K., 302
Hoggard, Walter C. Jr., 76
Hohmann, Bernard A., 114
Holaday, Mr., 9
Holden, Joan S., 60
Holman, Richard A., 62
Holt, Richard, 94, 292
Hooker, Ray W., 84
Hoover, Luther L., 62
Hoover, Richard A, 75, 162
Hopp, Harry, 81
Horner, Richard, 99
Horse Collar, 111–112, 124, 126
Horsman, Paul F., 69
HSFS. *See* High Speed Flight Station
Hubert, Wilbur, 81
Hughes, John K., 58
Humphrey, William R., 77
Hunter, Daniel, 81, 291
Hunt, Shirley A, 72, 235, 250
Huskey, Harry, 299
Huss, Carl R., 70, 235, 244, 251, 263, 273, 325, 327
Hyatt, Abraham, 302
Hydrogen peroxide, 50, 52, 117–120, 134

I

IBM, 16, 46, 82, 143, 237
IBM 704, 55, 69, 72, 82, 85, 109, 151, 250, 296, 297, 299
IBM 709, 151
IBM 7090, 72, 82, 151, 297, 330
Instrument research division (IRD), 40, 74, 83, 84, 179, 229, 235, 270, 287, 318
Intercontinental ballistic missile (ICBM), 2, 12–14, 21, 22, 26, 29, 96, 112, 115, 116, 131, 169, 211, 212, 233
Intermediate range ballistic missiles (IRBM), 13, 26, 28, 102, 169
International Geophysical Year (IGY), 10–12, 20, 22, 28, 31, 154, 310, 311, 327
International Latex Corporation, 165
Intrepid, 129
International Space Station (ISS), 172, 174
Iron Cross, 51, 117, 118, 300

J

Jackson, Bruce G., 66
Jacobs, Stephen, 67

James, John L. Jr., 62
Jenkins, Dennis R., 166
Jenkins, Morris V., 58, 69, 235
Jet Propulsion Laboratory (JPL), 29, 154, 170, 199
Jevas, Nicholas, 78
Jezewski, Donald J., 69
JF-104, 52, 119, 300
Jimenez, Jose, 326
Johnson, Bryant L., 78
Johnson, Caldwell C. Jr., 88, 220, 238, 244, 264–266, 276, 322
Johnson, Carol L., 77
Johnson, Elizabeth P., 72
Johnson, Harold I., 64, 74, 75
Johnson, John A., 99, 202
Johnson, John L., 99, 198
Johnson, Lyndon B., Pres., 170, 309
Johnson, Paul, 82, 143, 159
Johnson, Roy W., 28, 36, 198, 314
Johnson Space Center (JSC), 42, 57, 58, 111, 123, 231, 237–238, 242, 244, 246, 250, 253, 255, 259, 267, 287, 292
Johnson, W. Kemble, 59
Johnsville, 63, 74, 79, 103, 105, 106, 164, 216, 274, 316
Jones, Enoch M., 76
JPL. *See* Jet Propulsion Laboratory
JSC. *See* Johnson Space Center
Juno, 29
Jupiter, 2, 16, 26, 28, 29, 204, 297

K

Kadett, Heinkel, 117
Kano (KNO), 32, 155, 273, 324
Kansas Cosmosphere and Space Center, 107–109
Kaplan, Joseph, Dr., 310
Kaplan, Morton A., 327
Kapryan, Walter J., 65, 77
Karakulko, Witalij, 68
Karick, Francis S., 78
Kase, Louise E., 80
Kehlet, Alan B., 40, 66, 228, 233
Kennedy, John F. "Jack," Pres., 96, 112, 122, 170, 171, 218, 238, 314, 315, 328, 332
Kennedy Space Center (KSC), 150, 243, 279, 294
Khrushchev, Nikita, 13, 15, 22, 55, 326
Kidd, Louis M., 60
Killian, James R., 22–24, 27, 98, 169, 197
Kinzler, Jack A., 61, 83
Kistiakowsky, George B., Dr., 16
Klienknecht, Kenneth S., 60, 244
Kloetzer, Paul H., 78
Knox, Betty S., 60

Koch, Dick, 82
Kolenkiewicz, Ronald, 40, 65, 229
Koos, Dick, 64, 162
Koslosky, John T., 74, 330–332
Kovitz, Carl J., 76
Kraft, Christopher C., 40, 69, 96, 115, 120–122, 126, 128, 130, 143, 158, 159, 229, 236, 268, 317, 324–333
Kranz, Eugene F., 73, 74, 82, 121, 159, 244, 268, 324–326, 330–332
Krasnican, Milan J., 41, 68, 229
KSC. *See* Kennedy Space Center
Kuettner, Joachim, Dr., 82, 113, 114, 120, 121
Kyle, Howard C., 73, 74, 84, 244, 269

L
Laika, 5
Land, Edwin H., Dr., 16
Lane, William E., 68
Langley Research Center (LRC), 3, 33, 37, 38, 45, 50, 59, 61, 72, 73, 76, 82, 83, 86, 91, 99, 104, 134, 136, 137, 161, 170, 212, 222, 224, 225, 246, 277, 317, 329
Largent, Harold R., 68
Larson, Mark S., 62
Lauton, William T. Jr., 40, 77, 229
Lawrence, Ernest O., Dr., 14
Layton, Al, 82
Leatherman, Pattie M., 72
Lee, John B., 40, 68, 229, 238, 239, 322
LeMay, Curtis E., Gen., 27, 332
Leonard, Pauline O., 73
Lewis, Charles R., 270, 271
Lewis Flight Propulsion Laboratory, 169
Lewis, John H. Jr., 71, 73
Lewis Research Center (LeRC), 3, 41, 87, 110, 111, 134, 170, 204
Liberty Bell Seven, 106–109, 123
Lindell, Keith G., Lt. Col. Dr., 59, 63, 79, 256, 289
Lindley, Robert N., 56, 58
Link, Patricia D., 69
Link Trainer Company, 80
Linn, Kathryn E., 78
Little Joe, 43, 47, 65, 76, 84, 85, 103, 108–111, 164, 208, 220, 275, 281, 305, 323
Livesay, Norma L., 40, 60, 229
Llewellyn, John S., 67, 244, 272, 322, 328, 333
Lockard, Miles L., 62
Lovelace, W. Randolph II, 241, 257, 283, 302
Lovell, Jim, 163, 291
Lowe, Nancy C., 40, 73
Low, George M., 36, 41, 99, 198, 207, 209, 317, 322

LRC. *See* Langley Research Center (LRC)
Lunar Landing Research Vehicle (LLRV), 52
Lunney, Glynn S., 41, 64, 71, 152, 162, 229, 237, 239
Lytle, Carroll D., 77

M
MA. *See* Mercury Atlas
MA-1, 53, 66, 85, 111, 115, 123, 124, 165, 226, 276
MA-2, 112, 124
MA-3, 85, 124, 125, 146, 252, 324
MA-4, 125, 151, 152, 246, 263, 273, 291, 330
MA-5, 92, 125–127, 246, 263, 273, 291, 330, 331
MA-6, 92, 127–128, 130, 133, 139, 146, 246, 252, 263, 269, 272, 280, 290, 306, 324, 331, 332
MA-7, 116, 128, 130, 131, 134, 135, 137, 140, 252, 267, 271, 273, 277, 306, 332
MA-8, 67, 130, 131, 135, 140, 141, 271, 306, 333
MA-9, 67, 116, 125, 131–133, 136, 140, 151, 171, 252, 263, 264, 267, 271, 306, 333
MacDougall, George F., 40, 78, 87, 229
Magin, Betsy F., 40, 229
Magin, Roy, 60
Maloney, Philip R., 75
Malpass, Wenzell G., 62
Man-in-Space-Soonest, 27, 28, 294
Manned Ballistic Satellite Task Group, 36
Manned Satellite Panel, 28, 36, 197, 198, 208, 226, 294
Manned Spacecraft Center (MSC), 3, 61, 87, 90, 91, 93, 94, 133, 135, 138, 161, 168, 208, 213, 216, 218, 219, 221, 223, 231, 234–237, 239, 246, 250, 252, 255, 258, 260, 263–265, 267–269, 271, 273, 277, 279, 280, 282, 283, 285–287, 294, 317, 332
Mann, Ralph D., 62
Markley, J. Thomas, 65
Markos, Athena, T., 73
Marshall, Margaret, 78
Marshall Space Flight Center (MSFC), 29, 53, 114, 170, 200, 201, 279
Martin, Glenn L., 28, 30, 220, 226
Massachusetts Institute of Technology (MIT), 135, 169
MASTIF, 41, 42, 110
Mateo, Lopez, Pres., 32
Mathews, Charles W., 39, 69, 75, 82, 85, 103, 113, 123, 198, 218, 228
Matthews, C. Frederick, 55, 56, 58, 73, 74, 244, 273, 330
Mayer, John P., 40, 70, 161, 166, 168, 229, 241, 322
Maynard, Joan B., 73
Maynard, John W. Jr., 70

Maynard, Owen E., 58, 65, 110, 123, 244, 275
Mays, Jack, 291
MCC. *See* Mercury Control Center; Mission Control Center
McCafferty, Riley, 80
McCloud, Mary W., 69
McCraw, David L., 61
McDonnell Aircraft, 46, 56, 58, 63, 69, 76, 80, 87, 99, 105, 117, 119, 156, 213, 228, 234, 239, 246, 249, 268, 294
McElroy, Neil H., 26, 27
McRae, James, Dr., 16
Meadows, May T., 66
mechanical man, 124, 125
Medaris, John B., Gen., 28
Meintel, Alfred J. Jr., 75
Mengel, Jack, 53
Mercury Atlas (MA), 30, 66, 82, 103, 111, 126, 264, 277, 327
Mercury Control Center (MCC), 3, 51, 67, 69, 73, 79–82, 92, 93, 113, 143, 145, 146, 149, 150, 153, 161, 162, 166, 195, 219, 226, 236, 237, 240, 246, 254, 256, 264, 269, 270, 273, 274, 277, 280, 283, 284, 290, 291, 307, 328, 330
Mercury/Manned Space Flight Network (MSFN), 31, 32, 45, 46, 52, 53, 71, 77, 82, 124, 125, 131, 132, 153–155, 159, 160, 162, 254, 255, 261, 269, 330
Mercury Procedures Trainer (MPT), 64, 71, 74, 84, 130, 155–157, 161, 162, 277
Mercury Redstone (MR), 53, 77, 82, 103, 110, 113, 268, 270, 290, 326
Mercury Scientific Experiment Panel (MSEP), 133
Mercury Seven, 97, 104, 163, 257
Mercury Thirteen, 97
Meson, John K., 58
Messier, Roger, 62
Metcalf, George, 290, 331
Meyer, Andre J., 41, 76, 102, 229, 266
Miller, Harold G., 64, 71, 156, 162, 235, 244, 276
Mills, Anne K., xi
Minitrack, 31, 44, 52, 53, 82, 154, 271
Minneapolis-Honeywell, 69, 93
Minuteman, 26
Mission Control Center (MCC), 93, 94, 107, 120, 135, 143, 260, 271, 280, 291
Miss Sam, 47
Miss Sam (rhesus monkey), 110
Mitchel, Junior N., 62
Moore, Connie, xi
Moore, Thomas E., 69
Morris, Archibald E. Jr., 77
Morris, David P., Dr., 67

Moseley, William C. Jr., 66
Moser, Jim, 291
MR. *See* Mercury Redstone
MR-1, 82, 110, 120–121, 145, 159, 280
MR-2, 121, 122
MR-3, 110, 122, 125, 138, 272, 306, 307, 330
MR-4, 122, 123, 138, 166, 306, 330
MR-1A, 120–121
MR-BD, 122
MSC. *See* Manned Spacecraft Center
MSFC. *See* Marshall Space Flight Center
MSFN. *See* Mercury/Manned Space Flight Network
Muchea (MUC), 155, 271
Muhly, William C., 40, 78, 229
Munford, Robert E., 68
Muroc, 2, 50, 51, 117–119, 226, 282
Myles, David T. Jr., 74

N

NAA. S*ee* North American Aviation
NACA. *See* National Advisory Committee on Aeronautics
Nagle, Charles C., 62
NASA. *See* National Aeronautics and Space Administration
NASA Communications Network (NASCOM), 154
National Advisory Committee on Aeronautics (NACA), 2, 5, 7, 24, 27, 35, 55, 64, 86, 93, 95, 164, 169, 197, 300, 301, 309, 319
National Aeronautics and Space Administration (NASA), 2, 5, 7, 22, 25, 27, 36, 54, 59, 79, 83, 89, 95, 144, 159, 169, 179, 197, 295, 302, 309, 319, 329
National Aeronautics and Space Council, 28, 90
National Intelligence Estimate (NIE), 22
National Security Council (NSC), 10, 21–23
National Weather Satellite Center, 136, 137
Naval Research Laboratory (NRL), 2, 10, 11, 28, 30, 31, 44, 52, 53, 82, 98, 154, 170
Nesbitt, William J., 80
Newport, Curt, 107, 123
NIE. *See* National Intelligence Estimate
Nieman, Ed, 80, 332
Nimbus, 134
Noa (ship), 128
North American Aviation (NAA), 109, 113, 118, 119, 234, 239, 302
North, Gilbert, 64
North, Warren J., 36, 41, 64, 99, 198, 209, 213, 229, 256
Novaya Zemlya, 22
NRL. *See* Naval Research Laboratory
NSC. *See* National Security Council

O

OAD. *See* Office of the Associate Director
Oceaneering Expedition, 107
Oceaneering International Inc., 107, 123
Office of Naval Research, 82, 227
Office of the Associate Director (OAD), 39, 40, 228, 229
Office Services Division (OSD), 40, 229
O'Hara, Dee, 1st. Lt., 256
Ohnesorge, Thomas E., 68
Olasky, Charles C., 64, 74, 75, 157, 162
Olling, Edward H., 78
O'Loughlin, John C., 70, 235
O'Neal, Robert, 67, 272
Operation Paperclip, 52
Orion Multi-Purpose Crew Vehicle (MPCV), 172, 173
OSD. *See* Office Services Division
Osgood, Catherine T., 72

P

Packham, Leonard E., 74
Palmer, Carl B., 302
Pan American (PanAm) Airways, 80
PARD. *See* Pilotless Aircraft Research Division
Parsons, John F., 89
Patneski, Andrew "Pat," 91
Patterson, Herbert G., 40, 229
Pearce, Fred T. Jr., 69
Peebles, Curtis, xi, 120, 318
Peele, Theresa M., 60
Peenemünde, 13, 114
Perrine, Cal, 291
Pesman, Gerald J., 41, 229
Petynia, William W., 66
Philco, 80, 81, 88, 274, 291, 324, 330
Phobos, 174
Picard, Fredrick, 33
Piland, Joseph V., 78, 87
Piland, Robert O., 65, 234, 239
Pilotless Aircraft Research Division (PARD), 35, 60, 84, 101, 165, 179, 199, 305, 318, 321,
Piore, Emanuel, Dr., 16
Plaumann, Bob, 330
Point Arguello (CAL), 127, 131, 155, 246
Polaris, 26, 36, 220
Pollard, Richard, Dr., 67
Powers, Gary, 22
Powers, John A., Lt. Col., 79, 80, 93
Prandtl, Ludwig, 310
Prather, Victor, 165
Pratt, Billy, 279

President's Scientific Advisory Committee (PSAC), 16, 24, 27, 169, 197
Pressley, Will, 291
Preston, G. Merritt, 41, 69, 75, 110, 111, 143, 229, 242
PROCEDURES, 159, 161, 290, 291
Project Adam, 113
PSAC. *See* President's Scientific Advisory Committee
Purcell, Edward H., Dr., 16, 24
Purser, Paul E., 39, 56, 60, 85, 112, 179, 199, 220, 228, 233, 238
Pusey, Dr., 9
Pyle, Gov., 8

Q

Quarles, Donald Aubrey, 7–9

R

Rabb, Leonard, 41, 66, 229, 272
Radio Corporation of America (RCA), 46, 55, 56, 71, 81, 82, 93, 117, 143, 145, 162, 275, 290, 308, 325, 331
RAF. *See* Royal Air Force (RAF)
Raines, Martin L., Lt. Col., 59, 79, 82, 232
Rand Corporation, 29
Rann, John A., 78
Rayburn, Samuel T., 90
RB-57F, 165, 292, 344
RCA. *See* Radio Corporation of America (RCA)
RCAF. *See* Royal Canadian Air Force (RCAF)
RCS. *See* Reaction Control System (RCS)
Reaction Control System (RCS), 50, 51, 117–120, 130, 300
Redstone, 2, 16, 28–30, 43, 52, 53, 59, 64, 72, 77, 79, 82, 97, 103, 108, 110, 113, 115, 120–123, 145, 158, 159, 161, 164, 165, 208, 232, 236, 265, 268, 270, 277, 290, 326, 330
Reed, Carol C., 78
Reeves Instrument Corporation, 46
Reid, Henry J. E., 170, 198, 221
Reisert, Donald, 300
Rembert, Dick, 81
René 41 alloy, 103
RETRO. *See* Retrofire Officer (RETRO)
Retrofire Officer (RETRO), 71, 75, 82, 160, 162, 264, 273, 275, 325, 333
Rhodes, Richard, 112
Rhode-Worthman Committee, 112
Ricker, Harry H., 40, 68, 229

RKV. *See* Rose Knot Victor
Robert, Frank C., 40, 229
Robertson, Robby, 330
Roberts, Tecwyn, 58, 73, 74, 240, 244, 280, 328
Roberts, William, 74
Robinson, Sugar Ray, 327
Rochford, Alan M., 61
Rocketdyne, 113, 116, 302
Rodgers, William, 66
Roe, Alliot Verdon, 54
Roe, Humphrey Verdon, 54
Rogers, Charlie E., 62
Rollins, Joseph J., 40, 229
Rollins, Robert H. II, 68
Rose, James T., 65
Rose Knot Victor (RKV), 155, 305, 306
Rosenberg, David A., 327
Rosenbluth, Marv, 81
Rose, Rodney G., 58, 65, 78, 244, 276, 281
Ross, Malcolm, 165
Royal Air Force (RAF), 56, 248, 280
Royal Canadian Air Force (RCAF), 55, 56, 273
Rumbaugh, Sy, 81

S
Sachs, Ira, 82
Saecker, Albert J., 77, 78
SAM. *See* School of Aviation Medicine
Sam (rhesus monkey), 110
Samonski, Frank H. Jr., 77
Sanders, Newell D., 41, 198, 206, 208
Sartor, Ronelda F., 40, 229
Satellite Tracking and Data Acquisition Network (STADAN), 31, 52, 154, 330
Saturn, 29, 53, 71, 161, 172, 204, 275, 277
Saucer, Jean S., 68
Saunders, Acquilla D., 76
Saunders, James F. Jr. 68
Saunders-Roe, 54, 280
Sawyer, Diane F., 78
Sawyer, Ralph, 68
Schirra, Walter M., Lt. Cdr., 64, 106, 123, 128, 130–132, 135, 137, 140, 163, 291, 333
Schmitt, Joe W., 61
School of Aviation Medicine (SAM), 47, 97, 136, 251, 288, 323–324
Schriever, Bernard, Gen., 29
Schultheiss, Richard F., 74
Scott, David, 163
Scout, 77, 92, 125, 204, 208, 293
Seamans, Robert C., 99
Segment 51, 127, 134, 139, 290, 331

Seiff, Alfred, 46
Semipalatinsk, 22
Sharp, Dudley, 124
Sharp, Edward R., 170, 302
Shenzhou, 174
Shepard, Alan B., Lt. Cdr., 64, 96, 103, 106, 110, 122, 123, 129, 131, 138, 256, 272, 307, 330
Shewmake, Glenna, 62
Shields, Betty C., 60
Shoaf, Harry C., 7
Shoosmith, John N., 58, 72
Shortal, Joseph A., 27, 52, 84
Siegfried, Joseph E., 62
Sigma 7, 131, 135
Silverstein, Abe, 6, 41, 45, 56, 85, 97, 99, 170, 171, 198, 199, 207–210, 240, 243, 302
Simpkinson, Scott H., 41, 75, 103, 111, 143, 229
Sjoberg, Sigurd A., 244, 282
Skopinski, Ted H., 70
Slayton, Donald K., Capt., 64, 151, 307
Smith, Norman F., 76, 221
Smith, Richard F., 74, 78
Smith, Robert P., 66
SOFAR, 75
Solar Electric Propulsion (SEP), 172
Soule, Hartley A., 84
Space Launch System (SLS) 172, 173
Space Task Group (STG), 3, 4, 6, 7, 26, 29–31, 35–41, 43, 45, 47, 48, 50, 52–89, 91–93, 96–99, 102, 103, 108–111, 113, 115–117, 120, 122–125, 127, 133, 152, 154, 157, 159–161, 164–166, 168–196, 198, 199, 205, 208–210, 212, 213, 215–254, 256–291, 293–300, 304, 309, 318, 319, 322, 329–334
Space Technology Laboratories (STL), 96, 112, 114–116, 123
Spritzer, Edith C., 69
Sputnik, 2, 5–7, 10, 16, 21, 22, 26–31, 36, 40, 82, 95, 96, 101, 169, 170, 201, 217, 219, 220, 235, 297, 301, 318
SRD. *See* Structures/Stability Research Division (SRD)
STADAN. *See* Satellite Tracking and Data Acquisition Network (STADAN)
Stapp, John P., Lt. Col., 103
Stearn, Jacquelyn B., 40, 229
Stenfors, Harold, 81
Stephens, Emily W., 67, 272
Stever, H. Guyford, 301, 302
Steward, Harry D., 61
Stewart, Lester A., 78

STG. *See* Space Task Group (STG)
Stillwell, Wendell H., 300
STL. *See* Space Technology Laboratories (STL)
Stokes, Katherine S., 69
Stonesifer, John C. 76
Strass, H. Kurt, 65, 234, 276
Strategic Air Command (SAC), 13
Strickland, Jim, 81
Stromberg-Carlson, 45, 82
Structures/Stability Research Division (SRD),
    229, 286
Sturtevant, Paul M., 78
Suit, Kenneth L., 68
SURGEON, 160
Sutton, Dick, 292
Swigert, Jack, 163
SYSTEMS, 160
Syvertson, Clarence, 46

T
TAGIU. *See* Tracking and Ground Instrumentation
    Unit (TAGIU)
Taylor, Paul D., 40, 229
Teague, Olin W., 90
Teapot Committee, 29
Teller, Edward, Dr., 30
Tessler, Martin, 62
Texas Tracking Station (TEX), 154
Theil, Alphonse, 60
Theil, John, 60
Thibodaux, Joseph G., 198, 223, 322
Thomas, Albert, 90
Thompson, Dayle, 292
Thompson, Floyd L., 36, 83, 93, 170, 179, 198,
    222, 224, 316, 317
Thompson, Henry, 45, 53
Thompson, Robert F., 75, 76, 130, 245, 285,
    322, 323
Thor, 26, 27
Tiangon, 174
Tindall, Howard W., 287
Tiros, 134
Titan, 26, 161, 163, 219, 291
Titov, Gherman, 125
Tomberlin, Jim, 81
Towey, James E., 687
Tracking and Ground Instrumentation Unit
    (TAGIU), 45, 84, 85, 184, 261, 318
Tracking Systems Study Group (TSSG), 33, 84,
    85, 270
Trader, Arthur G., 62
Tsiao, Sunny, 33, 155

TSSG. *See* Tracking Systems Study Group
    (TSSG)
Tucker, Charles, M., 62
Tucker, Virginia, 304
Tynan, Charles, I. Jr., 76
Tyuratam, 21, 22

U
U-2, 165
University of Richmond, 86
University of Virginia, 86

V
V-1, 113
V-2, 30, 52, 117, 320
Vale, Robert E., 58, 112
Van Cliburn, 327
Vanguard, 2, 5, 10, 11, 16, 22, 27–29, 31, 44, 52,
    79, 80, 82, 84, 154, 297
Vavra, Paul H., 84, 261
V-2 Bumper, 320
Viking, 28, 30, 177, 305
Virginia Polytechnic Institute (VPI), 86, 206, 235,
    236, 238, 286
Voas, Robert B., Lt. Dr., 63, 64, 79, 213, 245, 256,
    288, 289, 294, 307
Vogel, Kenneth J., 78
von Braun, Wernher, Dr., 28, 29, 53, 122, 170, 302
von Ehrenfried, Manfred H., 74, 245, 290–293
von Kármán, Theodore, Dr., 53, 117, 204
von Neumann, John, Dr., 29
Vostok, 124, 125

W
Wafford, Larry, 81, 324
Walker, Joseph A., 27, 52, 119
Wallops Flight Facility (WFF), 50, 51
Wallops Island Station, 2, 35, 55, 84, 108, 201,
    215, 216, 220, 238, 265, 272, 274, 278
Walter, Hellmuth, 117
Walz, Joseph E., 327
Walz, Kenneth N., 327
Warren, James W., 62
Waterman, Alan T., Dr., 8, 9, 14
Watkins, Julia R., 40, 76, 229
Watkins, Shirley, 40, 229
Watts, George A., 58, 66
WB-57F, 165, 255
Weather Bureau, 97, 134, 137
Weather Service. *See* Weather Bureau

Webb, James E., 90, 197, 201, 238

Weil, Joe, 300

Western Electric Corp., 143

Weston, Kenneth C., 41, 67, 229, 272

Westphal, Ralph L., 78

West, Walter, 66

WFF. *See* Wallops Flight Facility (WFF)

White, Edward H. II., 107, 123, 163, 291

White, Lloyd, 81

White Sands Missile Range (WSMR), 30, 44, 45, 52–53, 154

White Sands Proving Ground (WSPG), 52

White, Stanley C., L. Col. Dr., 67, 79, 87, 143, 213, 245, 256, 289, 293, 294, 324

White, Ted, 81

Wikipedia, 176, 216, 295–298, 303, 306, 321

Williams, Thomas N., 69

Williams, Walter C., 36, 51, 76, 80, 97, 119, 121, 123, 128, 158, 199, 220, 225, 324, 332

Wilson, Ann F., 69

Wilson, Charles E., 26

Windler, Milton, L., 76

Winterhalter, David L. Sr., 68

Wobig, Orrin A., 61

Wolfer, Earnestine H., 78

Woodman, Ray S., 78

Wood, William H., 84

Woomera, 31, 129, 135, 155

Wright, Capt. 322

Wright, Orville, 309

Wright, Rebecca, xi

Wright, Wilbur, 204

WSMR. *See* White Sands Missile Range (WSMR)

WSPG. *See* White Sands Proving Ground (WSPG)

X

X-1, 51, 101, 104, 117, 126, 237, 241, 286

X-15, 27, 35, 50–52, 83, 97, 101, 117–119, 165, 205, 212, 215, 220, 226, 267, 300

X-1A, 117

X-1B, 50–52, 118, 300

X-Series, 27, 35, 50, 97, 99, 215, 226, 227, 267

Y

Yardley, John, 128, 332

Yates, Donald Y., Gen., 97

Yeager, Chuck, 117

Yorker, Lloyd D., 60

York, Herbert F., Dr., 16

Young, Edward F., 66

Younquist, Robertson C., 36, 198

Z

Zanzibar (ZZB), 32, 125, 155, 271, 272

Zarcaro, John, 291

Zavasky, Raymond L., 60

Zedekar, Raymond G., 63, 64, 74, 289

Zhukov, Georgy, Marshal, 15

Zimmerman, Charles H., 39, 56, 228

Zurakowski, Jan, 55

Printed in the United States
By Bookmasters